THE
OXFORD HISTORY
OF AUSTRALIA

THE
OXFORD HISTORY
OF AUSTRALIA

General Editor Geoffrey Bolton

Volume 1   Aboriginal Australia Tim Murray
Volume 2   1770–1860 Jan Kociumbas
Volume 4   1901–1942 Stuart Macintyre
Volume 5   1942–1986 Geoffrey Bolton

THE
OXFORD HISTORY
OF AUSTRALIA

VOLUME 3
1860–1900
GLAD, CONFIDENT MORNING

BEVERLEY KINGSTON

MELBOURNE
OXFORD UNIVERSITY PRESS
OXFORD   AUCKLAND   NEW YORK

OXFORD UNIVERSITY PRESS AUSTRALIA
Oxford New York Toronto
Delhi Bombay Calcutta Madras Karachi
Petaling Jaya Singapore Hong Kong Tokyo
Nairobi Dar es Salaam Cape Town
Melbourne Auckland
and associated companies in Berlin Ibadan

OXFORD is a trade mark of Oxford University Press
© Beverley Kingston 1988
First published 1988

This book is copyright. Apart from any fair dealing for the purposes of private study, research, criticism or review as permitted under the Copyright Act, no part may be reproduced, stored in a retrieval system, or transmitted, in any form or by any means, electronic, mechanical, photocopying, recording, or otherwise without prior written permission. Inquiries to be made to Oxford University Press.

Copying for educational purposes
Where copies of part or the whole of the book are made under section 53B or section 53D of the Act, the law requires that records of such copying be kept. In such cases the copyright owner is entitled to claim payment.

National Library of Australia
Cataloguing-in-Publication data:

Kingston, Beverley, 1941–
  The Oxford history of Australia. Volume 3, 1860–1900.
  Glad confident morning.

  Bibliography.
  Includes index.
  ISBN 0 19 554611 3.

  1. Australia—History—1851–1901. I. Title.
  II. Title: Glad confident morning.

994.03

Edited by Carla Taines
Designed by Guy Mirabella
Typeset by Asco Trade Typesetting Ltd, Hong Kong
Printed by Impact Printing, Melbourne
Published by Oxford University Press
253 Normanby Road, South Melbourne, Australia

CONTENTS

List of Tables *vi*
Publisher's Note *vii*
Acknowledgements *ix*
Map of Australia *x*
Note on Measurements *xii*
Prologue *xiii*
1 Materialism *1*
2 Belief *57*
3 Society *108*
4 Culture *174*
5 Power *237*
Epilogue *309*
Notes *319*
Sources of Illustrations *352*
Bibliographic Note *354*
Index *356*

TABLES

1.1 Principal articles of diet required per head, 1891 and 1901 *53*
3.1 Masculinity rates *114*
3.2 Age distribution of Australian population, 1861–1901 (percentages) *116*
5.1 Census listing of occupations, 1861 *284–5*
5.2 Census listing of occupations, 1901 *286*

PUBLISHER'S NOTE

The Oxford History of Australia covers the sweep of Australian history from the first human settlement down to the 1980s. It consists of five volumes, each written by a single author with an established reputation as a productive and lively-minded historian. Each volume covers a distinct period of Australian history: Aboriginal history; white settlement, 1788–1860; colonial growth and maturation, 1860–1900; the Australian Commonwealth in peace and war, 1901–41; the modern era, from 1942 to the present. Each volume is a work of historical narrative in its own right. It draws the most recent research into a coherent and realized whole.

Aboriginal Australia is treated in its entirety, from the dramatically recast appreciation of early prehistory to present-day controversies of place, identity and belief. Colonial Australia begins with the establishment of tiny settlements at different times and with different purposes on widely separated points on the Australian coastline. From these fragments of British society sprang the competing ambitions of their members and a distinctly new civilization emerged. As the colonists spread over the continent and imposed their material culture on its resources, so the old world notions of class, status and gender were reworked. The colonists came together at the beginning of the twentieth century and fashioned new institutions to express their goals of national self-sufficiency, yet they were tossed and buffeted by two wars and the dictates of the international economy. The final volume therefore reflects the continuity of Australia's economic and political dependence, and new patterns in the quest for social justice by women, the working class and ethnic minorities.

In tracing these themes, the Oxford History's authors have held firmly to the conviction that history needs to interpret the past as an intelligible whole. The volumes range widely in their use of source material. They are informed by specialist research and enlivened by vivid example. Above all, they are written as narrative history with a clear and dramatic thread. No common ideological orthodoxy has been imposed on the authors beyond a commitment to scholarly excellence in a form which will be read and enjoyed by many Australians.

ACKNOWLEDGEMENTS

In a world where knowledge is disseminated at a faster rate than it can be produced or recycled, and where the ownership of ideas is jealously guarded as intellectual property, one becomes very conscious of a bower bird habit. Intellectual debts of which I am aware are acknowledged in the Notes. Other influences are so deeply ingrained that only the ghosts of Australian history past in the old Mitchell Reading Room may recognize them. My acknowledgement to the trustees of the Mitchell Library is more than the conventional expression of gratitude. Since I was first introduced to its splendours more than twenty years ago, the Mitchell has been both refuge and source of renewal. There is nothing quite like reading David Scott Mitchell's own copy of a book and wondering who may have read it before. Technical assistance with the map and illustrations was provided by Kevin Maynard of the Geography School and Catherine Marciniak of the Audio-Visual Unit, both at the University of New South Wales.

This work has taken longer than it should because I like to do everything myself, but it would never have been done without J, W, PB and my PCW.

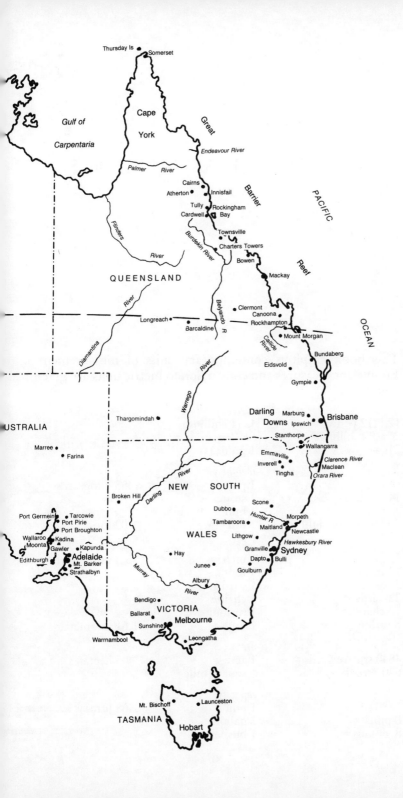

NOTE ON MEASUREMENTS

This book employs contemporary units of measurement. Equivalent measures and conversion to metric units are given below.

currency

12d (12 pence) =	1s (1 shilling)	
20s (20 shillings) =	£1 (1 pound) =	$2
21s =	1 guinea	

weight

	1 pound =	.453 kilograms
14 pounds =	1 stone	
8 stone =	1 hundredweight	
20 hundredweight =	1 ton =	1.02 tonnes

length

	1 inch =	25.4 millimetres
12 inches =	1 foot	
3 feet =	1 yard	
22 yards =	1 chain	
10 chains =	1 furlong	
8 furlongs =	1 mile =	1.61 kilometres

area

4840 square yards =	1 acre =	.405 hectares
640 acres =	1 square mile	

capacity

	1 pint =	.568 litres
8 pints =	1 gallon	
8 gallons =	1 bushel	

PROLOGUE

ON 10 DECEMBER 1859 George Ferguson Bowen, governor-elect of the newly created colony of Queensland came ashore from the government steamer *Breadalbane* at the botanical gardens in Brisbane.

Upwards of 4,000 persons were congregated on the banks, and the cheers that were given were worthy of an assembly of loyal Britons all the world over. His Excellency was received at the landing place by the Mayor and Corporation of the City of Brisbane; and as he stepped on shore a salute of twenty-one guns was fired. At the same moment a party of twelve young ladies, uniformly dressed in white, presented to Lady Bowen a bouquet of choice flowers. Passing under the triumphal arch, his Excellency and Lady Bowen entered the viceregal carriage, and a procession was formed to escort them to Government House. Along the whole line of route his Excellency was cheered after the genuine English fashion, and the procession that followed was of very creditable length and appearance. The banners carried by the various bodies of working men were especially noticeable for their appropriateness to the occasion, and the flag adopted as the Queensland ensign was frequently to be seen along the line of the cortege. The Union Jack, blended (out of compliment to Lady Bowen) with the Greek flag, waved on every side, and all the windows and balconies were filled with enthusiastic spectators.[1]

Later, in response to an address of welcome from 'the people of Queensland', their new governor told them that the name

QUEEN STREET, BRISBANE, IN 1860

selected for the colony was 'entirely the happy thought and inspiration of Her Majesty herself'.² This announcement, Bowen reported to his superior the Duke of Newcastle, Secretary of State for Colonies, was received 'with an emotion rarely witnessed in so large a concourse'. There were 'tears of joy, and shouts of "God save the Queen!"'³

'A new epoch in the annals of Australia has come to pass' proclaimed the *Moreton Bay Courier*. 'Our era has commenced.'⁴ While the *Courier* was thinking euphorically of the future of Queensland, the loyal ceremonies in Brisbane that day also marked a new phase in the history of this continent for they completed an administrative structure which had been evolving slowly, now to remain until federation in 1901. There were still disputed boundaries, especially between Queensland, South Australia, and the Northern Territory for which South Australia remained administratively responsible until 1911. There was also, from time to time, talk of new colonies, though nothing came of it. In Western Australia an executive council, only partially elected, continued to advise the governor and the colonial office until 1890 when there seemed enough suitable voters to hold elections for a representative legislative assembly. However, Western Australia's delayed self-government did not affect its existence as a significant entity among the Australian colonies. So, in effect, by the beginning of 1860 there were six colonies, each with its own government. During the next forty years this fact was basic to all other patterns that emerged.

1

MATERIALISM

THE SIX COLONIES of Australia constituted a rich resource for mid-nineteenth-century British enterprise. Early phases of exploration and pastoralism had revealed great wealth and vast potential, brilliantly confirmed by the gold discoveries of the 1850s. In 1860, with over half the continent but roughly explored and still unsettled, there was no need to think that what remained might be less valuable than what was already known. The mineral discoveries of the 1840s and 1850s on the fringes of existing British settlement gave rise to expectations of further discoveries as settlement advanced and, indeed, this proved to be the case in Queensland, the Northern Territory, and Western Australia, though the scale and richness of the golden decade was never repeated. Still, there was hope. William Westgarth, a Victorian colonist from the 1840s and a long-time enthusiast for all things Australian, on his return for the celebrations of 1888 was typically excited by the news of the Mount Morgan goldfield. It occupied a persistent place in the last of his many books about nineteenth-century Australia, though he had never seen it. He spoke for several generations when he wrote of his belief that

> beneath the varied surface [of Queensland] lay almost everywhere incalculable mineral wealth. Only some of this mineral wealth was

but just touched, Mount Morgan for instance. How many more Mount Morgans might turn up if the colonists could but get from the Home market the ready means to develop [*sic*] them!¹

In Queensland, the newest of the colonies in 1860, hopes for development and prosperity encompassed all that had gone before. A nucleus of old convict encampments and newer pastoral and farming communities in and around Brisbane was fired by the prospect of self-government, and especially by the prospect of winning control over the 670 500 square miles (1 727 200 square kilometres) of largely unexplored Crown land stretching to the tip of Cape York and into the Gulf of Carpentaria. By 1860 an area extending north to the present city of Rockhampton and west to the Warrego River had been declared open for settlement, i.e. the authorities were willing to accept applications for squatting licences in those districts. Beyond them was unknown, unexplored territory.

Symbolic of the contemporary mood, George Elphinstone Dalrymple, a well-connected young Scots immigrant, had set out with a small party of fellow land-seekers from the Darling Downs late in 1859 to follow the track marked a decade earlier by Ludwig Leichhardt en route for Port Essington. Dalrymple's party meant to locate land with squatting potential beyond the boundaries then open for settlement. Their journey took them to the fabulous valley they named Burdekin where they saw fine land, enough to satisfy all their requirements and those of their sponsors. It also spanned the months during which control over these lands was transferred from Sydney to Brisbane. Dalrymple arrived back in Brisbane on the last day of 1859 to find the new Queensland administration still in disarray—a conjunction of the summer season, Christmas festivities and the slow passage from Sydney of registers, documents and other paraphernalia of government. That evening, however, Dalrymple was invited, as befitted his status as a younger son, friend and relative of senior Darling Downs squatters, and a recently returned successful explorer, to *the* Brisbane New Year's party at Government House. He lost no time in pressing his claims and those of his friends for the lands of the Burdekin valley, for there they are, the first entries in the newly opened register of land claims for the Kennedy District, dated 1 January 1860.²

Where Dalrymple had been, others followed. During the next three years a steady stream of men and sheep made their way north through the coastal ranges and river systems, up the Belyando, along the Burdekin, and then toward Cape York or west into the waving acres of Flinders grass. The Gulf Country, watered by all those north and westward flowing rivers which had so confused Leichhardt became 'The Plains of Promise'.

Governor Bowen, himself a participant in a small way in the plans for a station in the Valley of Lagoons in the headwaters of the Burdekin and others further into the Kennedy, described the mood to the Duke of Newcastle, then Secretary for Colonies:

There is something almost sublime in the steady, silent flow of pastoral occupation over north-eastern Australia. It resembles the rise of the tide or some other operation of nature, rather than a work of man. Although it is difficult to ascertain exactly what progress may have been made at the end of each week and month, still at the close of every year we find that the margin of Christianity and civilization has been pushed forward by some 200 miles.[3]

By the late 1860s the optimistic sheepmen were discovering the truth about the Plains of Promise and the Cape Peninsula. The heat and wet were treacherous for sheep. So were the piercing seeds of the Flinders grass and the dreaded poison bush which grew along stock routes. The local inhabitants were no more inclined to welcome this invasion of squatters than they had Leichhardt, Kennedy, or the Jardine brothers.[4] Then a slump in the wool market and unpredictable seasons reduced many squatters to a diet of old mutton and pigweed. The monotony was broken only by tropical fevers. Sheep were replaced, optimistically, by cattle when northern gold discoveries and new meat-canning technology opened up viable markets for such remote beef, but although some spectacular mineral discoveries lay in store for the future, after the 1860s it was no longer the pastoral future of northern Queensland that sent investors scrabbling through their gazetteers.

Old-fashioned gold discoveries, a minor rush at Canoona near Rockhampton in 1861 and another on a useful scale at Gympie in 1867, were needed to confirm the newest colony as a serious option for intending settlers and a vigorous contender in the race for investment funds and economic development begun in the southern colonies some decades earlier. The gradual establishment of a chain of trading and mining settlements in bays and river mouths northward along the coast and the creep of railways westward to gather timber and wool sustained a sense of hope and feeling of promise. Indeed, Queensland's growth rate in the late nineteeth century outstripped the other colonies. Yet settlement was still sparse in 1900, the promise unfulfilled. Technology was increasingly seen as the key to both distance and climate. Artesian water would conquer dryness; refrigeration, heat.

Where Queensland led in the 1860s, the rest of northern and western Australia followed, not always knowingly or to advantage. In Western Australia the period of pastoral expansion was protracted and lacking the excitement of the 1860s in Queensland, but there, too, gold discoveries in the 1890s had a revitalizing effect, recreating something of the old optimism. Only those who had actually experienced the harshness of the north and west knew how precarious was the hold of the tiny white encampments along all those coasts. But from the mid-1860s it became increasingly difficult to evade the realities of the continent's dead heart. From Queensland it was possible to strike north at a tangent to the impenetrably dry country, keeping to the line of the Diamantina and eventually crossing, as Paddy Durack and Nat Buchanan did, through the Roper River lands to the downs of Victoria River and the Kimberleys.[5] From Adelaide John McDouall Stuart tried to push a path through the heart of the stony desert.[6] Intrepid young Western Australians set out in a clockwise direction across the continent, though they were not so much engaged in the scramble for land as in the search for knowledge of the value of the land and its increasingly elusive water supplies.[7] Belief in the potential of the arid interior which had mysteriously claimed Leichhardt and should have been buried with the sad remains of Burke and Wills persisted in the legend of Lasseter and his lost reef.[8]

There is no more striking evidence—export figures permit no other argument—of the continuing value of land and wool to the Australian colonists in the 1860s than their approach to claiming and dividing up the rest of the continent. Not only was Queensland seized and thousands of runs roughly marked in a scramble as dramatic as the 1850s rush to peg out claims on every new mining field. The Northern Territory came under the control in 1862 of the South Australian government in Adelaide where it was determined to sell enough land to make it a worthy investment according to the same principles used in the initial settlement of South Australia. It did not prove difficult to sell the promise of parcels of 160 acres (65 hectares) of country land plus a half-acre town allotment for £60 somewhere in the Northern Territory unsurveyed and unseen to speculators in London and Adelaide. The hardest part of the bargain proved to be surveying land to meet the sales. Real knowledge came later. It was still assumed in Adelaide in the late 1860s that sheep would make the first tracks across the Northern Territory as they had done elsewhere, that mineral discoveries and agricultural settlement would follow.[9]

In the end it was not the sharp hooves of sheep that made a path through the dry centre, but wagons bearing poles and miles of wire for the telegraph line. Instead of station homesteads, telegraph stations were built as markers and guardians of British settlement in the wilderness. The men of the telegraph discovered the gold and mined it too, creating at last a market for beef cattle in the north.[10] Subsequently these markets were extended by the technology which reduced beef cattle to Bovril and Camp-pie, and the advertising campaigns which made Bovril and Camp-pie regular features of the diet of millions of Britons.[11] They were also heavily dependent on lenient and concessional treatment from faraway governments in Brisbane, Adelaide and Perth, and some of the cheapest labour in the world—the Aborigines.[12] A high price was paid for sustaining the 1860s promise of thriving settlement and profits to match. Even so, as a cattle producer northern Australia could not meet Argentina or the American mid-west. Northern Australia's remoteness became an ironic advantage. Its obscurity conveniently concealed or obscured continual violations of practices and values the rest of the country held dear: a fair go, a decent wage, a white man's country. Old images connected with size, mystery, unexplored and unfulfilled promise enhanced its obscurity. A hundred years later it was still little known, sparsely settled and an earnestly misunderstood part of Australia, but by then it had been endowed with a new layer of symbolic meaning, for nineteenth-century enthusiasm centring on proximity to Asian trade had been disappointed and in its place had grown fear of failure, of emptiness, and of strategic vulnerability.

The end of the nineteenth century saw the end of exploration, the end of expansion, the closing of that chapter in which there had always been better land further out. What land there was became increasingly difficult of access. The north coast of Queensland retained its connection with the world because the steamships clung to the passage inside the Great Barrier Reef and made their landfall at the telegraph station at Somerset, Cape York, before striking out for the East Indies or the Indian subcontinent. The Northern Territory was more accessible by sea through Port Darwin than overland from Adelaide for bulky items such as personnel and food, but the north and west coasts of Western Australia, scene of the earliest European landfalls on the continent, could support very little in the way of regular shipping.

There had been no shortage of capital, both local and from overseas, to invest in the occupation of northern Australia. Dalrymple's backers were based mainly in southern Queensland and in Sydney.[13] Large sums of Melbourne money went into central Queensland and the district around Mackay,[14] while the prospect of both land and trading posts attracted established Pacific traders like the Sydney-based Towns and Co. to the far north.[15] Most of the risks of opening new country were borne by families and companies established in the older colonies. Some British capital arrived with the last wave of younger sons trying their hands at squatting in the colonies—the Hodgsons on the Darling Downs,[16] the Scotts in the far north at the Valley of Lagoons.[17] More came in the planned expansion of land companies in the earlier phases of pastoralism—the Scottish Australian Investment Company, for example, which retained William Landsborough to search for and take up massive tracts of central Queensland consolidated in Bowen Downs.[18] By 1870 the majority of pastoral holdings in Queensland were under the control of British and Australian banks and mortgage companies who had lent money for the purchase of stock, as an advance on the woolclip or to secure land against the threat of closer settlement legislation. Through the later decades of the nineteeth century it became more difficult for individuals to finance the larger and more arduous operations which were needed to carve runs out of the Northern Territory and the north-west of Western Australia and even more difficult for them to sustain them without the support of banks and finance companies. The days when a man could take his pay in sheep and set out for the bush had gone. The land 'further out' was likely to be semi-desert, impenetrable scrub or mangrove, the rivers dry beds of sand or raging torrents, treacherous with crocodiles. The men themselves were no longer the gentlemen amateurs or the sons of old colonists trained as stockmen and jackeroos as they had been, even in the 1860s. Professional managers appeared on the company-owned properties, men educated to make profitable decisions and to avoid unprofitable risks.[19] They no longer staked their claims and bargained later with the bureaucrats. Company policy was decided to take advantage of concessions that could be negotiated with governments, of services, labour, communications that would be provided. It was clear from the way in which Northern Territory lessees haggled over the amount of time they were to be given to stock their runs— three years? five years?—that fresh grass for hungry stock was no longer the imperative.[20] In the 1860s in Queensland Governor Bowen and Premier Herbert had generously offered to build harbour facilities and a township to assist their friends the Scotts in establishing their station in the ranges behind Rockingham Bay. Diplomatically they called the township 'Cardwell' for their colonial office patron. Where the land was good and the prospects of profits very high, private investors took the initiatives and government followed, as at Bowen, Townsville, Geraldton (Innisfail) and Cairns. But many promises and offers were needed to sustain settlement and investment later in the century in the far north and northwest. As the Queensland government had sent W. A. Tully, the government surveyor, to lay out Cardwell in 1864, so the South Australians sent G. W. Goyder to mark out a town to be named Palmerston as a base for southward settlement in the Northern Territory. Tully named his carefully laid out streets for the Oxford colleges of his patrons, but Cardwell survived only as a fishing village and a railway stop when at last the railway came. Its grandiose street plan and imperial names merged with the rainforest. Palmerston was swallowed eventually by Port Darwin, a name at least forward-looking and mindful of the concept of the survival of the fittest. Elsewhere along the farthest coasts, government residents, post and telegraph officers, and customs inspectors maintained illusions of settlement and a veneer of civilization over what had become a quietly desperate commitment to overinvestment in fantasy.[21]

Western Australia's most lucrative industries in the late 1880s after wool and sheep, were exotic: pearls and pearlshell found along the northern coasts from Broome to Somerset, then sandalwood, lumber, and horses, all of which were sent to India to supply both the Indians and the British army.[22] But everywhere north of the Tropic of Capricorn, the hypocrisy required to sustain settlement and civilization in this stark re-enactment of the harsh southern drama of first settlement highlighted compromises already enforced on imported ideals and cherished beliefs in those more benign southern areas.

Climate was generally advanced as the reason why there were so few women among the settlers in the tropics, though it was not climate that was crucial, or distance, but the cost of maintaining civilization where the profit margins where already low. White women could and did live in the north. Jeanie Gunn's account of her year in the 'Never Never' on Elsey station is only a little dramatized for effect.[23] But a wife required a house and a certain standard of comfort. She could not be expected to live or work, as the contemporary colloquialism put it, 'a black gin'. Further south it had not been unusual for wives and families to share the isolation and hard work of establishment. On the cattle stations of the north of Western Australia and the Northern Territory, however, it become common for the men to dispense altogether with the conventional notion of a wife. Quite unselfconsciously, Alfred Searcy described how:

nearly all of the drovers, cattlemen, and station hands had their 'black boys' [gins]. . . These women were invaluable to the white cattlemen, for, besides the companionship, they become splendid horsewomen, and good with cattle. They are useful to find water, settle the camp, boil the billy, and track and bring in the horses in the mornings. In fact, it is impossible to enumerate the advantages of having a good gin 'out back'.[24]

Besides casting doubt on contemporary dogma about the physical inferiority of the female sex, this revelation highlights the indispensible role of women in early settlement everywhere. However, as was also the case with Chinese and Aboriginal labour in areas unattractive to white unionists, isolation and climate encouraged discreet ignorance and the myths about tropical hazards and female weakness added their mite to fantasy, delusion, and the bigger balance sheet.

An early belief that white men could not work in the tropics was gradually modified by settlers determined to survive on co-operative-sized cane-farms and work them without assistance. The strength of the agitation against coloured kanaka labour came not from the north where failed goldminers were struggling to carve cane paddocks out of the scrub, but from southern Queensland, for example, around Bundaberg, where there was genuine competition for work. In the Northern Territory it was the same. Despite depression and unemployment, the contractors for the Palmerston (Darwin) to Pine Creek railway in the 1880s were permitted to use existing Chinese labour and to bring in additional Chinese, Cingalese, and Indian coolies. It was not the heat that made a difference to the white man's ability to work in the tropics, but the colour of the money. On the pearl luggers and *bêche-de-mer* boats, Aboriginal labour was welcomed, but mainly because nothing else was available.

Between 1860 and 1900 the value of wool exported from Australia rose 100 per cent. The quantity of wool exported during the same period rose from 100 million pounds to over 500 million pounds. Wool became more abundant, cheaper, and a popular fibre.[25] A cotton shortage, the result of the American Civil War in the 1860s, gave some incentive for cotton planting, particularly in those parts of Queensland perceived to have a climate similar to that of the southern states of the USA. Cotton, however, was expensive to plant and pick. Despite the view of its great advocate, Dr John Dunmore Lang, that it was the ideal crop for the family farm, since wives and children in Australia could do the stooping and intense cultivation which had required Negro slaves on American cotton plantations, Australian families seemed unenthusiastic about cotton growing.[26] In any case, the market for cotton began to decline. Several generations of British wives and children had worn cotton in the interests of cheapness and cleanliness. Improvements in the standard of living encouraged a move to higher status woollen fabrics.[27] If it was good enough for business and professional men to wear suits of best worsted, then why should their wives and daughters not have fine merinos, cashmeres, and any number of warm woollen petticoats. The woollen mills responded with new technology which extended the range and style of fabrics and reduced their cost. The demand for Australian wool grew and production costs were held down by continuous expansion. Sheep spread into every corner of the continent where they could survive and produce wool, and even where they could not. This expansion drew on reserves of hitherto uncropped grass at no greater cost than an initial annual licence fee of 10s for each 25 square miles. It also provided room for whatever natural increase might occur in the flocks. It didn't matter in the early years that the quality of flocks (and wool) began to deteriorate. A continuing demand for quantity as much as quality wool absorbed some very coarse fibres.

Expansion relieved some of the pressure on runs in the older districts and attention was then given to stock improvement and efficient management. With the possible exception of a few years during the late 1860s and early 1870s when the market was depressed, consistently good prices for wool encouraged improvements in fencing, watering facilities, and handling equipment as well as breeding programmes. The old established squatters began to reap the advantages of extending railways, regular communications, even of encroaching closer settlement. Free selection brought not only an opportunity to freehold choice portions of leased runs: free selectors themselves were a resident, captive, casual labour force. Canny squatters did well by consolidating their holdings, culling their poorer stock for sale further out, improving their facilities—these were the years when some of the grander homesteads were built—and adding to their financial reserves. Less capable squatters spent heavily and unwisely on freeholds which were not essential or worth the money, improvements which were unsuitable or unprofitable, or stock of inferior quality. They paid too little attention to management or went on expecting the old magic of natural increase and regeneration of pasture to work.

By the 1880s the effects of such bad or ruthless management were beginning to show in marginal pastoral districts. Natural increase without careful surveillance led to overstocking, disease, degeneration and the destruction of the natural grasses. Regeneration was hampered not only by poor seasons but by the spread of new pests such as rabbits, and the subtle though nonetheless substantial changes wrought in the natural vegetation by the chewing and tugging of all these imported animals. A succession of drought years during the 1870s was accommodated because there was still room on the frontier; but another sequence of dry years, beginning in the north in the late 1880s, and, during the succeeding decade affecting the whole of the continent, pressed unbearably on overstocked runs, ruined natural herbage and makeshift watering facilities. The obvious cost was seen in abandoned homesteads, drifts of eroded soil, and the collapse of many individual hopes. Less obvious was the destruction of natural resources, the loss of fragile native environments, the transformation of an ecology which had never been understood by its impatient invaders.[28]

Though the excitement of continuous pastoral expansion was more remote to the ordinary citizens of Sydney and Melbourne than it had become to the habitués of British boardrooms, the value of Australian wool underwrote all late nineteenth-century investment and development, especially in the wool-exporting cities. After hovering round a shilling a pound since 1872 the average price of wool fell to eightpence halfpenny in 1886, and to sevenpence three farthings in 1894.[29] The loss of even a farthing when multiplied by 600 million pounds of wool was enough to cause concern.

Thus declining confidence in wool triggered a series of other anxieties about the state of the economy leading to Melbourne's spectacular bank closures in 1893.

In comparison with the value of wool produced and exported after 1860, the value of gold and other minerals declined a little from their dramatic prominence in the 1850s. Gold and, for a few years in the early 1870s, tin, were, after wool, the next most valuable exports from the Australian colonies.[30]

It would be misleading, however, to think of the value of all minerals (and most other products, for that matter) only in terms of their export-earning capacity. Production, for example, of coal rose steadily after 1860, but as it was all consumed at home by growing industries, an expanding railway and shipping service, and rising domestic standards of comfort, its value may be overlooked. Likewise, though agricultural and farm produce did not occupy a notable place in the table of exports, their value in providing the bulk of food required for an expanding population, a population also with rising levels of consumption, must be borne in mind. As Anthony Trollope observed during his visit to South Australia:

the produce of a country which is exported always receives more attention than that which is consumed at home. Who thinks anything of the eggs that are laid around us, or of the butter made? In calculating the wealth of the country, who reckons up the stitching of all the women, or even the ploughing and hedging and ditching of the men?[31]

The real significance of mining exports and discoveries was for the image of Australia as a place where people invested money or thought of immigrating. The hope of mineral wealth was ever in the minds of ordinary men out searching for water or feed or a short cut. Colonial governments employed geologists to make systematic surveys and offered rewards for the location of worthwhile deposits. Each new discovery served to confirm the mythic ordinariness of all former stories.

A stockman with some knowledge of minerals kicked the rock used as a doorstop in the kitchen of Maryland station and set off the rush to the banks of Quart Pot Creek in 1872 which led to the rise of the last, richest, and most hopeful of the townships founded on alluvial tin through northern New South Wales and southern Queensland in the 1860s and 1870s. For Inverell, Glen Innes, Emmaville, and Tingha, a New England squatting heritage was a guarantee of respectability should it be needed.[32] But on Quart Pot Creek the town grew in weeks in a way that reminded old hands of Ballarat or Sandhurst in the 1850s. Its future was surely as promising. Searching for a name fit for a bishop they settled on 'Stanthorpe' (tin town) thinking that neither Maryland nor Folkestone, the names of the adjoining runs, conveyed the excitement and hope that was felt at Quart Pot. They were right to dream of greatness. Between 1872 and 1876, tin to the value of £715 000 was extracted, mainly by small-scale alluvial workings. With the output of these fields, those adjoining in the New South Wales, and those opened on the west coast of Tasmania, Australia was for a decade the world's leading tin producer. When news of a new El Dorado on a river called the Palmer (in honour of a recent Queensland premier, A. H. Palmer) somewhere beyond the Endeavour River, filtered down the tin races at Stanthorpe, the mines were already overtaken by mechanization, the town was in decline, and most of the work was done by speculators through the telegraph office.[33]

The quality of the stories was almost as important as the value of minerals in the ground, for the fortune often lay in company promotion and speculation. 'There is nothing finer in Australian history', wrote Randolph Bedford,

than the many times repeated adventures of the strong, young, hopeful men of Australia, hot foot to the new rush . . . the best gold is always furthest out. It is the spirit which has made the world, not the mere keeping of shops and the piling up of bank balances.[34]

A mere lad, Charlie Rasp, riding the boundaries of Mt Gipps station in 1883, collecting stones in his hat, brought to light the potential wealth of the 'broken hill' they called 'the hog's back'. Aboriginal stockmen led station owners and prospectors to the riches of Charters Towers in Queensland, and Parker Range and Mt Magnet in Western Australia. Bayley and Ford, the discoverers of Coolgardie goldfield, were tracked back to their claim which they had tried to keep a secret with the assistance of local Aborigines. The Morgan brothers listened to stories told by William MacKinlay's daughter of the mountain of gold her father had found and were rewarded by the fortune which now bears their name.

The entrepreneurial spirit thrived on distance and mystery. Nowhere can this be seen more clearly than in the persistent belief in gold in New Guinea. As early as 1864 a company was formed in Sydney with a view to prospecting in the eastern part of New Guinea, but disbanded when the imperial government refused support. A party of prospectors and adventurers set out to search for gold in 1872 in the *Maria*, but their brig was wrecked inside the barrier reef just north of Tully. Some of the survivors (including the young Lawrence Hargrave whose faith in the future led him eventually to experimenting with flying machines at Stanwell Tops outside Sydney) were rescued by Captain Moresby in the government survey ship *Basilisk*. They were the lucky ones.[35] Local Aborigines, already notorious for the rough time they had given explorer Edmund Kennedy, sustained their reputation as determined defenders of their home in the rainforest, and the nearby headland became 'Murdering Point'.[36]

Further attempts to claim the probable riches of New Guinea failed for want of official assistance, which had become increasingly important in providing the infrastructure of permanent settlement, or because there were difficulties in distance and climate. As happened in other unattractive or dangerous parts of northern Australia in the late nineteenth century—for example at Beagle Bay in the north-west[37]—the beach-head in New Guinea was secured by missionaries and ethnographers whose passions enabled them to overcome difficulties too great for the optimism of mere speculators and developers.[38]

Late nineteenth-century technology was greedy in its need for minerals. Copper, iron, and coal required serious investment and began to justify further expenditure on railways, shipping, and telegraphs. In contrast to gold, coal-mining, in particular, used traditional practices. It was also closely tied to the mainstream of industrial development. Whereas the never-ending rush to the newest goldfield added volatility, optimism, and even technical innovation to the mining industry, coal provided a sombre and steadying influence. The late nineteenth-century coal industry was concentrated, of necessity, on those fields close to the centres of settlement: Newcastle, Lithgow and Bulli in New South Wales, and after the 1870s, Ipswich and the Callide Valley in Queensland.[39]

The demand for coal rose steadily from 1860 and production expanded to meet this demand. New South Wales produced 368 000 tons of coal in 1860. There were some 979 coal-miners according to the New South Wales census of 1861, 900 of them in the Newcastle region.[40] By the turn of the century coal production had reached 5 million tons in New South Wales and about one-third of this was exported.[41] Increasing quantities of coal were required for railways and other steam-driven machinery, to produce the coal gas used for lighting, cooking, and heating, and then the electricity which supplanted gas. Coal was used too for simple domestic heating in the cooler southern cities, although as long as cheap or free firewood was available, wood was preferred.

While sailing ships could still carry heavy cargoes more cheaply than steamers, they carried Newcastle coal across the Pacific, picking up grain in San Francisco for Europe, or nitrates in South America. Coal was shipped to New Zealand until mines were developed there, to the islands of the Pacific which were becoming coaling stations for steam-powered European navies. It went to Chinese ports and to Singapore which became a major port using imported coal. Victoria and South Australia were important consumers of New South Wales coal for most of the century. Western Australia took a little before mines were opened at Collie in 1883. Tasmania and Queensland both managed to mine enough to meet their own requirements, but the steady procession of colliers from Newcastle south helps to explain why Victoria was driven to experimental economic policies in the 1860s.[42] Whereas New South Wales looked forward to steady industrial development based on its coalfields as well as a useful income from coal export, Victoria was forced to contemplate rigorous control of its export balances to allow for the continuing cost of coal. The brown coal deposits of Gippsland were known from 1857, but the technology to exploit them lay in the future. In the meantime, the cost of importing coal to develop and sustain Victorian industry as well as a modern urban lifestyle left little room for manoeuvre in the balance of payments. By the 1880s Melbourne's smog was an inescapable sign, not only of its advanced industry, but also of its dependence on New South Wales black coal.

There was some hope in water-generated electricity. As early as 1884 Tasmania's Mt Bischoff tin mine had its power supplied by a water wheel generator. From 1893 Thargomindah in south-west Queensland was supplied with electricity by dynamos connected to a wheel driven by artesian water. And when in 1895 the Launceston City Council began supplying power from a generator on the South Esk River, the foundations of the Tasmanian Hydro-Electric Authority were laid.[43]

Eventually the image of the lone prospector evoked only nostalgia amid the bustle and crowded conditions of most mining settlements. Individual miners or small groups of co-operators working early claims were superseded by wage-earning employees of mining companies. Deep-lead mining, quartz crushing and chemical extraction all required capital and organization. The experience of coal-mining became useful in other forms of mining for its underground working practices and employee organization, and men with experience in coal were preferred for employment. There was some direct migration, for example, from the Cornish tin mines, to Moonta and Kadina on Yorke Peninsula in South Australia. Both miners and managers were recruited for Australia by British-owned companies on British coalfields. Thus Newcastle seemed intensely 'English' and 'working class' socially and culturally. Those miners who could not, or would not become wage workers moved from one new field to the next. There were old-timers at Kalgoorlie who had seen service on half a dozen Australian fields and in New Zealand. (Though not exactly an old-timer, the 'roaring days' having passed before his childhood, Henry Lawson was one who had followed his hopes to the west in the 1890s via New Zealand.) The miners who moved to Broken Hill from

South Australia in the 1890s came from languishing copper mines, while Joseph Golding, father of early labour activists and feminists Kate, Annie, and Belle, was undoubtedly not the only old miner from Tambaroora on the Turon goldfield to end up in the mines at Newcastle. There was little growth in the proportion of the population engaged in mining from the middle to the end of the nineteenth century, but that workforce was recycled several times, rarely making much money, constantly hoping it would.

By the end of the century, the map had been filled in. 'Further out' there was only desert. The dream of a great new trading zone in South East Asia had faded as roofs rusted in Cardwell and Palmerston. Gold in the west kept something of the old enthusiasm for investment alive, but much British capital was moving away from Australia. True to the spirit of imperial optimism, local entrepreneurs combed the Pacific for exploitable raw materials and labour. An Australasian New Hebrides Company was formed in Sydney in 1889. New Guinea and Fiji continued to hold out promises of gold, land, or wonderful tropical products. And to the south lay Antarctica. An Australian Antarctic Committee was set up in Melbourne in 1886 under the joint auspices of the Royal Society of Victoria and the Victorian branch of the Geographical Society of Australasia.

In the long-settled southern half of the continent, pastoral and mining exploitation had settled down into a better appreciation of the problems of permanent settlement. The ruthless destruction of forests for no real purpose except that clearing was equated with development began to cause the occasional disapproving comment. 'Gum trees should be left standing in clumps,' observed George S. Baden-Powell in 1872, 'for if isolated they pine and die.'[44] Count Reinhold von Anrep-Elmpt speculated on the effects of clearing he observed in the upper Hunter and New England districts:

How much the consequences of such abnormal clearings have already affected the Australian climate is evident to the old settlers of the country, who claim that in earlier years the rainfall had a certain regularity. The soil then never dried out completely, the grass was fine and lush, birds enlivened nature and eradicated insects; there were few flies, only sporadic outbreaks of disease and no epidemics. The forest provided enough water for men and animals even at the height of summer; now the winter is colder and the summer hotter. Every year there are more bad harvests and a worse shortage of water; the rich grass of earlier years is replaced by weeds, cactus and thistle; flies and mosquitoes multiply to plague proportions and birds, like the Aborigines, disappear; epidemics break out among animals and men; all this is the result of thoughtless deforesting.[45]

Timber continued to be used extravagantly, especially on the fringes of settlement and where railways were being built.[46] Legislation which defined ringbarking as occupation of the land for free selection purposes came into operation in Queensland in the 1870s, though in 1881 the danger of wholesale ringbarking was recognized in New South Wales legislation to try to restrain ringbarking without a purpose or permit. In 1888 William Westgarth travelled from Melbourne north beyond Brisbane. He was so depressed by the sight of forests of dead trees that he suggested wholesale clearing and a programme of replanting with useful and tidy exotics.[47] In the 1890s, T. A. Coghlan noted the importance of the native flora, the tree and staghorn ferns, tiger lily, waratah, stenocarpus, flannel flower, star of Bethlehem, Christmas bush and bells, and wattle blossom to the 'metropolitan horticulturalist's business'.

It is a matter for deep regret that the state's beautiful native flora is being gradually destroyed. It now contributes at annual displays to the reduction of the debts on churches and other institutions; and the coast-slopes and the bush lands in the neighbourhood of every town are periodically denuded to furnish attractions to local flower-shows and charity bazaars, the well-meaning decorators of which, not content with merely plucking the blooms they require for their floral designs, tear them out by the roots, and create a waste where there was formerly a garden.[48]

Closer settlement when it took the form of agriculture was to blame for the most thoroughgoing destruction of the natural environment, though when neat farms appeared on the cleared land, no one seemed to mind. Agriculture contributed little to export income in comparison with the pastoral and mining industries. Nor did it attract significant or glamorous investment. But it did contribute substantially to the pace of industrialization and technological progress.

Without the enthusiasm of farmers for machinery and their willingness to experiment, there would have been less growth and investment in the workshops of South Melbourne, Sunshine, Granville, or many a small country town. The perishable nature of much food gave real meaning to innovations in canning and refrigeration. Meat-canning factories appeared for a season or two in improbable places—for example, on the banks of Henry Kendall's 'gloomy Urara' (the Orara River in northern New South Wales).[49]

Agriculture most felt the effect of the relatively full employment before the 1890s. Workers generally preferred urban employment to rural. Even in rural areas, government works on railways, roads and bridges, especially railways, competed with agriculture for labour. Government works paid better wages and were considered more attractive because of this.[50] Nevertheless, rural employment, some of it in the pastoral industry, more of it in agriculture, continued to expand. Having taken every possible step to exclude the 'unproductive' labour of housewives converting wild fruit to jam or children rounding up the cows for milking before and after school, McLean, Molloy and Lockett still find that rural employment expanded by a third in the 1870s, and by about 20 per cent in each decade following.[51] As always there were variations from colony to colony. In the 1870s expansion in rural employment occurred most noticeably in South Australia, Victoria, and Queensland, all colonies where large areas of farming land were pioneered in these years. In the 1880s New South Wales and Tasmania provided a greater share of opportunities for agricultural labour as transport and technology assisted expansion in labour-intensive industries like dairying and fruit-growing. In the south-west of Western Australia after the 1890s gold-rushes, there was a familiar demand for farm land involving, as in the 1860s elsewhere, and as A. B. Facey's recollections of his boyhood in the Narrogin district show, more hard work than economic reward.[52]

As well as providing increasing opportunities for employment, agriculture contributed more than is generally allowed to a rising standard of living in Australia during this period, and therefore to the attractive image of the Australian colonies as a place to live and invest. To support a growing population while at the same time developing exports of foodstuffs was impressive. Self-sufficiency in mustard, jam, sugar, olive oil and rum as well as in the more substantive breadstuffs and vegetables appears insignificant in aggregate statistics, but in particular districts these items were the basis of settlement and made the difference between survival and withdrawal. In the 1840s Caroline Chisholm had used the subtitle *Comfort for the Poor!* for her little book about the colonies *Meat Three Times a Day!*. H. Mortimer Franklyn indicated in his *A Glance at Australia in 1880* the change that had occurred by the subtitle *Food from the South*.[53]

The case for more farms to produce food for domestic growth as well as export was heard as frequently as the equitable argument in favour of free selection in the 1860s. Or it was asserted in the true spirit of intercolonial rivalry that it was a disgrace that Queensland, for example, was dependent on South Australia for wheat and could not even be self-sufficient in cabbages. The efficacy of South Australian policies on closer settlement and the subsequent success of South Australian agriculture were well appreciated in the eastern colonies. Some things it was necessary to import, but any colony of the size of Victoria or Queensland should be able to feed itself. The key to the successful development of agriculture in the 1860s and 1870s, just as it had been the key to the growth of the wool industry in the 1840s, was the appreciation and judicious application of new techniques and technologies to unfamiliar environments. Here South Australia led the way.

By the end of the 1860s South Australia was known as 'the Farinaceous Colony' and from the beginning of the decade was exporting wheat and flour. In the other colonies, despite the apparent failure of free selection legislation, agriculture continued to expand in accordance with the growth of markets, especially local ones, and of transport and storage facilities. Tasmania and Western Australia were generally self-sufficient in grain from 1860 and Victoria became so within the decade. As well as wheat, oats, potatoes and hay, both wheaten and oaten, were being produced, and in Victoria and Tasmania especially, the production of fruit and vegetables had begun on a commercial scale.[54] Most small settlements were reasonably provided with dairy produce, poultry and eggs, fresh fruit and vegetables. Where would the memoir of station life be without the Chinese gardener who made it nutritionally bearable? It might also be said that most households, except in the more densely populated parts of cities or in the most recently opened parts of the interior, were somewhere near self-sufficient in these foodstuffs. Markets for the sale and exchange of perishable goods operated in the major cities, but even in a town the size of Goulburn, for example, there was no need for a fruit and vegetable market.[55] Backyard gardens and the informal economy took care of existing supplies and needs.

Major developments in food processing and preservation during the later part of the nineteenth century made feasible the intensification of dairying in the areas remote from major centres. Fresh milk was brought in refrigerated wagons from central Victoria to Melbourne and from the south coast of New South Wales to Sydney.[56] Though the development of refrigeration was undoubtedly of great importance to the meat export industry after 1880, for the majority of ordinary Australians, its first impact was to improve the supply of fresh food, not only meat, but dairy produce and some fruit and vegetables. It meant ice-works in hot, dry, inland towns, and cool drinks, either beer, aerated waters or cordials in summer. Other technological advances expanded the range of comforts while strengthening the market for agricultural produce. Tin plate and canning technology, which had so stimulated tin mining in the 1870s, encouraged new markets for southern fruit and northern sugar. Australian labels like those of the Tasmanian Jam and Fruit Preserving Company, Rosella, IXL, and CSR treacle and syrup appeared on the shelves of country and suburban stores alongside the many varieties of Crosse & Blackwell.

By the 1880s there was even hope of controlling the water supply.

When the art of conserving the excess of wet seasons to balance the deficiency of dry years shall have been mastered and turned to account, phenomenal progress may be confidently predicted. Already very partial, but widely separated experiments have established the fact that copious treasures of fresh water underlie the arid plains, and need but the magic touch of the artesian wellborer's rod to gush forth in fertilising abundance on the thirsty land.[57]

Advertisements. lxvi

ICE! ICE! ICE!

IMMENSE ADVANTAGES OFFERED TO CONSUMERS
BY THE
VICTORIA ICE COMPANY,
LIMITED.

Office: Imperial Buildings, 77 Collins-st. West.
Works: No. 5 Franklin-st. West.

FOR THE PRESENT SEASON, 1880-1881.

This Company delivers Ice in Melbourne and Suburbs on the following favourable terms:—

60 lbs. per week, in six deliveries of 10 lbs. each, for four weeks ... ... 25s.
30 lbs. per week, in three deliveries of 10 lbs. each, for four weeks ... ... 15s.

For any large quantity notice must be left at the office before 10 a.m. on the day of delivery.

TERMS at the Ice Works, Franklin-street West. One Penny per Pound, or One Hundredweight for Six Shillings. Cases, Packing, and Cartage extra.

Special prices given for large quantities.

Orders from the Country for Ice should be forwarded the day before it is required, otherwise the delivery cannot be guaranteed. Large up-country consumers can arrange for a regular supply. No returned cases allowed for.

As the carriage per rail has to be prepaid, orders must be accompanied with cheque for same. Unless special arrangements have been made to the contrary, all sales must be paid for in cash. Country orders to be accompanied by a Post Office order or cheque, covering cost of Ice, case and packing.

Cost of case and packing, 28 lbs., 1s. 6d.; 56 lbs., 2s. 6d.; and 112 lbs., 3s. 6d. Packing of transit cases 9d. per 112 lbs.

Subscribers not receiving Ice regularly are requested to forward a memorandum of the default to the Secretary, as the Directors are desirous of making the deliveries as regular as possible.

C. W. UMPHELBY, *Secretary.*

Plans were already in existence for various irrigation schemes in northern Victoria and utilizing the waters of the Murray and Murrumbidgee rivers.[58] As farming land became more scarce, water was as gold, priming the pumps of optimism.

During the 1860s and 1870s pioneer farmers mirrored the pastoral settlement of earlier decades. They struggled with unsuitable seed and plant varieties, with farming techniques that were adapted only slowly to local conditions, with uncertainties in their marketing procedures and communications. Many of the interim solutions they adopted reflected directly the experience of the pastoralists. At the beginning, farmers too placed great importance on access to virgin land. Like the early squatters, farmers thought of moving farther out when farms seemed crowded or soil exhausted. Like the squatters they paid too little attention to replenishing or improving the soil. Just as wool suffered eventually from too little attention to stock quality and management, so wheat yields began to decline relative to area under cultivation in the 1880s. English seed varieties did not acclimatize satisfactorily. An inquiry into diseases in cereals growing in South Australia in 1868–69 revealed that although *lammas* wheats were most widely used, some fifty-two other varieties were mentioned by the 600 farmers questioned.[59] By then there was already a move towards *purple straw*, a locally culled quick-maturing strain of *red straw* which could be harvested before it was attacked by rust. Rust, which was a problem even in South Australia's relatively dry summers, inhibited the expansion of wheatgrowing in wetter areas. It was not until the end of the century that really suitable wheat varieties were developed by researchers like William Farrer and N. A. Cobb.[60] A minor irony of the movement of farmers into the drier parts of South Australia was that north of Goyder's line the risk of rust was lessened. But in really dry years, like 1888, nothing would grow.

There were comparable problems of finding suitable disease-resistant but reasonably high-yielding varieties in the east coast sugar industry. Highly developed West Indian canes began to fail, and the prospects for the industry, intensified by labour and other technological problems, looked doubtful until sturdy semi-wild strains were introduced from their home in New Guinea. Like wheat, sugar needed technology,

GENERAL VIEW OF IMPLEMENTS AND STOCK ON THE FARM OF MR. JOHN RIGGS, GAWLER PLAINS

especially if growing, cutting, milling, and refining were to escape the plantation system.[61]

Many of the farms established in the movement towards closer settlement in the 1860s were either too small, or undercapitalized, or both. A clear trend in this period was the amalgamation of small farms and the creation of better capitalized, better managed operations. This rationalization conflicted with the ideology of the free selection movement but it allowed crop rotation with fallowing instead of clearing yet more virgin land.[62] It gave better protection against drought, and space for more effective water conservation. It made the use of machinery more economic. Suitable machinery and allied techniques were as important for developing the less tractable acres of South Australia, the Wimmera, the Mallee, the western plains of New South Wales and the Darling Downs of Queensland as the use of suitable varieties of wheat. Names like Ridley, Bull, Mullens and Mackay have become legendary as the inventors of machinery and processes, but many a small town had its inventive blacksmith, its innovative farmer, its own machinery workshop producing pieces custom-built to local specifications.[63] Some of them, unique in design or manufacture, may still be seen as exhibits in local museums or rotting in paddocks and yards.

Wet seasons in the 1870s helped to sustain the transition to farming begun in the 1860s. In New South Wales wheat-farmers abandoned the coastal valleys to more suitable forms of agriculture and moved to the highlands, the northern plains and southern slopes. The good years raised hopes for all that land beckoning beyond the fifteen-inch (37.5 mm) rainfall line since 1867 declared unsuitable for farming by South Australian surveyor-general G. W. Goyder.[64] By the end of the 1870s wheatgrowers were settled so far to the north of Adelaide and west of Port Augusta that the terminus of the Great Northern Railway line at Government Gums was renamed 'Farina'. Facetious suggestions that the next stations on the line should be called 'Bran' and 'Pollard' became irrelevant after more ordinarily dry years returned and it was established that there was indeed some meaning to Goyder's line.[65] Despite enthusiastic adoption of modern theories about tree planting and deep ploughing as a means of attracting and conserving moisture, it gradually became apparent that rain did not necessarily 'follow the plough'.[66]

At Tarcowie, well within Goyder's line, William Stagg, aged eighteen, his father's only help on their 199-acre block, was ploughing in November 1885. 'I have not ploughed it over and above deep as Father does not hold with deep ploughing, though I do,' he wrote in his diary. But the purchase of a second-hand mowing machine for £3 12s 6d and a horse rake for £8 10s enabled them to get their hay off in half the time it usually took with scythes.[67] The next year they acquired a 'stump jumping' plough though William thought it would be better called a 'stonejumper because it is very little better than any plough among roots, but a great deal better among stones'.[68] His story of ploughing, haymaking, and waiting for rain was interspersed with the problems of finding money for the land tax and the interest on their mortgage. Both his sisters, who were younger, were in service in nearby townships, and William himself contributed to the family cash supply by selling wood he cut, or birds he caught, in Tarcowie. Most of his work simply took lots of time, energy, and suitable weather. On 4 August 1886, for example, he wrote,

We began to build a water-closet down by the shed. It may seem strange to some people that we should be here between 10 and 11 years, and had none up, but ours is by no means an isolated case where there is plenty of trees.[69]

For all the hard work involved, farms round Tarcowie were viable if not always profitable, and by the end of the century, William Stagg had acquired his own block, though, with the long drought of the 1890s, he still could not afford the luxury of marriage. By the end of the century, however, the famous Farina was nothing but a dusty, almost deserted railhead, bustling only on rare occasions when mobs of cattle arrived to be carried south. The train came once a fortnight. To the south lay abandoned farms and crumbling chimney stacks. Northwards the land was considered too poor to support further extension of the railway, too dry for horses. Camels were the preferred mode of transport. Yet half the land cultivated for wheat was still in South Australia, and South Australian wheat led wheat export tables. It was shipped directly through railheads at Port Germein, Port Pirie, Port Broughton, Edithburgh (surveyed 1869), Wallaroo, all neat gulf-side towns surveyed and symmetrically laid out according to Adelaide's larger order. Bound mostly for Britain, the wheat was carried like wool in clippers sailing the roaring forties route via Cape Horn. Towards the century's end markets for surplus South Australian and Victorian wheat were also opened in South Africa and in the Pacific (Guam and the Marianas).[70]

Dry summers, interlocking railways and shipping, the ease with which mechanization was applied in the flat landscape were important ingredients in South Australia's preeminence as the farinaceous colony. In Victoria, and to a lesser extent in southern Queensland, railways and mechanization went side by side over the black soil plains. Difficulties encountered in crossing the Great Divide, also the absence until 1889 of a bridge across the Hawkesbury River, hampered the transport of New South Wales wheat. There too, frequent summer rains discouraged mechanical harvesting of damp and flattened wheat. Labour shortages at harvest time, however, were an incentive to mechanize. By the time the limits of land available for agriculture were established, subsistence farming was giving way to mechanization and cash cropping wherever possible. Even the dairy as a sideline on many a mixed farm became an important source of income with the advent of refrigeration.[71]

Though there was a drift away from old ideas about the farm as a self-subsisting economy, the railway station and the local township had never been more significant.

Interior towns or 'townships', as they are at first called, have here as elsewhere much similarity of look. The streets are always rectangular, and usually conveniently wide. The first and most conspicuous edifice is the hotel or 'public' as it is summarily called, and which is not seldom a solid stone or brick edifice. The next is the blacksmith's forge. Then cottages, mostly of wood, straggle like pygmies along the edges of the wide grassy streets. Then a little church or two, which have, at least, a better chance than the cottages of being stone or brick, like the 'public' though of far inferior dimensions. The public school early rears its modest head.[72]

These towns could, and did provide a kind of self-sufficiency which was strengthened by distance and which complicates the assumed dichotomy between urban and rural life during this period. The railway, the postal service, and local ingenuity bridged the gap between the country and the city. When William Stagg required a rifle he bought one by mail order from Adelaide.

Perhaps as a legacy of the popularity of Steele Rudd's 'Dad and Dave' stories, we tend to see the selectors and small farmers of the late nineteenth century as deluded, ignorant, and poverty-stricken.[73] Such judgements are crassly based on technologically produced affluence. They ignore the importance of food production in a world more precarious than our own as well as the value of the ancient knowledge of agriculture to society's collective wisdom and morality. (They also assume a level of affluence and sophistication in the cities that must be regarded as dubious.) Compared with farming conditions in the Old World, those in Australia were harsh but open to change. Clever or knowledgeable farmers survived. Successful farmers needed the skills of smart businessmen. Though farming seemed a link with traditional practice, as did coal-mining, or many urban trades for those who followed them, unhappily it was not. There was no room in Australia for old-fashioned or peasant-like attitudes to agriculture. And while no farm could be considered a model of industrial relations—on family farms, especially, long hours were frequently worked by all—there was little evidence of the servility and deference which characterized relations between farmers and their labourers in the Old World.

Defying expectations of the 1860s that a class of yeoman farmers would unsettle the squatters and provide conservative political ballast, farmers themselves became a more demanding and troublesome group. In many districts they began to organize to considerable advantage, lobbying for railways, or concessions, just like city businessmen.[74] Were they then to be classified as landowners or capitalists? Were they still deserving of sympathy or special treatment? The 'plight' of the selector had been a hardy perennial in parliament and the press, a useful foil for the 'plight' of the workingman. In time that plight turned to sturdy independence, even confident modernity. Yet on the margins of the good earth, in areas of low rainfall, poor transport, and recent settlement, it was remarkable that farmers could survive, let alone produce enough to require storage at railheads. In this lay the root of urban ambivalence towards them, the reluctance to admire, the need to patronize. Farming was fundamentally about survival, not only the survival of the farmer and his family, but increasingly also of city folk who were learning to consume the wealth earned by rural exports. Farming, even precariously, conferred a sense of independence and self-worth which the urban working class was striving to find. Here the farmer and the suburban homeowner had much in common, though when they saw themselves as rural landowner and urban worker, it might seem that they had nothing in common at all.

It was still possible in the 1960s to stand at the intersection of Swanston and Collins Streets in Melbourne and by squinting, see views very like those in the illustrated papers of the

1880s and 1890s. It seems amazing that in a town which was so new and in a community which was still quite small, buildings of such stature, such solidity, and such costliness had been envisaged and built, buildings which must have been far in excess of the requirements of that time, for they showed a magnificent capacity to accommodate the needs of a fairly expansive century.

Melbourne was, though it is no longer so clearly visible to the passerby on Collins Street, the most dramatic example of confident, extravagant public building programmes in the late nineteenth-century Australian colonies. A similar sense of solidity and scale was demonstrated by many of its private citizens in their bulky and ornate mansions and villas. Elsewhere, beyond Melbourne and Victoria, there was also evidence of building for the future. Elaborate parliament houses and town halls, court houses, government office buildings, post offices were thrown up in every major colonial city and large country town. Adelaide's town hall, commenced in 1864, and opened the next year, cost £25 000 to build; Brisbane's, commenced 1865, cost £28 000; Melbourne's commenced in 1867 and opened in 1870, cost £100 000. This town hall included the largest organ in Australia at a cost of £7000. Sydney Town Hall, commenced in 1868 and opened in 1875 cost £80 000 to build.[75] At the same time post offices were built, in Adelaide and Brisbane in 1872, Melbourne in 1867, and Sydney 1874, though some of the decorations and clock towers were not added till the 1880s.[76] As well, ambitious companies, banks especially, built elaborate palaces or temples for their head offices and solid neo-classical double-storied premises appeared in multiples wherever money was changing hands. When contemporaries marvelled at the evidence of progress and civilization that had appeared in such a short time in Melbourne (and to a lesser extent, everywhere settlement warranted it), they were tacitly paying tribute to the value in labour and materials that it had been possible to find and spend.

Even without the extent of investment in buildings for government and private business purposes, the sheer volume of housing built between 1861 and 1900 was enormous. At the beginning of the 1860s, the available housing was overcrowded, and much of it was temporary or of very poor construction. N. G. Butlin has estimated that about one-fifth of the population of the eastern colonies (Queensland, New South Wales, Victoria, and South Australia) of just over one million at that time lived in tents, bark huts, or other temporary dwellings (Queensland had the highest ratio of temporary dwellings, South Australia the lowest). As settlement stabilized, housing became more permanent. In the cities, however, there were fewer rooms than people needing them. By 1891 this situation had so changed that in the same four mainland colonies, a population of 3 003 692 had access to 3 058 000 rooms in permanent houses. By the end of the century there were 100 000 more rooms than there were people.[77]

This newer housing was also of more substantial quality. The average house grew from three rooms in 1861 to five in 1900, and in Sydney or Melbourne, it was more likely now to be of brick or stone.[78] In Adelaide it would almost certainly be of rubble and stone, but in Brisbane weatherboard houses raised on stumps prevailed. With the shift to more substantial and permanent housing went a fall in the number of owner-builders and owner-occupiers. In the 1860s it seems that many temporary homes were 'knocked up' by those who lived in them, though not necessarily on land to which they had legal title. By the end of the century it is thought that between 60 per cent and 70 per cent of houses were rented.[79] The effect of increasing growth and regulation in the building industry was greater uniformity in the standard and design of houses, regardless of whether they were for owner-occupation, sale, or rent.

The owner-builder of the 1860s might cut his own timber, collect his stones, and mix a mud mortar from a nearby creek. But brick houses with timber joinery, plaster ceilings, and iron roofs required outlay on materials, if not always on labour. Building societies which organized savings for the purchase of a ready-built house, as well as some banks and insurance companies which became involved in house and land mortgage business, grew greatly in number and popularity. Their willingness to take risks rose as it became clear that all circumstances in the Australian colonies were tending towards widespread ownership of small parcels of urban real estate, whether for owner-occupation or rent. By far the greater part of the money invested in housing came from local savings. After the 1860s the small savings of ordinary people were earmarked for the eventual purchase of a home.

Taken together, the building industry and the production of building materials employed about one-third of the Victorian male workforce in the 1860s. Although the relative proportions of building and construction to other forms of manufacturing employment declined by the 1890s, they were still providing about a quarter of the jobs for men. Outside Melbourne and Victoria the building industry may have been a little less dynamic, but the volume of work accomplished speaks for its overall importance.[80] Given the proportion of the urban workforce engaged in the building trades and/or saving for or paying off a home, it is no wonder that the building industry collapse of the early 1890s was widely felt.

There were not only public buildings and houses to be constructed. The increasingly complex urban infrastructure also demanded attention—roads, harbours, bridges, water supply, waste disposal, public transport, lighting. Initially interest in these services was mainly selfish. Property owners organized to protect themselves against fire which could wipe out the value of their buildings, so early building regulations were chiefly concerned with fire prevention.[81] Street cleansing had a higher priority than the provision of fresh water, sewerage, or garbage collection. The responsible citizen who could afford to do so took precautions so that his own home was adequately drained, had a good well and enough space for an earth closet and waste disposal. (This may not have been as serious as it sounds in the days of the perpetual kitchen fire, the compost heap, fowl-yard, and pig man, and when tin cans and bottles were almost invariably recycled.) The chief offence usually came from other people's drains and cesspits and whatever overflowed into the streets. The best answer to this was a larger house in a semi-rural suburb like St Kilda or Woollahra where the problems were simply solved by space.

The late nineteenth century, however, saw a growing awareness of the commercial value of much of the waste created by society and industry. A society which had been both too affluent and well-served for space to care what happened to its scraps and cast-offs gradually became conscious of the possibility of turning some of the rubbish to profit and of needing more of the space. A plant was set up by the Yarra River in 1885 to transform wastes from the abattoirs, tanneries, etc., into blood and bone.[82] Formerly much of this matter had simply been washed into the river to rot. Unfortunately similar profitable schemes for dealing with sewerage could not be found, though many were tried.[83] Both Sydney and Melbourne set up government inquiries into drainage and sewerage problems in the 1880s resulting in the early stages of the comprehensive systems we know today. However, a comparison of the colonial capitals as congenial places to live in 1892, found that though it was

with the single exception of Buenos Ayres . . . the largest and most populous city in the Southern hemisphere . . . the financial, artistic, commercial and manufacturing metropolis of Australia . . . well lighted and paved [with] an excellent water supply . . . the great drawback to Melbourne, is the defective drainage and sewerage.

Brisbane had 'good drainage into the river'. Hobart was 'a cheap and agreeable place of residence', well lit and drained, with tramways to the suburbs, while Perth could boast a main street nearly two miles long planted with Cape lilacs and mulberry trees; Adelaide was declared 'the best drained city in Australasia'. In addition it was lit by gas and electricity, well paved and served by tramways.[84]

In 1860, there were perhaps 340 miles of railway in Australia.[85] By 1900 there were 10 566 miles. Even more than building the cities, this massive investment of money, labour, and materials spoke to the spirit of late nineteenth-century Australia. Land was a vital resource because it brought forth wealth and the promise of more. Shelter was essential, but could also easily be left to the initiative of individuals. Railways were not essential. They were visionary. They promised to secure, if not to tame the continent. Both profits and promises were re-invested in railways. Railways were the miracle technology of the 1860s. In the Old World they defied climate. Despite ice, rain, and snow, trains steamed past on iron rails, keeping timetables with novel and awe-inspiring precision. In Australia, railways defied distance as well as climate, spreading a thin web of settlement to places which were otherwise inaccessible or insecure. The shadow of the train against dry grass under a relentless sun, the snorting engine, the plume of steam, the smell of coal in the air were not then romantic reminders of civilization but reassurance that settlement could be sustained. 'The flaunting flag of progress is in the west unfurled' wrote Henry Lawson. 'The mighty bush with iron rails is tethered to the world.'[86] Perhaps only in the USA were railways used on a grander scale proportionate to area and population. Most transport development after 1860 was in railways. There were heroic attempts to utilize the Murray–Darling river system with paddle steamers.[87] Elsewhere, the real hope for sustained settlement or intensive use of resources more than a couple of days by coach or dray from the coast rested with the iron rails.

Symbolic of the hope invested in railway building were the rival schemes floated during the 1870s for a north–south transcontinental line. By then it was possible to envisage a grand design, a real network of railways covering the continent. Neither of the proposed transcontinental lines, one extending the existing line north from Adelaide to the Arafura Sea, the other running through western New South Wales and Queensland linking western lines and carrying them north to the Gulf of Carpentaria, was ever built. Nor were equally grand but slightly more practical suburban circles planned for either Sydney or Melbourne. Though technically viable, and of unlimited potential for attracting future growth, their immediate cost was far beyond any likely return. The sheer size of even modest railway-building projects, their fascinating modernity, their need for labour, and stimulus to all kinds of other services, and the willingness of colonial governments and overseas investors to put money into them, gave railway building glamour such as ordinary road building or river dredging never had.

Before the days of rail, transport facilities just grew, rather like houses, as they were needed. Indeed the badness of roads which passed as coaching routes and the brilliance of coach drivers who negotiated them, repairing damage to their coaches as they went, became legendary.[88] It was more efficient to build railways than trunk roads. The quantity of labour required just for paving relatively short sections of city streets was prodigious. Railways made political sense. As soon as they were built, transport improved dramatically. They multiplied the distances over which perishable foodstuffs could be transported to market. They multiplied the distance over which workers could travel each day, extending the range of suburban housing development. Intercolonial trade, however, continued to prefer coastal shipping lines. Breaks of gauge and missing links in the railway system were inconvenient, especially where bulky goods required trans-shipping. But the railways, with their neat, impersonal carriages, were a boon to ladies who suffered from seasickness, and so different from the undignified jostling of coaches.

Though they failed to deliver the profits governments and overseas investors had hoped for, the railways were a constant source of income to townships and districts through which they passed. Whole towns grew up (and disappeared as quickly) round construction camps as the railway advanced. Difficult tunnels, cuttings, and bridges sometimes took years to build—though the invention of dynamite in 1867 greatly speeded excavation work formerly done by navvies with picks and shovels. At major junctions, and at ends of lines, towns like Junee, Dubbo and Wallangarra owed their existence largely to railway workshops, accommodation for crews and many other workers, most of them well paid, most of them in skilled or secure jobs. In the cities, similarly, whole suburbs lived beside and off railway workshops: Parramatta–Granville, Williamstown–Newport, South Brisbane.

Between 1860 and 1890 railways were the fastest-growing industry in the colonies. Then came manufacturing, with a rate of growth also greater than in the pastoral, building and construction, or distribution industries.[89] These high growth rates, both in railways and in manufacturing, were largely a product of their relatively undeveloped state in 1860. At that time, what manufacturing there was, concentrated on providing materials for building and construction, processing food and drink, and the provision of clothing and footwear. Local demand for machinery to assist in gold and other mining, in agricultural expansion, and in the construction of railways led to the growth of metal manufactures, which were sustained increasingly by the need to repair and replace existing machinery. Throughout the period, manufacturing for the building and construction industries maintained its position as the largest area of activity, though growth in metals and engineering was more dramatic. Processing of food, drink, and tobacco grew in proportion to the population and was more susceptible to variations in the supply of raw materials than to the availability of funds for investment. In the manufacture of clothing and textiles especially, large numbers of women workers meant a low wage structure, and therefore a different kind of significance for these industries in an economic environment devoted to growth, high wages, and modernity.

Except in New South Wales and Tasmania, concentration of industry in the metropolitan areas was marked. The characteristic manufacturing establishment was small, often little more than a shed in a backyard, employing a handful of workers. Some of the complications of regulating factory work in the 1870s and 1880s were a result of its barely visible nature. It was necessary first to define what constituted a factory. Contemporary statistics are erratic and unreliable, an indication in itself of the 'relatively minor place that manufacturing industry held in relation to the total activity of Australia'.[90] Aggregate figures, whether they be N. G. Butlin's 4–5 per cent of GDP in 1861 rising to 10–11 per cent in 1891 or Allan Thompson's 15 per cent of GDP in 1891,[91] still disguise the ephemeral nature of much late nineteenth-century manufacturing activity in Australia. Processing plants lasted only a season. Production runs were short, and there were workshops whose only claim to continuity was that a building existed on the site.

The boasting publications of the second half of the nineteenth century with their many illustrations of large, tidy factories billowing smoke from numerous chimneys evoked an impression of advanced industrialization.[92] So too did the dark sketches of narrow crowded streets, with their barefoot children and untidy lean-tos, but the reality was something less than these hints of comparability with contemporary industrial Britain. For one thing, bare feet were not such a bad thing in most parts of Australia. For another, the streets shown were most commonly inner-city areas whose inhabitants survived in the host of urban service industries and casual trades, while the most gruesome streets in the pre-industrial dockside areas and in the shadow of abattoirs, fellmongeries, and tanneries defied accurate depiction of their filth and stench.

In places like the Hunter Valley of New South Wales where coal, water, and raw materials were all found in convenient proximity, there were conscious attempts at industrialization on a more coherent scale. Newcastle was frequently envisaged as the Liverpool of the south. But even here, activity was fitful, a little reminiscent of eighteenth-century industrialization in the valleys of Shropshire and Derbyshire. Maitland and Morpeth were the really thriving centres, profiting from the conjunction of easily available coal for steam-driven machinery, a rich agricultural hinterland, and water-borne transport. (There was a regular steamship service from Sydney to Morpeth every day; fare in 1873, 20s cabin, 12s 6d steerage, meals included.) Newcastle, though it had coal and transport, was situated clumsily halfway between the Sydney markets and the meat, wheat and wool which required processing, a most reliable and lucrative activity. It was an obvious site for iron foundries, smelting works, chemical plants, railway workshops, shipyards, but all came and went, rarely with much profit, always at the mercy of the limited colonial market with its sporadic demand.[93] To an extent distance created its own market. For a long time, for example, local potteries commanded a market for everyday ware otherwise not worth transport and loss through breakage. They survived, not only at Newcastle, but at Lithgow and Bendigo. Soap manufacturers, on the other hand, failed regularly despite the proximity of their raw materials. Their product was neither fragile nor perishable. In some cases it was cheaper to bring manufacturing to the market as the Arnotts found in moving their biscuit factory down the river from Maitland to Newcastle, and eventually to Sydney. That was bad for employment in the valley, as indeed it was every time an industry closed down or moved, but it worked for Arnotts, which became one of the nineteenth-century manufacturing concerns to survive.[94]

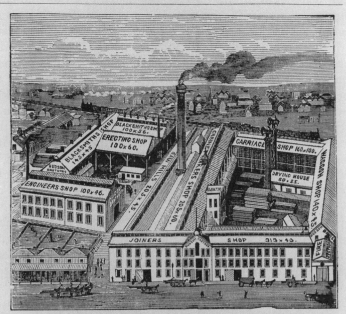

HUDSON BROTHERS,
Timber Merchants, &c.

SAWMILLS, REDFERN,

Manufacturers of every description of Joiners' and Cabinet Work,

ALSO MANUFACTURERS OF RAILWAY CARRIAGES, WAGGONS, COLLIERY PLANT AND TRAMWAY ROLLING STOCK,

Branch Mills, Wharf and Timber Yard, Murray St., Pyrmont,

AND HARDWOOD MILLS AT THE MYALL LAKES.

Sleeping Carriages and every description of Specialties of Railway Rolling Stock manufactured to designs adopted by the various Colonial Governments.

Our Sleeping Carriages are running on the New South Wales Railways, and are highly approved by travellers in this and the neighbouring colonies.

HUDSON BROTHERS.

In Victoria, where the obvious resources for industrialization, water power, coal or iron were lacking, tariffs on many imports were introduced to create a protected environment favourable to local industrial development. The Victorian tariff worked as effectively against manufactures from the other Australian colonies as against Britain. It also seemed irrational at times, as when it worked against the import of special machinery or particular components like calfskin needed by otherwise thriving and self-sufficient shoe manufacturers. Industry and employment were created, as in the manufacture of dynamite from 1876 for the construction industries, but not as much as Victorian publicists imagined.[95]

Approximately 50 per cent of gross domestic capital formation in Australia in the 1860s and 1870s was British in origin. Such high levels of investment, combined with continuous migration, expanding markets for British manufactured goods, the export to Britain of wool, minerals, and meat, ensured close and continuing links with what most Australians at this time knew as 'the mother country'. Yet in relative terms, the Australian colonies were much less significant fields for British investment than the Americas, particularly the USA and some of the South American states.[96] During the late 1870s and 1880s, Australia began to attract more private speculators, especially in mining, whose interest shifted quickly and uncomfortably for the eastern colonies during the 1890s to Western Australia. There the whole cycle of mineral exploration and discovery, subsequent immigration, demand for housing, water supplies, railways, and employment which had been so important in promoting economic growth in the eastern colonies began again.[97]

The leading Australian borrowers in London were the colonial governments, and they raised funds mainly for railways and associated public works. Banks and land mortgage companies borrowed extensively in Scotland (many, indeed, were founded in Scotland and subsequently moved their operations to London or to the other colonies), or outside the London money market, for example, in Liverpool, to invest in land, housing, and pastoral development. Mining companies also borrowed in London, but most of the funds

A GLIMPSE OF SYDNEY FROM DARLINGHURST

for other kinds of activity, such as the establishment of manufacturing enterprises or the building of privately owned utilities like gasworks or tramways came from local savings. Government loan issues were conducted first by Anglo-Australian banks, so that in 1866, the Queensland government found itself practically penniless when the London banking house of Agra and Mastermans, through which a loan had been raised, failed. Later the greater certainties of the Bank of England and the London and Westminster Bank were preferred. Before 1870 and during the early 1870s, government issues were slow to sell, with most investors preferring land and mining stock. Larger issues, greater faith in the reliability of colonial governments, and a tendency to identify all colonies as 'Australian' rather than as individuals made government borrowing easier. Further, a strong Australian lobby, composed largely of semi-retired bankers and politicians, began to promote Australian issues in London with skill and knowledge. Finally, the changing habits of British investors who found they had increasing access to a more easily manipulated stock market helped to explain the immense popularity of all Australian stocks in the 1880s.[98]

In addition to overseas borrowing for public works, colonial governments raised revenue for general purposes from a number of other sources. Chief of these was revenue from the lease and sale of Crown lands, revenue earned by government enterprises, like the post office and telegraph as well as the railways, and that which came in the form of indirect taxes such as customs and excise duties, stamp duties and other taxes on the operation of business and commerce. Further, all colonies introduced some form of probate or succession duty during this period as the easiest form of indirect tax to impose and administer. South Australia was the first colony to introduce a tax on income (1884) followed by Tasmania (1894), New South Wales and Victoria (1895). Only the residents of Queensland and Western Australia were still free of direct income tax at the end of the century.[99] (Income tax almost certainly hastened votes for women. Unless they were taxed, wives would be tax shelters. Unless they voted, the claim 'no taxation without representation' could be raised.[100])

As a source of revenue, land was obviously a diminishing

one in the small colonies of Victoria and Tasmania, and ceasing to be lucrative in South Australia where the poor quality of much land as well as the policy of selling it for farming meant that after the 1870s there were few expectations from that source. Western Australia continued to receive a steady income from land. Queensland's income from land in 1900 was £573 754 8s 9d. Over half of this was from rent of leased runs.[101] A falling away of land income was noticed in both Queensland and New South Wales in the 1880s and gave rise to reviews of leasing arrangements in the hope of raising more revenue from direct sales. In other colonies, land taxes were introduced to bolster land revenue.

Both New South Wales and Victoria derived good incomes from the effective use of their public utilities—in 1886 the income from this source in New South Wales amounted to £3 3s per head of population compared with £2 13s 4d from various forms of taxation, and £1 13s 7d from land. A small colony, with a small population and fewer and less effective public utilities, Tasmania received only 15s 7d per head of population in revenue from these sources, 9s 2d from land, and £2 14s from taxation in the same year.[102]

Collecting income tax was a complicated business, especially in a scattered and remote population. This was, indeed, one reason for the reluctance of Queensland to impose such a tax. By contrast, South Australia's relatively orderly urban and farming communities facilitated the collection of income and land tax. Direct taxation began to produce statistics from which fairly precise calculations of individual income could be made. Formerly what was known of wage rates, investment patterns, probate figures, etc., gave very rough indications. The earliest estimate of private wealth was made by T. A. Coghlan, in a paper at the Australasian Association for the Advancement of Science in Hobart in 1892. Coghlan calculated subsequently that in New South Wales in 1898, the average male wage earner had an income of £98, the average female £36, 'but to this sum should be added the approximate value of board which servants receive'. As he noted, 'there are no data on which could be founded an estimate of the amount of income derived by persons superintending the employment of their own capital'.[103]

However, for some time, colonial statisticians had been

monitoring the aggregate figures on wealth and property-ownership as well as those on wages and prices. They were generally in agreement that their information showed the wealth of individuals in the colonies to have been rising fairly steadily since the 1860s. By the end of the century, private wealth in New South Wales, Victoria, and South Australia was said to average about £260 per head. This was somewhat behind Britain where the comparable figure was computed to be £302, but ahead of France (£252) and the USA (£246) which were the next most wealthy countries according to the same criteria.[104]

The meaning, or lack of meaning, of such statistics can quickly be illuminated by reference to what we know of poverty and inequality in that most wealthy society, Britain, and by the knowledge that Australian statisticians were criticized in Britain for their optimistic interpretations of their own material.[105] Coghlan especially was in danger of reading too much into the averages. Working from probate figures he calculated that half the wealth of New South Wales was in the hands of a mere 2367 persons. Even so, the proportion of the population owning enough property to make a will necessary for probate purposes had risen from 5.7 in every hundred who died in 1865 to 16.2 in every hundred dying in 1900. At the same time, the average value of these estates rose to the mid-1880s, after which it began to fall— the average was £2053 in 1876, £3173 in 1886, and £1959 between 1896 and 1900. Coghlan was surprised to discover that by 1900, 25 per cent of estates being valued for probate were left by women. It had been assumed earlier 'that few women held property worth taking into consideration'.[106] Indeed, research on Victorian probate records suggests that in that colony in 1860, only 1.38 per cent of wills valued for probate were those of women.[107] Besides again emphasizing the problem of reading too much into all these statistics, the rising tide of wills (growing at a faster rate than the population as a whole as the population aged) suggests a society in which people were becoming more settled, a society, even, in which the accumulated wealth of an earlier generation was beginning to show. It suggests a society in which small inheritances were easing the way for an increasing number of people. Such an optimistic view, however, should be balanced by the more sober fact that well over half the male population and almost all the female population left no significant property. As Rubinstein points out, their standard of living might have been quite satisfactory while they were alive, but their widows and orphans were left with nothing when they were gone.[108]

The power to transmit advantage across the generations is one of the keys to a class system. The traditional aristocracy was composed of those whose landholdings, 'blood', and therefore political and social position, all of which usually ensured wealth, were handed on from one generation to the next. The middle classes could hand on little by way of land, title or privilege, but they could teach their offspring their skills, whether as professionals, in trade, or simply in managing money. A business or a professional name handed on from father to son may have been a small inheritance, but it was a beginning.

By British or American standards there were no really great fortunes made in Australia before the end of the nineteenth century, but there were many modest fortunes derived from land ownership or trade. They formed the basis of an ever-widening class of smaller wealth holders. Undoubtedly the short span of Australian history had something to do with this pattern. Perhaps more significant were the relatively larger roles played by overseas investment capital and colonial governments in initiating and underwriting the development of Australian resources. The great entrepreneurs and robber barons lived somewhere else.

More than half the total wealth of New South Wales in the 1890s was made up of the value of land, houses, and improvements.[109] The evidence relating to home building during the course of the second half of the nineteenth century suggests that probably 30 per cent, perhaps 40 per cent, owned their own home or were in the process of paying it off.[110] C. W. Dilke wrote in 1890 of the 'considerable body of small proprietors, and of house-owning workmen [who] have become sturdy supporters of the present order of society'.[111] Though it is suggestive rather than conclusive, such evidence points to a more subtle kind of class distinction than is usually made. Those who were able to translate their good luck, their skills, their management abilities, or the prevailing economic circumstances into secure material possessions, most typically land and housing, drew away, though not yet noticeably, from those who remained at risk in both the housing and employment markets. Increasingly, as such security was transmitted to the next generation, the difference between those who had it, and those who were without, became more obvious. This was not exactly a capitalist class, though certainly the creation and manipulation of capital in a small way was involved. Rather this group had a vested interest in security and stability. It cared little who managed the national economy, or by what principles it was managed, so long as such management was both steady and reliable.

The orthodox economic view of the second half of the nineteenth century is of a period of steady growth followed by collapse in the 1890s.

The Australian economy between 1860 and 1891–2 may be described as one in which rapid growth of output was accompanied by declining external prices, slowly rising domestic prices and money wage rates, relatively stable expansion of employment and incomes and the absence of exchange rate fluctuations or severe balance of payments disequilibrium. The growth of output, employment and incomes coupled with a gradual but discontinuous diversification of activity is explainable primarily in terms of expanding British markets, rapid technological change (including changes of taste) in certain lines in Britain and Australia, heavy inflow of British capital and large investment outlays on 'productive' and 'developmental' assets; broadly, these outlays were divided into two distinct sectors, public and private, in the economy. Other factors, population growth and concentration, a high marginal propensity to import, a low marginal propensity to save and 'exogenous' factors of climatic events and conditions played relatively subsidiary roles.[112]

By the late 1880s 'a tendency to stagnant productivity in leading sectors' along with a 'sectoral disequilibrium' in the demand for capital[113] brought about the 'depression', the 'downturn', the 'slump' of the early 1890s, as well as a tendency to dramatize the contrasts between earlier growth and subsequent decline.

There was no need for more building. Railways had been built everywhere they could conceivably be useful, and in quite a few places where, frankly, they were not. Houses had been built for all the people who could afford them and many who were scratching for the deposit. There were enough government office buildings and other public buildings to supply the requirements of decades to come. All this building had been necessary, obviously. The gradual accumulation and recycling of human effort which building represents had barely begun in Australia in 1860. Furthermore, the activity had created its own momentum, its own prosperity in the demand for labour, materials, equipment, people to service the workers, and others to service them. Then suddenly nothing more was needed.

From mid-1891 depositors began to show their lack of confidence in building societies by withdrawing funds. By mid-1893 most major banks had been forced to suspend payment at some stage. In the six weeks between Easter and the end of the financial year in 1893 twelve banks closed their doors. The crisis was aggravated by other crises on the London money market, by the huge interest bill falling due at mid-year, by falling wool prices, and by a creeping drought. Most of the banks reopened after they had had time to sort out their affairs, but many of the building societies were utterly bankrupt. Savings disappeared. Business and personal failure was widespread. Employment levels reached their lowest point by about 1896. Rabbit trapping, possum snaring, all manner of rural self-sufficiency plans were an answer for some. Surreptitious sewing, laundry work, and miracles of making do kept other families going. There were many moves and broken families, especially in Victoria where the delicate economy felt the loss of confidence most severely.[114]

The records of the Scottish Australian Investment Company, one of many investment and management agencies operating in the pastoral industry in eastern Australia in the late nineteenth century, show that during the 1860s and 1870s they were returning 20 or 25 per cent profit on their invested capital.[115] The balance sheets and annual statements of innumerable other English and Scottish firms in pastoralism, mining, or general banking and agency business in Australia tell a similar story.[116] Contemporary estimates, which were probably better informed about what went unrecorded than modern calculations based purely on available statistics, show overseas interest and dividend payments of about £33.5 million in the quinquennium 1881–85, rising to £50.1 million 1886–90, £63.7 million, 1890–95, and falling to £60.6 million 1896–1900.[117] That so much was exported and that it was not begrudged is best explained by the fact that its real extent was not understood. Australian investors, of course, were doing equally well, or almost, and regarded the extent of British investment as a guarantee of their own prospects. They may even have believed that British investors were carrying a heavier share of the risk in the least likely ventures. This may not have been so. One of the many pieces of interesting information revealed by the 1915 Census of Private Wealth in Australia was that there were then 3120 non-residents in receipt of income to a total of £1 073 348 from Australia. Not only were there more women than men (shades of the Scottish Widows and Orphans Investment Fund), but the average income of female non-residents was £409 compared with £256 for male non-residents. The comparable averages for the resident population were £48 for a woman and £146 for a man. The total income of 2 191 945 residents was £240 163 204—a third of the income derived in Australia went to people outside the country in 1915.[118]

Investment created work, of course, and distance ensured a limited labour supply for all of this period. So wages remained at a satisfactory and fairly steady level in most trades until the 1890s. But because of the pattern of investment, increasingly the work was government work, government buildings, railway workshops, all the thousands of supervisory and administrative jobs which resulted from government borrowing and government developmental projects in support of private enterprise.

Before the 1890s there had been seasonal and local fluctuations in the availability of work, some on quite a large scale, as for example, when a major piece of railway construction was completed. In Queensland in 1866 navvies brought to the colony by the contractors Peto, Brassey and Betts, were paid off with no prospect of further work at the end of construction on the Ipswich–Toowoomba railway. They marched on the government store in Brisbane demanding 'bread or work' and stormed the building. Instead of either bread or work, they were given free passages to the northern goldfields—neither the first nor last time that this most convenient form of government relief was deployed. Large-scale unemployment occurred regularly towards the end of the shearing season, though gradually a system evolved whereby shearers followed regular routes from shed to shed, and sometimes to New Zealand, thus extending the season to its utmost limit. Small selectors surviving from year to year on the income from a burst of shearing mopped up work nearby. Knowledge of seasonal labour patterns evolved and was effectively handed on and developed. One of the strengths of the union movement in the bush was its capacity for disseminating information about the availability of work. The union informed its members where the work was and what kind of work to expect. Thus the efficiency of the available labour force was maximized, its mobility enhanced by communication and organization. Wasteful searching for work was minimized; labour was available when and where it was needed; wages and conditions were kept steady and under surveillance. Rural employers, in the pastoral industry in particular, benefited from the unionization of the itinerant workforce, though, as the strikes in the early 1890s showed, they also feared the potential strength of the unions and their own loss of bargaining power should union membership become universal.

Instability and insecurity were typical of most jobs in processing, building and construction in the towns as well. Unionization was seen as a means of regulating the level of employment in a trade or industry, and also the flow of new workers into it. Sustained levels of skill, experience, efficiency, and organization to prevent oversupply of labour meant high wages. In Victoria, the success of the tariff in creating and expanding markets for protected goods encouraged the government to enter the field of wage regulation as well, through the establishment in 1896 of wages boards. A tariff system keeping out unwanted imported goods was widely reinforced by the exclusion of unwanted workers. Non-white labour was easily barred from union membership or by so regulating employment, especially of Chinese in the furniture-making trades, that it was not worth an employer's time to take them on. Sex segregation further restricted the deployment of labour. So did active union involvement in the regulation of apprenticeships, which were limited with a view to controlling the number and quality of trained workers in each trade.

In comparison with those industries involved in production, construction and processing, the service industries were the least well paid or organized, though they were also least subject to seasonal and regional variation in availability of work. Cooks, laundresses, and housemaids were always in demand, as were farmhands, stablemen, and wharf labourers. Service industries and professions connected with banking, trade, and government, expanded to match the growth in economic activity. There was a notable expansion of white-collar jobs offering security and promotion prospects, though only at the cost of utter devotion. Clerks were needed in banking, the law, accountancy, insurance, retailing, agency work and, of course, government offices. Here too, distance played a part in creating and protecting jobs, since an ordinary consignment of wool, for example, might pass through several representatives' hands before reaching Sydney or Melbourne, which were themselves only staging posts on the way to London. The same process in reverse was required to bring marble fireplaces from Italy to Western District homesteads, or magazines and dresses from London to Rachel Henning on Marlborough station in central Queensland, or consignments of pianos from Germany to Paling's music warehouse in Sydney. Banks, agencies, and government departments all found it necessary to maintain branch offices in many small towns. In a more densely settled country they would have been unnecessary or done twice the business.

The relatively large numbers of people engaged in service and clerical occupations (by 1891, perhaps one in four of the workforce) undoubtedly had the effect of demystifying or democratizing the management of money and the processes of administration. In 1860 clerical workers were a privileged group living close to the bourgeoisie. By 1900 they, along with all the others who depended on their education and good manners for their jobs were more respectable but not necessarily as well paid as the working classes. Through them, knowledge of the ways of capitalism became commonplace. Clerks in banking and insurance were not intimidated by the workings of these institutions or overawed in their precincts. Domestic servants took their well-learned notions of respectability and polite behaviour back to their own homes or to the homes of their husbands. Among the skilled working classes petty capitalism was readily accepted. High or steady wages made saving possible. From mid-century, savings banks sponsored by the government through the post office, friendly societies and life insurance companies had become part of the ambitious working-class culture in Australia.

Regular saving and, to a lesser extent, borrowing became common. By the turn of the century, three out of every ten people had savings accounts. The average value of an account was £32, to a total of over £30 million held in savings deposits.[119] Life insurance, membership of friendly and building societies were popular destinations for savings, with homeownership the main object. By the 1890s 30 to 40 per cent of houses were in the process of being purchased over a long period by the families who lived in them. These people were among the least seriously affected by the Depression. As there were more houses than people able to buy them, there was little point in hounding those who were slow with their repayments. At the same time, homeowners were more likely to keep their jobs, because of the kinds of people they were, and because they had so much to lose. Meanwhile secure occupancy of a house and garden brought other advantages, a fowl-run, a vegetable garden, fruit trees, space for a small business, like the laundry work undertaken by Katharine Susannah Prichard's mother, or the freedom to take in a lodger.[120] Those who suffered most in the collapse of the housing market were people saving the deposit for a home with one of the defaulting building societies or banks, speculators who had overstretched their commitments, and all whose jobs depended on continuing buoyancy in the investment industry, from builders' labourers to shop-assistants.

Between 1860 and 1900 concepts of the nature of work and the relationship between hours worked and wages received changed greatly. In traditional and rural occupations, a high proportion of remuneration for labour or service was still given in board or keep or some other form of exchange, for example, sheep or cattle which a stockman received in lieu of wages, the proportion of a crop allowed to a share-farmer, cream from the milking or the drippings from the kitchen, traditional perquisites of the dairywoman and the cook. Here a precise cash award related to the number of hours worked was scarcely relevant.

The early unions began to keep records of hours, wages, conditions and levels of employment for the benefit of their members. This led slowly to greater comparability from one workplace to another, and eventually to a desire for systematization and regulation. But in many jobs, for most of this period, it was common for workers doing the same job to be receiving different rates of pay based on the employers' estimate of their worth, or because of their length of service, loyalty, family responsibilities and so on. In the 1860s, the concept of the eight-hour working day—eight hours on six days of the week, or a forty-eight hour week arranged some other way—seemed a humane though ambitious goal for those still working seventy hours or more. As industries moved towards this goal, not to be working at least forty-eight hours came to be seen as less-than-full employment. To be unable to work forty-eight hours was to risk unemployment altogether. While unskilled, seasonal, or casual labour were still widely available, unemployment had only a temporary meaning. There were always people out of work, on the move in search of work, or who worked casually as jobs turned up. Unemployment in New South Wales and Victoria in 1891 just before the bank crashes was calculated at about 7.5 per cent. It rose to nearly 11 per cent in 1896, and fell by 1901 to 6.5 per cent.[121] Australia-wide unemployment at the end of the century was still 4.8 per cent of adult breadwinners. Coghlan thought this needed explanation amid the general return to pre-Depression conditions of wealth and prosperity. It had become clear through the preceding decade that

> an increasing ratio of wealth accumulation and an increasing distribution are consistent with a dearth of employment amongst a certain class of the community. It is a sound sentiment that a condition of regular employment of the whole population willing to work is more conducive to real prosperity than a condition, which, while affording increased opportunities of saving to a large proportion of the community, leaves a not inconsiderable section uncertain as to their means of support during a large part of the year. The

condition of affairs to which allusion has been made is one of the inconsistencies of modern progress by no means peculiar to Australia, and to remedy it is perhaps the chief aim of modern social legislation.[122]

Before the 1890s Australian society set high standards for individual self-sufficiency, at least for men, and took a fairly harsh view of those who failed to meet them. The Depression had a very unsettling effect on the belief in individualism, exposing as it did the pathetic vulnerability of savings or lifelong reputations for steadiness and hard work. Because of the greater stress on opportunity in the colonies, there was little sympathy for failure through lack of application. Only South Australia and Western Australia had state-subsidized benevolent asylums. In all other colonies, poverty was punished by the way in which it was relieved. Only fatherless families, chronic invalids, or women for whom insurance against sickness or provision for unemployment was impossible, could expect sympathy if they became destitute. In this group poverty was both expected and tolerated. Women, after all, were known to be weak, dependent, and not entirely to blame for their plight, on a par with the victims of industrial accidents or chronic illness, but even they were expected to work to contribute to their own support.[123]

'I think this is a frightful country, with nothing to do but look after these tiresome sheep and cattle in good years, and to stand by and see them die in bad ones', said Miss Possie Barker, part-Aboriginal daughter of the owner of Boree station.

'Still it's a good country to make money in,' I said.

'Perhaps it is,' Possie replied. 'I don't care much about money. You see, people have to wait so long before they get any. What's the use of having money when you're old.'

Possie didn't live to be old. It would not have done in 1891 for Jesse Claythorpe, the hero of Rolf Boldrewood's *A Sydney-Side Saxon*, to marry her. But Jesse, the poor English immigrant, made money, and in that he was not unusual. Judged by such a simple measure as the amount of food consumed in the colonies, people at the beginning of the last decade of the nineteenth century were 'remarkably

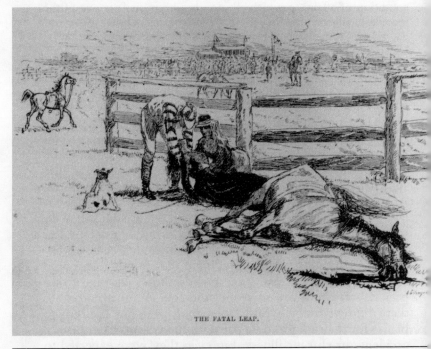

The death of Possie Parker, an illustration from the original serialized version in The Centennial Magazine, *1899*

prosperous'.[124] Coghlan calculated that the people of New South Wales in 1891 consumed more energy in the form of food than anyone else in the world. In fact, they consumed about twice as much on average as was thought necessary for the amount of work they did. 'Though the excess may be looked upon as waste,' wrote Coghlan, 'it is none the less evidence of the wealth of the people whose circumstances permit them to indulge in it.'[125] As early as 1860 Archbishop Polding thought 'worldly prosperity appears to have produced its mischievous effects with peculiar rapidity and certainty'. He was concerned that Catholics were already spending more on their homes and their families than on their church.[126] Another churchman, Congregationalist R. W. Dale, thought that there was not enough want in Australia when he visited in 1889 to keep church people adequately employed in 'good works'. The first volume of economic

Table 1.1: Principal articles of diet required per head, 1891 and 1901

	1891	1901
Flour	260.0 lb	238.2 lb
Oatmeal	4.2	7.0
Rice	11.8	9.7
Meat	291.5	297.2
Beef	177	166.5
Mutton	105	118.8
Pork, etc.	9.5	11.9
Potatoes	206.5	197.7
Sugar	93.5	107.8
Butter	16.7	19.6
Cheese	5.2	3.7
Tea	7.8	7.9
Coffee	11.5 oz	8.0 oz
Cocoa and chocolate	none quoted	8.7 oz

Sources: W and P, 1892, p. 837, *W and P*, 1900–1, p. 765

theory written in the Australian colonies (by W. E. Hearn, professor of history and political economy at the University of Melbourne) was called *Plutology, or the Theory of the Efforts to satisfy Human Wants*. It was a study of the theory of wealth and consumption.[128] Though it made no particular reference to Australia, one wonders whether Hearn would have been moved to write it had he remained in Ireland. Another economic optimist, H. Mortimer Franklyn, noted with concern in 1887 the growing Australian propensity for conspicuous consumption and sheer waste and also a fondness for gold, jewellery and diamonds.[129] By the end of the century, however, the colonies had fallen well behind Belgium, Germany, and Switzerland in the table of consumption, and Denmark, Sweden, and France were pressing close behind.[130]

Some interesting changes had occurred in the pattern of food consumption between the beginning of the 1890s (the earliest available statistics) and their end. Table 1.1 suggests a decline in the consumption of cereals over the decade and a rise in the consumption of sweet things, including cocoa and chocolate, which evidently had not been thought worth inclusion in the calculations in 1892. It should also be noted that although no fruit or vegetables other than potatoes were included in these tables, total expenditure on fruit and vegetables was a little less than on bread in 1891 and somewhat more than on bread in 1901. (£2 693 000 compared to £2 781 400 in 1891; £3 155 000 compared to £2 561 000 in 1901.) These figures are difficult to interpret, partly because they take no account of home-grown vegetables, partly because prices reflect seasonal variations. Salt consumption was specially noted by the statistician as excessive, perhaps 20 pounds per head per annum used in every home, although much of this it was hoped went into meat preserving.

There were some notable variations in consumption patterns from colony to colony. Queenslanders were the greatest meat eaters in 1891. Western Australians seemed to have an unusual weakness for sweet coffee and tea, also rice, though the likely explanation is that Aboriginal workers received rice and sugar and little else as food handouts. Tasmanians ate the most potatoes. By 1901 Western Australia had emerged as the most prosperous colony judged simply by its eating habits, consuming 443.9 pounds of meat per head as well as more wheat, rice, tea and coffee than anywhere else. Queensland had taken the lead in sugar consumption, and Victoria had become the colony where the least meat was consumed (209.3 lbs per head). This may have been a consequence of the Depression. It was certainly related to the loss of so many male meat eaters to the west.

If the workers thought they were working harder at the end of the century than they had a decade earlier to maintain their purchasing power and standard of living, they were right. There was some bitterness, a sense of betrayal at the loss of optimism and opportunity. Tension between capital and labour became more evident. The search for economic security intensified. Both employers and employees accepted the idea of arbitration because it protected them against loss of income during strikes. Governments were expected to protect investors against risk. Workers preferred fixed wages to incentive, bonus or profit-sharing schemes.

Despite her sympathy for the working class and wish to see them living easier lives, Beatrice Webb, on her visit to the colonies in 1898, thought 'a rather gross materialism' was one of the worst characteristics of the Australian people.[131] More ominous for the future was an observation made by Alfred Deakin in one of his notebooks in 1892:

It is depressing to note how little real love of literature, art, or ideas has been fostered in our seasons of plenty among the well-paid and reasonably leisured artisans and business people generally. Selfishness and shams, cant and materialism rule us, up and down and through and through.[132]

The most common question asked about the proposal for a federation of the colonies was 'what shall we get out of it?' Some of the answers were simple: a consolidated approach to customs and tariffs, a more coherent and impressive image in world money markets. This last had been urged for decades by hard-headed brokers and bankers. But who would own assets like railways? Who would take responsibility for the unequal debts and liabilities of the colonies? With the utmost ingenuity formulae were devised to ensure that no colony lost materially by federation. Though 'Commonwealth of Australia' invoked a unified perception of our 'geographical dowry second to none the world over',[133] there was little in the Constitution itself to suggest positive intentions to share resources.

2
BELIEF

THE MENTAL WORLD of late nineteenth-century Australia rested on two assumptions neither of which was usually subject to examination. One was the idea of progress, so skilfully delineated in its British context by J. B. Bury in 1920, 'an idea so eminently colonial', according to William Westgarth who described the colonies as 'a sort of apotheosis of progress'.[1] The second assumption was that civilization was basically British. There were plenty who did not like Britain, but they did not envisage a world in which Britain was not a powerful presence, nor British values the standard of behaviour. The very existence of the colonies was evidence of progress and the power of British civilization.

These beliefs were sustained and encouraged in the decades after gold by the marvellous growth of the cities, Melbourne in particular, and the spectacular impact of technology on that growth. Cable trams and railway systems, gas-lit streets and public buildings, hydraulic lifts, telephones and telegraphs were built into the urban fabric. In 1879, and again in 1880, first Sydney, then Melbourne, were venues for a World's Fair or International Exhibition, the ninth and tenth in a series descended from the great Crystal Palace Exhibition in London in 1851. In Sydney 1 117 536 people came to see 9345 exhibits; in Melbourne there were 12 792 exhibits and attendance was 1 330 279.[2] Though these were small in comparison with previous exhibitions in Paris, Vienna, and Philadelphia, that a World's Fair could be held in Australia at all confirmed the progress the colonies were making. Self-congratulation and self-advertisement reached new heights in 1888, however, when Melbourne staged another exhibition to mark the centennial of white settlement. The Victorian premier, Duncan Gillies, promised 'a public occasion which not only sums up the past, but also prognosticates the future of Australian achievement'.[3]

Summing up the past was already an accepted method of extending and elaborating the successes of white settlement while at the same time pushing the unsavoury aspects of the story into a progressive perspective. On 8 May 1885 the Historical Society of Australasia held its first meeting with David Blair as president and decided that its aim was to study 'the authentic history of the British Empire in the South'.[4] An analysis of the histories written in Australia between 1860 and 1888 reveals at least twenty-eight accounts, including half a dozen or more which were published in conjunction with the centennial celebrations. In addition there were many volumes planned for 1888, but late in appearing, also less ambitious ephemera, souvenirs, etc. As critics, H. G. Turner and Alexander Sutherland, commented:

> The period of the Australian Centenary was responsible for the publication of a large number of books of a professedly historical character, but they contain little original matter, many of them appropriating hundreds of pages of previous volumes without the faintest trace of acknowledgement.[5]

Optimism, faith in the future, enthusiasm for economic and political opportunities and personal achievement were also characteristic of most of the literature written for potential immigrants, which in practice meant most of the memoir and travel books as well as the more obvious handbooks and guides. When occasionally a critical account did appear—like Anthony Trollope's not always flattering description of his visit, in part to see his not-very-successful squatter-son— nowhere was the outrage greater than in the colonial press.[6]

Exhaustion following the excesses of 1888 and uncertainty arising from the economic and labour troubles of the 1890s brought a brief pause in the chronicles of progress. However, the idea of federation and the prospect of a 'nation for a continent and a continent for a nation', gave fresh inspiration. 'She is not yet', sighed poet James Brunton Stephens in 1877. Six years later he was urging the 'Sisters Seven' (for New Zealand was still then assumed to be part of 'Australasia') to go forth to meet their 'fair Ideal'. And in 1900:

We cried, 'How long!' We sighed, 'Not yet.'
And still with faces dawnward set, 'Prepare the way,' said each to each.[7]

As the native-born population became statistically dominant, there was new interest in the future of 'the Australian type'. The knowledge that the Australian population had been self-selected since mid-century by the act of immigration gained significance as the convict generations receded. Hopes for social progress built into the plan of South Australian society emerged in the other colonies and were elaborated under the influence of Charles Darwin's message about natural selection and survival of the fittest. It was believed that the 'unfit', whether physically or morally, would have difficulty in reproducing. They would not find mates or enter stable marriages. Thus their types would wither away. Those unfortunate children they did produce could be institutionalized, or better still, placed with good families whose influence and training might overcome defects of birth. Such arguments at least balanced the idea that a higher than average capacity for roguery and social irresponsibility had evolved through natural selection from the original convict settlers.[8]

The symbols of British civilization were more specific. The union jack flew from flag-poles. 'God Save the Queen' was sung on official and patriotic occasions. Her Majesty's mails carried her crest and likeness into the narrowest streets and remotest settlements. Though it was possible to scorn the symbols, the Queen's English was hard to avoid either as the medium of formal communication or informal exchange. Behaviour began with language and was defined in basically British ways. Though by the end of the century 'typically Australian' social attitudes and forms of behaviour seemed to be emerging, many of these were more British than British. That typical Australian phrase, 'fair dinkum', for example, first recorded in the 1880s, is completely English, originating as Lincolnshire dialect, 'dinkum' meaning work.[9]

The dark side of this faith in progress and British civilization was fear of the primitive or the elemental, whether in nature or in human beings. The Australian bush, the natural environment, was being tamed, not only by sheep and settlement, but by the workings of the English language and the romantic imagination. Flora and fauna were tamed as apple, honeysuckle, oak, emu-wren, kangaroo-rat, or native dog. Systematic classification by botanists, geologists, chemists, and zoologists revealed the simple constituents of a landscape which once occasioned fear and frustration. Ghosts, spirits, supernatural creatures were all banished by rationality before they could be known or remembered. Katie Langloh Parker's *Australian Legendary Tales* were dismissed as 'the prattlings of our Australia's children', an Aboriginal equivalent of Rudyard Kipling's *Jungle Book*, or fairy-stories deemed suitable mainly for children.[10] A landscape richly peopled by spirits disappeared almost before their presence was known. The recent past, redolent with tragedy, guilt and revenge, began to slip into sentimentality ('the ghost you may hear as you pass by the billabong') with scarcely the rattle of a chain.

The Aborigines had become less troubling than the climate's unpredictability. But in their role as primitive human beings, they excited more interest and sympathy than as displaced and diseased victims of progress. The study of anthropology as it emerged in the second half of the nineteenth century cast the Australian Aborigines as the archetype of a primitive society. Christian missionaries might pity them as people living in ignorance of the Word of God and therefore in need of salvation, but anthropology turned them into living fossils. Even so, the discovery of more colourful and incomprehensibly primitive tribes in New Guinea too quickly relegated the Aborigines to collections of bones and artefacts gathering dust in museums.

The most disturbing manifestations of primitive or elemental behaviour, however, were found within the skins of 'civilized' man, sexual needs to be fully acknowledged before the century was out, and an urge to violence hitherto controlled by the combination of law, religion, and exhausting labour. Late nineteenth-century Australian society, with its limitless desire for growth and its attachment to ideals of British civilization, posed new difficulties for the understanding and control of those primitive instincts.

Anxiety about the quality of the population, the purity of the race, the perfect fragility of civilization were not unique to Australia. Thomas Malthus's exposition of the relationship between war and want, Charles Darwin's devastating vision of the human as animal, Benjamin Disraeli's relentless advocacy of the civilizing mission of British imperialism defined a late Victorian pride in British civilization and the dread (or dream) of undermining it. In Australia the paradox of the primitive in a civilized setting was sharply experienced. On 'Kanga Creek' (Sparkes Creek near Scone, NSW) the young Havelock Ellis became so intensely aware of his humanity, his sexuality and his loneliness that he was driven to devote the rest of his life to understanding them.[11] William Howitt went to the heart of Australia to search for the remains of Burke and Wills and came back determined to understand the qualities by which the Aborigines survived.[12] At Ooldea waterhole, Daisy Bates smoothed on her gloves, raised her parasol, and stepped out of her tent to visit 'her people'.[13] The Australian continent seemed to be crying out for people, yet the trend in civilized society was toward repression or management of the birth rate. The circumstances of Australian settlement initially rationed opportunities for sexual activity, but when at last there were enough women to permit 'normal' sexual behaviour, the sexual instinct, it seemed, had become too civilized.

Life on the frontiers of settlement contained, even glorified the urge to violence. While there were still rewards to be had, the work ethic soaked up latent energy. More was consumed by the sporting life. Work was thought to have an 'ennobling' effect. It produced self-respect, a sense of responsibility, a healthy mind in a healthy body. 'Nowhere', wrote B. R. Wise,

> whether in public or private affairs, does the individual count for so much as in Australia. There is no helpless fluttering against the iron bars of class or tradition. Every stroke of work tells. A man can use his strength in Australia, whether it be strength of muscle or of brain. The daily victory over the forces of Nature in the material world gives confidence in other directions.[14]

Such faith in the value of work was at odds with faith in the capacity of technology to lighten the burden of human labour. Contemporary definitions of what was good or right rested on the premise of hard work. When a man could no longer use his strength, his alternatives were limited and dulling. A man was his work. Masculinity was defined by vigour and energy. Idleness suggested femininity and weakness. In this atmosphere of restless energy, women learned to endure the frustrations of idleness, though at some cost to their health and sense of self-respect.[15]

It might be thought that a belief in God should be added to the idea of progress and the importance of British civilization as one of the basic assumptions of Australian society in the second half of the nineteenth century. However, this would overlook changing attitudes towards Christianity. It was not that people ceased to believe in God, or even that there was a marked decline in formal religious observance. Rather, the idea of God began to change. Individualism began to appear in both religious belief and practice.

Perhaps only two people in every hundred in the Australian colonies were non-Christian in 1860, not counting the Aborigines. Among them the Chinese predominated. Another two in every hundred would or could not be specific about their religious position. Most of the non-Christian groups had a spiritual, religious, or cultural home, somewhere they turned for inspiration or reassurance. The Chinese and the Jews were also able to sustain places of worship. From the early 1880s there was even a mosque in Marree. For the Aborigines, however, Australia was their home, their only source of spiritual life. Since the early days of white settlement, missionaries and others had sought to penetrate the belief systems of the Aborigines, mainly in the hope of discovering keys to an easier and more efficacious conversion to Christianity. By 1860 understanding and knowledge of the Aboriginal spirit world, such as it was, was closely related to the models and perceptions provided by Christianity. Did the Aborigines believe in a supreme being, in life after death? Did they have concepts of sin and redemption? For their part, the Aborigines attempted to incorporate the white men into their system of spirits as departed relations or visitors from the dreamtime.

During the second half of the nineteenth century two distinct intellectual movements had the effect of further separating the two worlds of black and white religious belief. In the southern parts of the continent, orthodox missionary work, which had become dominated by evangelical attitudes, increasingly identified with (and was disheartened by) the manifest record of failure at conversion or even 'civilization'. On the northward-moving frontier, the same evangelical or muscular version of Christianity espoused by white settlers showed a diminishing level of tolerance for the elusive quality of Aboriginal belief. There was loss of patience with the slow assimilation of the Aborigines to Christian concepts of property, work, and morality. Meanwhile, among scholars and more civilized folk in the southern cities, the conjunction of the new atheism with anthropology began to devalue all forms of supernatural belief. Further, pioneering anthropological studies of religion such as James Frazer's *The Golden Bough* set the framework for subsequent research in religious belief and behaviour in a compellingly Eurocentric manner.[16] By the time Baldwin Spencer and Francis Gillen were ready to publish their accounts of central Australian Aboriginal culture, the Aboriginal spirit world was already defined as mere magic, an example of totemism. Spencer and Gillen's picture of the Aborigines stressed their Stone Age primitiveness, their savagery.

> We had been living amongst and witnessing the daily life of savages who were yet in the Stone Age, and whose manners of life, customs and beliefs were akin to those of the early ancestors of mankind—just as the quaint, egg-laying and pouched mammals on which they fed were akin to, in fact the surviving relics of, ancient groups of primitive animals which have elsewhere been replaced by higher forms![17]

The anthropologists claimed as 'fully initiated members of the Arunta Tribe' to have witnessed the 'most sacred customs and beliefs', yet they hesitated to use the term 'ceremonies' for what they had seen, for

> though the native ceremonies reveal to a certain extent, what has been described as an 'elaborate ritual', they are eminently crude and savage. They are performed by naked, howling savages, who have

no permanent abodes, no clothing, no knowledge of any implements, save those fashioned out of wood, bone, or stone, no idea whatever of the cultivation of crops, or of the laying-in of a supply of food to tide over hard times, and no words for any number beyond three or four.[18]

If this was not a completely effective appeal to the materialist sensibilities of the likely reader,

the way in which they draw and use their own blood, smearing it over one anothers' faces and bodies as a gum with which to fix the colour down [in the] simple but often very decorative designs drawn on the bodies of the performers[19]

was bound to impress those who were struggling with their own primitive desires or who had already rejected the Christian blood sacrifice as crude and barbarous.

Appropriate comparisons, between, for example, the significance of ritual objects, the 'churinga' of the Aborigines or their ceremonial use, and the ritual objects of Christianity or Judaism, seemed blasphemous. Nor would they appeal to a Protestant ascendancy striving as it was still to stand apart from such superstitious practices. To compare Aboriginal male initiates with the traditional Catholic priesthood would be offensive or aggressively sectarian. And while the 'Dreaming' and the spirit world of the ancestors were not incompatible with concepts of heaven and the resurrection, the symbolism of the Eucharist had become so refined that comparison with the ritual significance of drinking blood and eating flesh was unthinkable.

'In the beginning', said *Genesis*, 'God created the heaven and the earth.' Aboriginal belief in the origins of earth and life differed from those of their invaders mainly in the more humble role they saw for themselves. They did not assume they had 'dominion over the fish of the sea, and over the fowl of the air, and over every living thing that moveth upon the earth'. They were just as inclined to regard inanimate objects as endowed with spirit life as animate ones, and far from taking their dominion for granted, devoted much time and energy to increase and fertility ceremonies. This seemed quaint to a society which agonized over whether it was sacrilegious (or efficacious) to pray for rain, but put its faith in the slogan 'rain follows the plough', and in superphosphate to restore ignorantly farmed or exhausted land.

It should be possible to outline the basic elements of Christian belief which dominated the Australian colonies in the 1860s, but since there were so many attempts in the succeeding decades to define 'common Christianity' for use in state-run schools, and so much acrimonious dissension, it is perhaps unwise. As those who thought the Bible, the Ten Commandments, the Apostles' Creed and the Lord's Prayer formed a basis for instruction in Christianity discovered, such simplicity produced nothing but scorn. Two facts, however, were very influential on the religious belief of the majority of the population nominally Christian. Most were able to read, and did read the Bible for themselves. What they made of it can be judged from the proliferation of sects among the most avid of the Bible-readers and from the drift to non-adherence or free thought. Secondly, material security changed many of the questions about human existence formerly answered by Christianity, though it did not alter questions about death. So the belief in heaven and hell and the possibility of a life after death persisted where the Ten Commandments had been forgotten and the Creed had become a meaningless gabble. Heaven was losing its attraction as the only place where there was no want or fear and where all men were equal, but it was gaining, logically enough, as the scene of family reunions, while wives, husbands, and children still died young. The gospel vision of Armageddon when the last trumpet would sound and the dead would awake and rise from their graves still had literal meaning. It was not until 1895 that the first cremation could be arranged (when the body of Mrs Henniker was burned on a pyre on the beach near Brighton, Melbourne). Much of the resistance to cremation came from a belief that it was safer to await the day of resurrection in a proper cemetery, with or without garlanded urns and marble headstones.

Tension between traditional religious belief and the rising expectation that science or rational investigation could solve the problems of mankind dominated the conscious belief of the late nineteenth century. The Christian churches perceived it in the apparent decline in the numbers of worshippers taking their places regularly and the subsequent diminution of the church's moral authority. They tried to combat this decline. Naïve victories were won when alternate Sunday activities were banned. But frequent exhortations to return to the simple and innocent ways of an earlier age, when the churches were powerful and the people poorer, sounded hollow, especially when the unfortunates of society, the drunkards, the shiftless, the recalcitrant, the foolish were also exhorted to give up their wicked ways, work hard, conform, and be prosperous. The challenge of affluence, itself a direct consequence of the successful application of reason and science, produced a new willingness to search for moral values appropriate to the changing economic and social conditions. But the churches clung fast to their old moral values, so that, for example, the practice of equality began to surpass their teaching on the subject. Their response to this as to other forms of challenge was a sectarian one. Nonconformity, itself a product of earlier attempts to modernize the church, came closest to an understanding of what was needed to maintain a balance between body and soul in the late nineteenth-century world though its emphasis on an individual and personal kind of religion made it appear weak in contrast with the authority of Catholicism, or the traditionalism of Anglicanism. In so far as these strengths could command numbers, they were still forces to be reckoned with, especially in alliance with the forces of democracy. Some such hope led the Catholic church in particular to concentrate on numbers at the expense of theology, notably in its widespread but poorly equipped education system. In the competition for believers, the strongest animosities were not between those who believed and those who did not, but between the different Christian denominations. Their common plight did not produce a united front or lead them to examine those aspects of their problems that were shared because they were specific to Australia.

'Australia', Mary McKillop told the Pope, 'is in every sense a dangerous place for Catholics', and she went on to spell out the dangers—mixed marriages, 'money, and the comfort it brings', the ability of the state to provide both education and charity along secular lines. The few initiatives devised to meet Australian circumstances raised little interest in faraway Lambeth or Rome. 'It is an Australian who writes this', Sister McKillop added in explanation and apology.[20] Most church leaders were not Australian during this period. They were trained elsewhere and best informed about larger problems confronting the churches, often therefore misperceiving the particular problems of organized religion in Australia. Thus too the church acted as one of the most important institutions maintaining the structure of colonialism, confused though its position was by the presence of Rome and the Irish question. The slow appearance of Australian-trained priests and clergy did nothing to convince anti-clerical nationalists otherwise.

In the 1860s pioneering work, and the sense of mission this aroused, obscured some of the more fundamental problems the churches were facing. The Church of England in Victoria and South Australia had achieved some measure of independence from Britain and a viable legal status in the 1850s. Fear of being implicated in problems beyond its knowledge and influence, as well as the desire to protect its accumulating property, led the Church of England in New South Wales to establish its own government through an act of the New South Wales parliament in 1866.[21] Thus a degree of local autonomy was secured for administrative purposes. Administration seemed to be of greater concern to the major churches than matters of doctrine and theology. For most of this period, church leaders were almost wholly preoccupied with funding and staffing, not only churches, but schools. Meanwhile in learned and expensive quarterlies, both local and imported, in pamphlets, newspaper articles, letters to the editor, speeches to self-improving and socially interfering societies, in debating clubs, at dinner parties (till religion was banished as detrimental to civilized conversation), and, if Joseph Furphy is to be believed, in drovers' camps on the banks of the Murray, thinking people taught themselves versions of a debate which was taking place throughout the Christian world.

That debate at its simplest was about evolution, leading to complex arguments about scriptural interpretation and authority. Dr Bromby, headmaster of Melbourne Grammar School lecturing on 'Prehistoric Man' in 1869, gently suggested a possible line of human development to set against the case for creation described in *Genesis*. When popular writer Marcus Clarke wrote of 'the dogmas of the priesthood' and the outworn creeds of Christianity as explanations for rising indifference, or when the Chief Justice of Victoria,

George Higinbotham, spoke (in 1883 at the invitation of Charles Strong) to the Scots Church Literary Society on 'Science and Religion', the growing gulf between 'institutionalised religion bound by creeds and dogmas' and the recently rediscovered humanity of Jesus, it was as if the publishing industry had nothing else to print. The Anglican Bishop of Melbourne, James Moorhouse, most ably responded to both Clarke and Higinbotham with a liberal orthodox theological view, but there were few churchmen of his calibre in the colonies.[22] Charles Strong's interpretation of the doctrine of atonement was such, however, that he was driven from the Presbyterian church to set up his own Australian church, which was characterized by the absence of both creed and dogma.[23]

Debate on the latest developments in theology was most forthcoming in those churches with the best educated clergy and congregations. The Unitarians, notable in Melbourne in the 1870s for their chief preacher, Martha Turner, the eminently respectable and thoroughly well-educated sister of Henry Gyles Turner, banker, littérateur, sportsman, man about town, and also a leading Unitarian, with their alternative to the old trinitarian notion of God, commanded more space and attention than their numbers warranted.[24] Congregationalists responded with intellectual enthusiasm to the challenges of the time. Presbyterians and Anglicans had plenty of scope as leading churchmen in each denomination took both very advanced and extremely conservative theological positions. The antithesis, for example, to Charles Strong could be seen in Adam Cairns, first principal in 1866 of the Melbourne Presbyterian Theological Hall, 'a fearless preacher and denouncer of everything he thought to be evil' including the running of trains or opening of public libraries on Sundays and attempts to legalize marriage with a deceased wife's sister. He remained 'bound to the doctrine of verbal inspiration of the Bible'.[25] Alongside Moorhouse's sophisticated and elegant expositions of the Church of England's position, Frederic Barker, Bishop of Sydney from 1854 to 1880, seemed uninterested in science or the higher criticism, though his evangelicalism and abhorrence of Catholicism endeared him to many conservatives. Mervyn Archdall, long-time vicar of Balmain and a leader of the Australian

Protestant Defence Association, never let his 'doctrines sag or melt like lollies in a bag'. Frequently polemical, always scholarly, he criticized Darwinism, attacked Catholic ritual, deplored Anglican latitudinarianism, and upheld the scriptures as the source of all authority in religion.[26]

Least conspicuous of the major Nonconformist churches in the theological debates of the 1870s and 1880s were the Wesleyans or Methodists, though they contributed their mite to sectarian discussion. At least four different branches of the Methodist connection thrived in Victoria, New South Wales, and in South Australia where, taken together, they were the largest Nonconformist group. Methodists in general had less formal education than other Nonconformists, though like the Unitarians and Congregationalists, they were not trammelled by ancient and hierarchical structures of government. Nor did they suffer from a sense of self-importance or responsibility for carrying on the dignified traditions of the Churches of England and Scotland. In Victoria, South Australia, and New South Wales during the last decades of the nineteenth century, Methodists continued to add to their numbers and to sustain their record for attendance where other denominations lost ground or stood still. This may have been because of the absence of theological controversy or because church self-government was effective. Yet the Methodists were not without their internal debates, chief of which was whether the weekly class meeting should remain obligatory. Those Methodists whose steadiness, industriousness, and sobriety had made them prosperous, found weekly attendance at class meetings increasingly onerous and came to prefer regular attendance at Sunday worship.

There was much in the relatively autonomous nature of Methodism that coincided with the needs of the religiously inclined members of the late nineteenth-century lower middle class, also the earnest self-improving farming communities who were its other source of strength. The emphasis on Bible reading and the simpler forms of early Christianity were in sympathy with critical rediscovery of the historical Jesus. Its participatory structures made Methodism attractive to the active enthusiast for self-improvement or community involvement. As Renate Howe's analysis of the social composition of Victorian Methodism shows, there were plenty of examples of successful self-improvement and effective participation in local government among the members.[27] Yet Methodism also held out comfort and moral guidance to people cut off from their roots or inclined to view opportunity fearfully. As well, it provided one of the better (and much appreciated) outlets for shameless commitment, for enthusiasm amounting even to passion.

The great achievement of Methodism in the colonies at the turn of the century, the symbol of 'the Australianisation of an English-born religious tradition' was unification of various groups of Wesleyans, Primitive Methodists, and Bible Christians.[28] Unification demonstrated maturity and confidence, though it may also have been a sign of loss of variety, of rigidity or respectability. Indeed, the gap thus created began to fill with smaller, newer enthusiasms, the Salvation Army, Seventh-day Adventism, Christian Science, as well as the many varieties of unbelief or passionate belief in secular forms of salvation.

By the end of the century, most Protestant clergy had 'modified their views on biblical inspiration and had come to terms with modern science'.[29] Not many went as far as the Rev. Robert Potter BA of Melbourne whose eight sermons with notes on the scientific aspects of Christian doctrine, *A Voice from the Church in Australia*, were published in 1864. There he speculated on the relationship between matter and consciousness suggested by the resurrection, and on the new absolutes being created by science in a world still full of sin: 'as humanity becomes more intellectual, humanity does not become more moral... We see moral depravity and corruption flourishing in Paris and London amidst the brightest glare of modern intellect.'[30] He went on to argue that there was no difference between the intellectual condition of 'savage' and 'modern' man. Progress came with the accumulation of moral wisdom.

Years later Potter was revealed as the author of an immensely popular adventure story called *The Germ Growers* supposedly written by Robert Easterley and John Wilbraham.[31] In this story the two authors become close friends while at Oxford, travel to Australia, fall in with some Aborigines near Daly Waters on the overland telegraph line which is being constructed. The Aborigines are afflicted with some peculiar ideas about white men and eventually the source is revealed—a colony of strange beings who are culturing lethal germs with which they are waging war on the rest of humanity. The germs are distributed about the world in flying 'cars', invisible to the human eye, but casting a shadow on the ground as they pass overhead. Their leader (who has 'As Gods' worked in ancient Hebrew on his headdress) is called Niccolo Davelli. The two heroes become caught in 'this world of the ether', and learn about germ culture and the laws of light, gravity and motion which they must overcome to escape, which they do. In the end they marry sisters and live side-by-side, and happily ever after, a perfect tale of mateship.

Potter's science fiction suggests a more sophisticated appreciation of the relationship between science and religion than we have come to assume. Well before the end of the century, the problem for the church ceased to be the threat of mass desertion to free thought, spiritualism, or rational modern alternatives to religion. Instead it became apathy or disdain, especially among poorer people. A timid or censorious response to social problems did not improve the image of the church either, for people resented the possible inference that their attendance was only insuring charity in time of need. At the same time, there was a danger of the church being identified with the established order.

Between the curious few and the uninterested many were the mass of ordinary church-goers who sought security or preferred to ignore the possibility of change. The available figures show a rise in church attendance from mid-century to a peak of 38 per cent of the population attending regularly in 1870. Thereafter there was a decline till about 1886, when attendance again began to rise, though it never again reached the level of 1870.[32] By the end of the century attendance was lower than it had been in 1860 but not as low as might have been expected. Many clergy continued to preach about holding to the old paths while in their heads they sympathized with the new critical attitudes. Dr Thomas Roseby at Camden College argued that 'Unless the whole scheme of nature had first existed in the thought of God, it would never have been *realised* in the facts of experience.'[33] Roseby felt that God's will was easier to live with than free will.

COLLINS STREET EAST ON SUNDAY MORNING

Almost certainly those who took refuge in the will of God found the orderly and hierarchical structure of the church a comfort. The church was a social institution that had recognizably survived translation to the colonial environment. By the second half of the nineteenth century individual clergymen may have been despised as effeminate or impractical. Yet church-going conferred status. It was an important adjunct to upward mobility. In a society where it had become difficult to differentiate people from one another on the basis of speech or dress, for example, church-going was a useful part of the ritual of fashion and status. Within denominations there were higher and lower status churches as well as popular or fashionable preachers who merited review in the daily press.[34] Though fashion-conscious wives were sometimes said to be the only reason their husbands were seen in church, it may have been the conversations and the contacts which were really the attraction. On more than one occasion Henry Parkes's finances were rescued as a result of Sunday morning consultations outside St Mark's, Darling Point. Church building kept pace with the growth of the suburbs, as with the expansion of settlement, suggesting that the churches filled a social or community need. Fashionable city churches did not suffer from suburban expansion, but poorer inner-city churches missed both the money of those who had moved to the suburbs and their bourgeois skills.

In 1892 T. A. Coghlan noted that there had been a small increase since 1872 in the proportion of weddings celebrated by registrars in New South Wales. Church weddings were most popular in 1873 when only 4.08 per cent of those married went to civil registrars, and least popular in 1884 when 7.18 per cent were performed by registrars. An average of 6.09 per cent of the marrying population went outside the church between 1872 and 1892.[35] From the majority who chose a church wedding, there was a slight tendency for the more permissive churches and clergy, like the Congregationalists, to attract more than their obvious share of weddings. It would be unwise to extrapolate much further about the role of the churches from this or other rites of passage. State registration of births, marriages, and deaths certainly worked against the authority of the churches in the long term, but the immediate impact of some state measures, such as the regulation of burial grounds, was to enhance the role of the church as the body designated to carry out administration.

In rural Australia Sunday retained its traditional aspect as a day of rest, though people sometimes travelled considerable distances to attend worship and to socialize. 'To put on a clean shirt and to camp on Sunday is the stockman's open profession of allegiance to a Higher Being', Rosa Praed observed.[36] Mary Mcleod Banks remembered Sundays at Cressbrook in southern Queensland as

happy in spite of the restrictions which hedged them round; we might not run, nor gather fruit, nor sing songs, nor read anything but books on religion or the magazines laid out on the verandah table for Sunday reading. But we had our friends to talk to, and took our meals with our elders. All work had ceased, and the observance of things above the common routine brought a sense of dignity and restfulness into our lives. We enjoyed helping to arrange the chairs in rows in the hall for the morning service and to lay out the books. The short sermon, read from a book and probably more delectable than many another preached throughout the world on that day was no burden to us; we often heard one by Charles Kingsley. We sat, looking out on the wide, quiet bush, and listened—the earth and its promise were fair in our eyes.[37]

In Victoria, British statutes dating from the seventeenth and eighteenth centuries prohibited Sunday trading, most public transport, organized sport and other recreations, and the publication of newspapers. The Lord's Day Observance Society maintained a sharp watch for any breaches of the sabbath and, before the 1880s, was not successfully challenged. Then the argument that leisure might be beneficially employed in educational pursuits such as visiting museums, art galleries, and libraries or even in innocent and healthy excursions to the country and the seaside penetrated the self-improving side of dour Presbyterianism, or was it the chink of loose change in too many pockets? Sunday newspapers, however, remained anathema, banned decisively by the Victorian Sunday Newspapers Act of 1889.[38] In Sydney too, there was a flurry of sabbatarian activity in the early 1880s. There since 1878 the museum, the art gallery, and the public library had opened on Sunday afternoons and trains ran. Prosecutions against newspaper sellers, barrowmen, and others

SELECTOR'S HOME—SUNDAY MORNING

for Sunday trading were petty harassment more than religious zeal. Arguments which carried weight in Melbourne—such as that the quiet church- and home-centred Sunday was essentially British and preferable to the rowdy public entertainments of the continental Sunday, favoured, it need not be said, by French and Italian Catholics—were less compelling in more Catholic Sydney. Only flagrant abuse of the law against revenue-raising public entertainments by free thought lecturers, who also made their meetings attractive with musical items, caused Henry Parkes to prohibit the use of public halls in 1887. Church-sponsored entertainments, crusades, revivals, missions, tours by famous preachers, monster outdoor meetings were tried to lure the urban crowd.[39] From the Salvation Army's brass and timbrel bands to the Melbourne Methodist Church's Pleasant Sunday Afternoon—a family outing designed to compete with the attractions of the botanic gardens and the museum—the range was enormous, but the churches were losing their traditional claim to the undivided attention of the people on their day of rest.

The church, however, continued to exert its influence on daily life through widespread acceptance of Christian values as the basis for morality. Though this period saw a subtle shift towards a secular/legal moral code, unconscious habits of mind were steeped in generations of Christian teaching. The search for a secular morality grew out of the primarily Protestant socialization of those who led the way out of the old religion.

In *National Life and Character: A Forecast* (1893), C. H. Pearson described the puritan tradition of family life by which he had been raised:

> The mother who almost doubted if it was not sin to love the babe that smiled up in her face; the children who spoke with bated breath and were trained to orderly composure on Sundays; the belief of young and old that they lived in a world whose amusements and thoughts were irreverent and grotesque by the side of life with its awful duties, even as laughter above a deathbed would be; the conception of marriage as indissoluble; the recoil from libertinage of thought or of moral tone as from shame and death, are all parts of a system that could only be maintained while the New Testament was believed in as something more than the best possible moral code,—as the actual word of God.[40]

Pearson believed that these attitudes were being replaced by others more 'genial', 'charming', and 'natural' especially in family life. The belief that all children were conceived and born in sin, and that only severe discipline and training would transform them into passable adults was fading. Sexuality was shedding its Christian condemnation as a necessary evil, the human weakness enslaving the soul. Pearson's optimism in writing this epitaph on puritanism was calculated. The New Testament was still the moral code, though now it was interpreted by 'public opinion'. These words, 'genial', 'charming', 'natural', heralded the morality of public conscience and personal insecurity. Sexual practice was transformed from instinctive behaviour governed by religion and folk wisdom to almost a science. Though 'experts' did not emerge until later, this period saw intense debate over the formulation of a new secular puritanism. 'Wowsers', the anxious guardians of traditional views, set 'the tone in politics and the press on moral issues'.[41] Some support for the word 'Commonwealth' to describe the federation of 1901 was based on its allusion to Cromwell and more especially, seventeenth-century Puritanism.[42]

The new puritans saw the family rather than the church as the guardian of moral responsibility, with women and children as main agents and beneficiaries. Campaigns against prostitution, theatres, gambling, and drink, or in favour of censorship were all aimed at protecting the otherwise vulnerable position of women within marriage. Women still inhabited a world dominated by age-old fears and practices. They were still regularly exposed to the dangers of pregnancy, childbirth, venereal disease. Their survival as well as that of their children depended on the goodwill and economic skills of others. The loss of traditional moral sanctions emphasized their vulnerability. It is not surprising therefore that they became fearful or evasive about sexuality.

Despite the efforts of reformers like Henry Keylock Rusden, who in 1878 published the first 'Australasian edition' of Charles Knowlton's 1832 American pamphlet on sexuality and birth control, *Fruits of Philosophy*, ignorance, repression and self-control continued to be the dominant methods of dealing with sex.[43] A distinction emerged between sexual activity which was basically a male preserve and a private matter (women who initiated sexual activity were 'bad', prostitutes), and reproduction, which was of legitimate interest to women, but also a matter for public policy. Birth control, however, had long associations with libertines and prostitutes, radicals and free thinkers. It was private, mainly of concern to men, and therefore suspect with 'respectable' women. Even though W. C. Windeyer's judgement in the Supreme Court of New South Wales in 1888 made legal the sale of Annie Besant's pamphlet, *The Law of Population* with its birth-control advice, this was not the sort of thing that found its way into the hands of young women.[44] Birth control became acceptable for married women after the experience of one or more pregnancies had given them the necessary experience, courage, or determination to seek out ways of controlling their own fertility. Such information was provided by chemists, nurses, midwives, friends, or by discreetly baffling advertisements for mail-order devices and preparations, or by doctors or husbands.[45] Experience was crucial. None of the literature made much sense without it. As menstruation was also regarded as an unmentionable female illness, the whole business of reproduction seemed distasteful anyway. Unmarried women did not need much persuasion to ignore the subject. Their sexuality was an inconvenience, best translated into charm while they concentrated on the outward and visible securities of love and marriage. After marriage, young wives expected the knowledge and experience of their husbands to deal with their well-repressed instincts. Thus gratitude and dependence were deepened by ignorance. In his essay on '"Civilized" Sexual Morality and Modern Nervous Illness' (1908), Sigmund Freud might have been describing these women.

Their upbringing forbids their concerning themselves intellectually with sexual problems though they nevertheless feel extremely curious about them, and frightens them by condemning such curiosity as unwomanly and a sign of a sinful disposition. In this way they are scared away from *any* form of thinking, and knowledge loses its value for them. The prohibition of thought extends beyond the sexual field, partly through unavoidable association, partly automatically. . . I think that the undoubted intellectual inferiority of so many women can. . . be traced back to the inhibition of thought necessitated by sexual suppression.[46]

Young men, on the other hand, were free to read, to experiment, though the risks were not fully understood. In Henry Handel Richardson's portrait of Richard Mahony's disintegration, guilt overshadowed the physical effects of venereal disease. William Chidley's memoirs provide striking evidence of contemporary views of sexual practice, both safe and acceptable, corrupt and depraving.[47] Sexual guilt, in particular about masturbation (which was the forbidden key to experience for both sexes), began to compete with more conventional religious guilt, not only as a reason for admission to insane asylums, but more generally as a source of anxiety.[48]

These changing ideas about sexuality reflected changes

occurring in Britain and the USA at the time, though it is possible that in Australia some emphases were different. Good health, youthfulness, affluence and leisure all sharpened interest in sex. Tension between the desire for respectability and the search for modernity may have been greater here, as were sexual insecurity and segregation. Certainly, official concern with the quantity and quality of the population arose early, as Neville Hicks has shown, and this heightened debate between those who desired population growth and those who were interested in the freedoms conferred by scientific birth control.[49]

'Cleanliness is next to Godliness' was probably the most frequently used aphorism of the late nineteenth century, accompanying, as it usually did, the unwelcome application of soap and water to juvenile dirt. At a practical level, cleanliness became the cornerstone of public health. It also symbolized important aspects of the new puritanism—a clean mind in a clean body. Public cleanliness was a masculine matter, for medical officers and local government; private cleanliness was mainly a female responsibility. Standards of hygiene, both in public and private, were authoritatively laid down and actively pursued, especially in the later years of the century, by doctors, scientific societies, health and sanitary associations, all claiming professional knowledge or special expertise. Nor did they confine themselves to ways of purifying the water supply, eradicating outbreaks of disease, or advising on the frequency with which floors should be swept or baths taken. Their views were sought on such things as the amount of air necessary for factory workers or miners, on the length of time that shop-girls could stand without straining their reproductive organs, the best kind of artificial food for a new baby.[50] Inevitably questions of public well-being shaded into moral ones. In the public sphere cleanliness was enforced. Tiles bearing the message 'Do not Spit', for example, were set into railway station walls and fines imposed on offenders to guard against the spread of tuberculosis. Women, who bore private responsibility for the physical as well as the moral well-being of their families, were not easily supervised or regulated. So they were taught to believe that it was impossible for them to be too fastidious about cleanliness. The douches and syringes which became fashionable may have

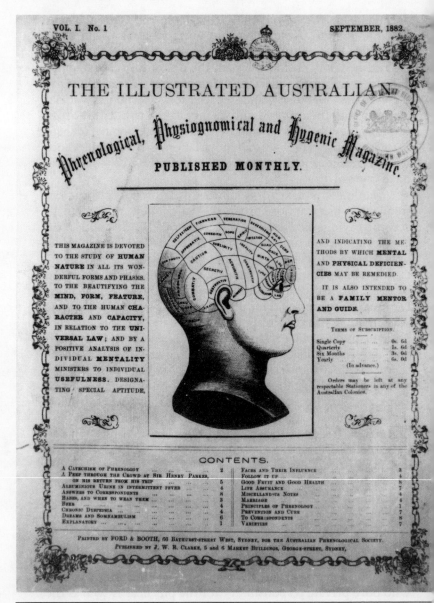

Only the first issue of this magazine has survived but such eclecticism was not unusual

had contraceptive intentions, but their great appeal was as an aid to cleanliness. Cleanliness became associated with purity, then with femininity, then with lack of curiosity.

Dr Phillip E. Muskett, author of several late nineteenth-century manuals on health, lifestyle, and child-care under Australian conditions, was of the opinion that accuracy, punctuality and economy came next after cleanliness in the litany of desirable personal characteristics. Few would have disagreed, though Muskett was careful to point out that though these virtues were of obvious value to the young man seeking material advancement, they were equally relevant to his wife in running the household.[51] As a mother she was responsible for inculcating the right habits in her children by precept and training. Thus women were uneasy agents of the new secular morality. They were believed to be weak, and their weakness was essential to control the waywardness of men. But they were also known to be strong because of their adherence to traditional knowledge. They could be despised for their conservatism, yet respected for maintaining standards.

This new puritanism with its Protestant undertones was particularly severe on Irish Catholic mores.[52] Catholics had to contend not only with their traditional religious beliefs, some of which adapted badly to rapid modernization, but also ideas about them which were both inhibiting and demoralizing. Repressed Presbyterian imaginations, for example, relished the diet of sadism and kinky sex suggested by convent walls and neighbouring presbyteries. Semi-pornographic revelations of the 'Maria Monk'[53] variety were reinforced by the somewhat sinister appearance of nuns in their medieval habits and priests in their soutanes. Colonial anti-Catholicism fell readily into abuse of 'the scarlet woman' and the 'whore of Rome', repeating scurrilous stories culled from the international literature. Perhaps the mere idea of voluntary celibacy in a society where there was already too much that was involuntary was upsetting. Certainly Irish Catholic links with gambling, horse-racing, and drink were seized upon by the wowsers.

It was the Catholic church, moreover, which bore the brunt of anti-authoritarian and nationalist attacks on religion. In the absence of an established church, and with more than half the population already adopting a relaxed attitude towards the authority of the existing churches, the Catholic church was the obvious representative of traditional church authority. Catholicism in the 1860s was perceived as Romish and sinister. The later Irish version was less interesting though more troublesome. Events such as the tour of the Redmond brothers in 1883 to raise money for Fenianism aroused noisy demonstrations and rabid prose which were more about the threat of Irish disloyalty to Britain than the authority of the Church of Rome.

Cardinal Moran in his *History of the Catholic Church in Australasia* (1896) first formulated the idea that the Catholic church in the colonies during the preceding fifty years had undergone 'Hibernicization'. Most of the bishops appointed since the middle of the century, with a few notable exceptions, had been Irish. The exceptions were significant—the Spaniards, Serra and Salvado in Western Australia setting forth to christianize the Aborigines like their countrymen in Mexico and California two hundred years before, the English Benedictines, Polding, then Vaughan building an empire from Sydney. The Irish bishops were less than glittering, though individuals like James Quinn in Queensland and Matthew Gibney in Western Australia attracted dubious attention. As well as the bishops, a majority of the clergy were Irish, though there was a scattering of Spaniards, Italians, and Frenchmen. The impact of Ireland was further felt through the orders of nuns imported initially to minister to the need for charity but, with the implementation of compulsory education, to provide teaching staff in hundreds of parochial schools needed to counteract state secular education.

Irish priests, bishops and teachers had great difficulty in adjusting to the facts of religious plurality in the colonies. They were accustomed to a high degree of social and religious uniformity which led to a much easier imposition and acceptance of the teachings of the church. Mixed marriages, for example, were rarely encountered in Ireland, whereas they were common here. A high degree of authoritarianism seemed necessary simply to maintain a recognizably Catholic environment.[54] Before the 1870s the Catholic church was more permissive than it became later. Marriage outside the church was permissible. Education outside church schools was acceptable. The church was less legalistic and intrusive.[55]

From the 1870s Catholicism everywhere felt itself under siege, in Italy from the Risorgimento, in Ireland from Irish nationalism, in the Australian colonies from the movement for secular education. The Catholic bishops in Australia generally welcomed conservative developments in church doctrine such as the Syllabus of Errors and the decree of Papal Infallibility. This may have been, as John Molony has argued, because they were closely in sympathy with the ideas of the Roman Curia at the time, or more precisely because so many of them were protégés of Paul Cullen, Rector of the Irish College in Rome and Bishop of Dublin.[56] Or it may have been just practicality because to dispute those papal pronouncements was to provide additional ammunition for the enemies of the church and the papacy in Australia, and because, being men of only ordinary intellectual ability, they had no ideas of their own.[57] In any case statements emanating from the embattled church in Rome, threatened as it was by Italian nationalism, were sympathetic to the church in Australia struggling also against nationalism and secularism.

For the parish priest, the doctrine of Papal Infallibility was less puzzling than the daily problems of guiding a flock which was no longer submissive, but confident, no longer pious, but disrespectful. Irish immigrants who had responded well to space and opportunity had also adapted their religious behaviour to colonial necessity and discovered that the neglect of ceremonies and duties enforced by a shortage of priests brought no discernable response from on high. Nor were many Catholic clergy equipped to win respect in the wider theological debates of the time.

Bishop Polding had thought it best to accept colonial realities and to trust the eternal verities till the church was really needed. More recently arrived bishops, such as the Quinn brothers in Brisbane and Bathurst, and Murray in Newcastle, were not so patient. They began to impose greater discipline, first on the priesthood, then on the laity. Priests were forbidden to drink or gamble and their daily timetables were strictly laid down. The 1869 stand against mixed marriages was a dramatic example of tightening control on the behaviour of the laity. Whereas under Polding, marriage whatever its complexion was considered a worthy achievement, now there were strict definitions of acceptability. This may have reflected an easier marriage market with a greater choice of partners, but it was also a far-reaching assertion of church authority. Whether it worked is another question. There was no notable decline in the number of mixed marriages, nor a rise in the number of applications for special permission. It probably did sharpen awareness of Catholic authoritarianism, and cause some who might have stayed to drift away. The church began to expect a high standard of piety from its ordinary adherents. O'Farrell has argued that under the leadership of Cardinal Moran pious practice was substituted for thoughtful or reasoned understanding of church teaching.[58] 'Yes, dearly beloved, indifferentism is the deadliest of all errors.'[59] This had the effect of keeping both clergy and laity continually busy with religious observance, though there was a danger that if or when the action ceased, so too did the meaning of religion.

Those who could accept the church's prescriptions on marriage and its busy ordering of their daily lives were unlikely to reject the values which came to represent Catholicism to the rest of the society during the late nineteenth century. These included a timid rejection of much behaviour suggesting modernity; an inward-looking self-absorption so that Catholics mixed as little as possible with the rest of the community, keeping to their own schools, sporting and social organizations. The church became a world in itself. There was a tendency to confuse the Australian present with an Irish otherworldliness which was absorbed from teachers and clergy, and which in its very fantasy seemed more spiritual and emotionally satisfying than the commonplace problems of becoming that still-uncertain creature, an Australian.

Not all Catholics were given to dreaming of Ireland. Many who accepted and tried to live according to the church's prescription of what a devout Catholic man or woman should be were intensely engaged in the battle to win their share of the good life and to bring up their children decently within the church. This Moran recognized when he spoke in sympathy with the strikers in 1890 and when he acquiesced in the involvement of Catholics in politics on behalf, mainly, of the labour movement.

However, that sentimental piety and acceptance of authority which made good Catholics had less fortunate consequences when carried into the politics of the labour movement. Ideas were very important as a stimulus to early working-class politics. Many of the early labour leaders were addicts of the printed word, incessant debaters, and of the generation which, having discovered self-education, began to demand a better education for the next. Among other things they brought to politics the latest, most modern, most heretical views of what was wrong with government, religion, society.

The simple gospel of the Christian church with its emphasis on equality in death, the virtues of poverty and humility, and the social benefits of altruism was entirely compatible with, maybe even the source of the simpler versions of socialism current towards the end of the nineteenth century. 'Socialism, of course, simply aims at realising the Christian's provision, per medium of an endowment policy endorsed by the whole community.'[60] It was easy for good Catholics to fall in behind the advocates of what was essentially a secular version of Christianity. And that, to the chagrin of church leaders, is what they did. Working-class leadership came from the intellectually stronger and more rigorously exercised minds in secular revolt against superstition and the power of priests. The Catholic working class in Australia responded in politics as they had been taught to respond in church. They accepted authority. They attended piously, carried out their duties carefully and without question. When challenged they responded protectively and with great emotion.

Though only about a third of the people regularly attended church, at census time an overwhelming majority nominated a faith to which they assumed some connection. In the Australian census of 1901, of the 3 773 801 counted, only 133 292 were classified as non-Christian, or as having no definite religion, or said they did not wish to state their religion. Of the acknowledged Christians, about one-third were given as Church of England, with slightly more of them men than women. Roman Catholics accounted for a little more than a fifth of the total population with roughly equal numbers of women and men. Methodists outnumbered all other Protestant denominations, thus confounding Marcus Clarke's 1877 prediction that the religion of the coming Australia would be a form of Presbyterianism.[61] New South Wales was predominantly Anglican and Roman Catholic, but in Victoria, Presbyterians and Methodists together outnumbered the Catholics, and there was a strong Baptist church. In South Australia, Methodists alone outnumbered the Catholics and there were more Methodists and Baptists together than Anglicans. Victoria and Queensland were the strongholds of Presbyterianism. In both Presbyterians outnumbered Methodists. The highest proportion of Catholics was to be found in New South Wales, followed by Queensland, Western Australia, then Victoria. The numbers professing adherence to other denominations were relatively small, although there were some differences worth noting—for example, both South Australia and Queensland had strong Lutheran communities. In South Australia, Lutherans made up the largest group after Anglicans, Methodists, and Catholics. In Queensland, Lutherans came fifth after Anglicans, Catholics, Presbyterians and Methodists.

Altogether 42 131 people said they objected to stating their religion. In New South Wales and Queensland this was roughly 1 per cent of the whole population. In Western Australia and Tasmania a slightly higher proportion rising to 2 per cent in South Australia objected to stating their religion. The people of Victoria were least reluctant to state their religion, though there more people than anywhere else admitted to having no denomination, no religion, or gave themselves as 'undefined Protestant' or 'other Christian'.

Women outnumbered men as Methodists in both South Australia and Victoria, and as Catholics in Victoria. Despite the fact that there were still approximately 200 000 more men than women in the population as a whole, Congregationalists, Baptists, the Church of Christ, and the Salvation Army all showed more women than men in most colonies. Western Australia remained a male stronghold, even in these denominations.

The religions where men plainly predominated were non-

Christian. Of 35 500 Muhammadans, Buddhists, Confucians and Pagans, nearly 35 000 were men, and nearly 17 000 of these were in Queensland. Men were also most noticeable among the decidedly non-religious and those who objected to stating their religion. Of 9182 self-acknowledged free thinkers, only 1319 were women. Free thinkers were disproportionately in evidence in Queensland and Western Australia. There were 834 male agnostics, 137 female, 245 male atheists, 29 female, 5149 men who said they had no religion, 1333 women. And whereas 28 443 men objected to stating their religion, only 13 688 women did.[62]

Certain social characteristics are illustrated by these rough statistics, for example the loss of male population from Victoria in the 1890s and the corresponding gain to Western Australia, or the tendency of men who were working class, mobile, restless, to leave their religion behind. The established communities of the south-east had settled into nonconformity; frontier communities were more likely to contain nominal Anglicans, Catholics, or self-conscious atheists and agnostics.

Historians whose idea of what constitutes a religious society has been shaped by the Christian tradition have seen nineteenth-century Australia as 'the kingdom of nothingness' or the 'first genuinely post-Christian society'.[63] Yet a contemporary observer, Charles Dilke, who visited the colonies twice, thought that the people had become more sober, industrious, and God-fearing by the time of his second visit in 1889 than he had found them in 1867.[64] For, even among the majority who were not regular church-goers, there was indifference rather than outright rejection of religion.

Such ambivalence or nominal adherence was reflected in the Constitution. Religion was first thought irrelevant to federation, and better left alone as a likely source of unnecessary and damaging sectarianism. But the preamble to the Constitution now states, 'the people . . . humbly relying on the blessing of Almighty God, have agreed to unite'. And section 116, while forbidding Commonwealth legislation for the establishment of any religion or the imposition of any religious tests, leaves open the possibility that any of the states might so legislate. There was no reference to God in early drafts of the Constitution. After all He was not invoked in the constitution of that most religious country, the United States of America. But an effectively orchestrated church campaign in which Protestants and Catholics combined without difficulty—it was certainly impressive to see Catholic lawyer P. M. Glynn acting as advocate at the later conventions for rigid Victorian Presbyterian J. A. Rentoul—brought God into the preamble. Section 116 seems nothing more than an electoral bet each way for Christians and non-Christians.[65]

In comparison with the forces of organized religion, the forces of organized unbelief were minute. A new popular religion, the Salvation Army, gathered supporters more easily than any of the organizations advocating, for example, atheism, free thought, or spiritualism.[66] What was really missing from religion in the colonies was grandeur or style built on the inheritance of the ages in older societies, or that intensity of belief, evident in societies less secure materially, and less committed to the use of reason to solve human problems. As Joseph Furphy wrote to C. H. Winter, 'a man doesn't need to be a psalm-singing sneak to admit that beyond the material elements of life, beyond the minerals, plants and animals, beyond the changing seasons, and so forth, there lies a purpose.' He went on, 'My own Sunday-school conceptions of heaven, hell, the judgement day, etc., have long ago gone by the board, and without affecting my sense of right and wrong in the slightest degree.'[67]

Dilke had noticed that people in the colonies thought for themselves with an unusual level of confidence, and were unwilling to tolerate ideas as dictated to them by clergymen.[68] Some of religion's most careful students were those who were learning to reject it, like Melbourne pieman E. W. Cole, who became a bookseller in his determination to publicize his own version of the story of Jesus and St Paul.[69] The scriptures were read carefully, interpreted literally. After stripping away all belief in the supernatural, what was left was a simple system of ethics, which reappeared in several guises—in the bushman's creed of mateship or W. G. Spence's idea of the New Unionism.[70] It was not necessary to be a regular church-goer to have firm views on morality. Those who consciously did not adhere to formal Christianity were often very demanding in their attitudes to both public and private morality. Most outspoken radicals were deeply moral. Their criticism of religion was of its hypocrisy, of the failure of those who accepted it to live up to its teachings, of the comfort and power of its position while it preached the virtues of poverty and self-discipline to the masses. They rejected the hierarchical structure of Christianity, its arbitrariness, its irrationality. They believed that society could be made more humane, more equitable, more efficient, if only people would throw out their superstitions, banish the irrational, take control of their own lives.[71]

Alongside the churches and those small organizations committed to opposing organized religion, quasi-religious organizations flourished. In 1890, for example, it was estimated that in New South Wales alone there were 10 000 Freemasons in 185 lodges.[72] Having moved away from his father's uncompromising Congregationalism, S. W. Griffith followed his legal mentor Arthur Macalister, a leading Queensland Freemason, and joined the Brisbane Victoria Lodge of the United Grand Lodge of Ancient Free and Accepted Masons of England in 1865. He continued a devoted Freemason all his life.[73] Freemasonry allowed ceremony and ritual observance, performed not by priests, but by the meritocracy of the membership. It thus reflected the goal orientation and success ethic of male society, preaching brotherhood but practising exclusivity. Memberships were probably overlapping, but taken together, the numbers accepting the rules and principles of Freemasonry, the friendly societies, and trade unionism suggest that the male population had begun to develop organizations which either augmented or substituted for the traditional churches.[74] Brotherhood, self-help, mutual responsibility and protection of the weak were values compatible with both Christianity and democracy, but they were more suited to egalitarian than to hierarchical organization.

Women remained within the churches where their importance as agents passing on basic Christian moral values to their children was enhanced. The world they inhabited was still irrational. Domestic life was barely touched by the scientific and technological innovations which transformed the work of so many men. Women might learn to use the unseen forces of electricity, but could not yet expect safe or painless childbirth.

In the 1860s and 1870s tiny coteries of sympathetic souls gathered in Sydney and Melbourne (Stenhouse's Balmain circle, for example, and the Yorick Club to which Marcus Clarke belonged) to talk, to read and publish each other's work.[75] Most aspired to be writers or sought something of the bohemian lifestyle such associations suggested. More earnest literary/intellectual associations appeared devoted to the study of almost everything from Shakespeare to social questions. By the 1880s there was no shortage of venues for the young man who wished to pursue his self-education in literary and philosophical areas. Bookshops were remarkably well supplied with the most recent publications from London, and to a lesser extent, the rest of the world. Public libraries provided access to the classics of the European intellectual tradition. David Stenhouse pursued his passion for Greek philology in Balmain. In Brisbane S. W. Griffith translated Dante. J. F. Archibald with his mixed Catholic and public school education from Warrnambool, and Christopher Brennan with his classical/Irish/Australian scholarship from Riverview College and the University of Sydney, both thought they were more at home in the mental world of the French and German symbolists than in the Australia of John Shaw Neilson or Henry Lawson. For them, as for many of the educated or would-be educated, a kind of internationalism was the best antidote to the inevitable provincialism of ideas in Australia.[76] Professional associations of doctors and accountants tried to keep their members in touch with developments both in Australia and overseas through their journals which were sometimes surprisingly eclectic. The editor of the *Australian Medical Journal*, Dr Patrick Maloney, was known about Melbourne as a poet. The *Australian Insurance and Banking Record* carried articles on literary and social matters. The founders of *The Australian Economist* (1888–98) were Walter Scott, the professor of Greek at the University of Sydney, A. C. Wylie, a teacher, and Arthur Duckworth, an accountant with the AMP Society.[77] It was frequently remarked, however, that most successful men, whether in business or the professions, were either uneducated or uncultured or both. The facilities for intellectual development were excellent for those with wealth, leisure and stimulus to study the larger questions of the time, but those most likely to take advantage of them were workingmen. Making money seemed a study in itself, or else it produced arrogant contempt for all forms of knowledge categorized as 'useless'.

The universities were most respected when they aided the solution of practical problems, as when Professor David Orme Masson of Melbourne and Professor John Smith of Sydney were able to advise on water supply questions.[78] Professors like Badham, Hearn, or Pearson, whose contributions to knowledge and ideas were recognized beyond Australia, found it difficult to sustain their work at such a distance. (Both Hearn and Pearson, in fact, deserted their lecture rooms for the more pressing challenges of colonial politics.[79]) A few happy appointments gave youthful unknowns the opportunity to build a life's work and a reputation out of research they began in Australia—in biology, for example, Baldwin Spencer.[80] But the very ease with which scientific 'discoveries' were made as the continent's resources were mapped and catalogued discouraged the thoughtful or reflective scientist.

Scientific research was aided by royal societies in all colonies except Western Australia. These pre-dated the universities and were mixed gatherings of gentlemen amateurs and serious scholars. Wives and daughters collected botanical specimens and made intricate drawings, which, in the case of Ellis Rowan for example, reached the level of high art.[81] Astronomy, geography, ornithology and botany had their own societies before the end of the century. The Australasian Association for the Advancement of Science was founded in 1888. Public lectures, meetings, and conferences disseminated scientific information to a curious general public. Literate and adventurous scientists found an easy place as newspaper columnists. For example, John Smith, professor of Chemistry at the University of Sydney, contributed accounts of his study-leave trips to the *Sydney Morning Herald*, Baldwin Spencer partly financed a trip to central Australia by writing columns for the *Australasian* and the *Leader*, and on his return, giving public lectures illustrated with carefully censored lantern-slides.[82] The highlight of a family outing in the second half of the nineteenth century might be a visit to a museum. Collections of zoological, palaeontological, geological and anthropological materials were established in each of the major cities during this time. They provided entertainment as well as instruction. Gerard Krefft, curator of the Sydney museum, sometimes concealed himself behind a screen to observe public response to displays he had organized. He wrote to Charles Darwin in 1873:

I got a splendid youthful Chimpanzee skeleton which is greatly admired by visitors as the skeleton of a 'poor baby' & several foeti in bottles, near it (dogs, lemurs, monkeys etc) which are very seriously ogled at by any number of women. All these people believe they are human relics in an early stage of development . . . there is nothing so instructive than to watch a number of women young or old contemplating the human foetus. You cannot *hear* what they say, but you can always guess; and how they edge near to this interesting case keeping their eyes ahead so that male visitors shall not detect their very natural curiosity, is truly wonderful. I left the collection of 'human foeti' exposed until I heard one day a little girl, with a baby in her arms, call to some companion, 'come on Fanny and see the babies in the bottle'. Since then the case has lost its attraction as I removed the objects though thousands of women still look at the early stage of dogs and cats with a puzzled air.[83]

Professor Frederick McCoy, the energetic director of the National Museum of Victoria from 1858, built up a collection of mining and economic exhibits which became the basis in 1870 of the Industrial and Technical Museum (later the Melbourne Museum of Applied Science). Similar economic and mining museums were established in Sydney (1880) and Adelaide (1893) partly instructive, partly nostalgic for the gold-rushes, always in hope.[84]

At a popular level, botany, biology, and geology reinforced the message of Darwin and fostered independent approaches to the teachings of religion. The idea of evolution became the modern model of progress. So society was seen evolving from its master–slave stage, through capitalism with its worker-servants to eventual partnership or co-

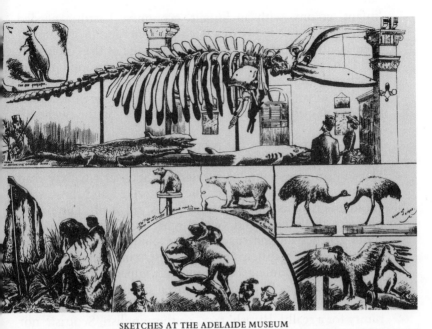

SKETCHES AT THE ADELAIDE MUSEUM

Among the native animals is an exhibit of 'The Primitive Child of the Forest: the Noble Aborigine in his Native Lair'

operation.[85] At a more complex level of political, economic, and social thinking, Herbert Spencer's idea of society as analogous to a biological organism was acquiring the status of an unexamined belief. In Australia, as in Britain, economic, demographic, and social 'laws' derived from the observation of 'nature' became self-fulfilling, as in Pearson's attitudes to family life. A simple biological organism in comparison with the complexities of society as a whole, the family became a model of 'natural' interrelationships. The biological family itself gained from the organic analogy: each member had 'natural' rights and duties according to his or her position within the whole. The family with its male 'head' became a 'natural' unit of economic individualism.

Arguments for the introduction of tariffs to protect local industry justified departure from the teaching of Adam Smith and the 'natural' interplay of market forces by reference to J. S. Mill's qualifications on 'natural' liberty. As an 'infant' colony with an 'infant' economy, Victoria needed protection of her industries as children needed protection of their health, rights, and morals. The image of the colonies as children and of Britain as mother gained through usage. Much of the debate about political independence was couched in restless adolescent terms, a fitting prelude to the translation of Empire into 'family of nations' in the twentieth century.[86]

The experience of a fair proportion of the population during this period was optimistic 'proof' of theories of social Darwinism as expounded in Herbert Spencer's widely read volumes of early sociology. The fit survived and those who could adapt to colonial circumstances did well. The native born on the other hand, were living examples of theories about genetics and inheritance. The evolution of an 'Australian type' was much anticipated. Optimists concentrated on the effects of environment in producing a superior physical and moral specimen. Pessimists pounced on lingering traces of criminal heredity.[87] All were agreed on the splendid opportunity for a really 'British' race, neither English, Scottish, nor Irish, but a mixture of all three, and its challenge to the Asiatic Chinese and Japanese. Such ideas were understood easily alongside the imperative for stronger breeds of horses, sheep, and wheat, though the temptation to concentrate on quantity rather than quality was evident among human population theorists as well as livestock breeders.

F. W. Eggleston was in many ways a typical product of the mentality of late nineteenth-century Australia, though in some respects he was also unusual. Born in 1875, the son of a Melbourne boom-period lawyer who was himself the son of an immigrant Methodist minister, Eggleston's childhood was clouded by his mother's illness and death when he was barely nine years old. She died at twenty-nine of tuberculosis compounded by pleurisy, having borne five babies (four survived) in eight years, to a husband eight years older. By the time he was an adult, Eggleston had broken with the strong Methodism of his father and stepmother, though the Depression kept him from the university and forced him into his father's practice. In his quest for self-education he joined an evening debating class taken by Alfred Deakin, whom he had met and admired through his legal work. Thus he became acquainted with the Victorian liberal tradition. Friendship and intellectual company came through a group of similarly placed young people who in their Gladstone Debating Society argued an understanding of the unsatisfactory world they were about to inherit.[88] Here he met his future wife, Lulu Henriques, and the St Kilda Jewish bourgeoisie which produced among others, Isaac Isaacs, John Monash, and the Phillips, Fox, Ellis, and Goldstein families.[89] The conjunction was fruitful for Eggleston. His horizons were extended well beyond the normal Anglo-Australian ones, and he was shielded by Lulu and her friends from the cruder aspects of Australian anti-intellectualism.

By the 1890s when Eggleston encountered them, the ideals of Victorian liberalism were dominant in Australian politics. The unchanging objective was 'to assert the sanctity of the human personality and to promote the free creative activities of individuals'.[90] Once privilege was eradicated, the free spirit of humanity should assert itself, and government would become rational, orderly, and minimal. Though there was a tendency for this *laissez-faire* theory to be reduced to simple economics—that is, to be understood merely as the function of the market—experience in the Australian colonies provided ample evidence of the effectiveness, even necessity for human intervention in managing market forces. The fragile relationship between colonial sources of wealth and the availability of labour strengthened awareness of human as well as financial capital. Arguments about the 'freedom of the individual from exploitation' were more easily accepted. The same imbalance between resources and labour supply encouraged democratic procedures and collective involvement in development which became known as 'state socialism'. Politicians did their best to maximize differences of opinion on the rate at which democracy should be extended, or the extent to which government should involve itself in economic management and protection of individuals from exploitation, but there was little dissent from the underlying principles.

Some dissent came from acknowledged conservatives

like James Macarthur[91] or Henry Wrixson, who consistently opposed any extension of democracy and asserted that wealth and standing were better indicators of ability than mere opportunity. They also believed that wealth conferred the duty of social, moral, and national leadership, and that neglect of responsibility would result in a loss of wealth. These essentially eighteenth-century views took on new meaning with the rise of a newly wealthy class, not yet habituated to their traditional duties and responsibilities. They naturally preferred to think that wealth was a reward for hard work and superior talent, not a result of good luck or extreme ruthlessness. As Wrixson said in an 1868 election address, 'Wealth is the only badge of our aristocracy, but it confers a nobility neither exclusive nor enduring.'[92] He emphasized duty and responsibility, not only for himself, but for others. Economic self-sufficiency was the first duty of every man. To seek the benefits of family life without being able to afford them was irresponsible, but no restraints or disincentives should be placed in the way of the man who accepted his responsibilities. Wrixson supported the Victorian Charity Organisation Society because it sought to relieve hardship without creating dependence on charity, but he also supported legislation to compel factory owners who had been negligent about safety to provide compensation for injured workers. Privilege, in his view, carried an equal measure of responsibility which should be enforced legally if the moral imperative failed.

A similar sense of responsibility can be seen behind conservative opposition to free selection legislation. In his study of the Monaro, Sir Keith Hancock has shown how squatter hostility to the new selectors was in part dismay at the destruction of the environment which they as prior arrivals had begun to understand and manage conservatively.[93] In the absence of an authoritative conservative tradition in the management of nature, most instincts for traditional practice focused on the management of labour.

Archibald Forsyth was an employer who relied on the skill of his employees. He worked with them to adapt their industry (rope-making) to new market conditions and technology, converting successfully from seafaring rope to wheatgrowing binder-twine. Conservative in his attitudes towards animals—he didn't like to see them exploited or cruelly treated—he might also be regarded as paternalist in his treatment of his employees. Yet in his utopian novel *Rapara*, an imaginary island somewhere in the Pacific which is a 'model Radical State', the gifts of nature are open to all. Wealth, the result of labour, is secured to individuals in proportion to their enterprise, energy, and thrift, while indolence and extravagance produce poverty. Inherited wealth is taxed.[94] Another employers' representative, Bruce Smith, son of Howard Smith, founder of the coastal shipping firm, and generally regarded as an arch-conservative, attacked what he saw as the coercive aspects of colonial liberalism and advocated 'equal opportunities'. 'Every man and every woman must be allowed to "unfold" as he or she may think fit.'[95] This required paying fair wages for work well done with a view to redistributing wealth in the interests of order and stability. (It also meant votes for women.) Smith believed that the redistribution of wealth was preferable to the redistribution of power, because ultimately, 'the workman possessed of his own freehold' developed a strong 'regard for the rights of property'.[96] He was in favour of unions because they brought order into industry and advocated organization among employers as well, though he was opposed to coercive industrial legislation, since, as he argued, there was no means of enforcing such legislation.[97]

Writers like Forsyth and Smith appealed to the idea of opportunity and the sense of fair play in Australian society. In this they were echoing the values of classical liberalism, notions of individualism and justice and the importance of property which were pre-industrial in their origin. Conditions in late nineteenth-century Australia nourished these ideas, especially the reversion to a rural or semi-rural existence experienced by so many. Distance protected and elevated skilled workers in the most necessary and traditional trades and crafts. Radical artisan ideals deriving from the early effects of industrialization in Britain were reinvigorated by working conditions in mid-nineteenth-century Australia. But both classical liberal and radical artisan ideals seemed old-fashioned in their emphasis on individual rights, slow in producing change and negative in advocating the exercise of moral restraint. In comparison, collectivism, whether liberal or socialist, promised faster change, and impressive action, the most modern methods applied with the greatest efficiency and goodwill.

One of the most famous (and until its translation by Russel Ward in 1977, least read) accounts of the Australian colonies in the late nineteenth century has been Albert Métin's *Le Socialisme sans Doctrines*, first published in Paris in 1901.[98] The mild experiments in collective enterprise and social reform Métin described to his European readers were essentially products of advanced colonial liberalism or 'state socialism'. Victoria with its semi-governmental authorities to manage such enterprises as railways, irrigation, water supply, posts and telegraphs, its wages boards, its charity commission, led the way. By the end of the century all colonies were committed to publicly owned and managed enterprises, and some state intervention in labour relations.

Russel Ward has traced Australian collectivism to honour among thieves or mutual support among the convicts.[99] State enterprise was justified by economic conservatives as socially responsible, though W. K. Hancock described it as 'collective power at the service of individual rights'.[100] Queensland liberal S. W. Griffith thought 'it is only the State, i.e. the community in the aggregate, that can enforce the rule of freedom', and prevent 'the complete domination of the weak by the strong'. (This article on 'Wealth and Want', written in 1888 for the *Boomerang*, was attacked as 'communistic'.[101] C. C. Kingston told a public meeting in Adelaide in 1891, 'I wish to be classed as a State Socialist—as one who recognizes it is right for the State to interfere for the good of society.'[102] The difference between liberal state socialists and labour socialists lay in their view of human nature, and therefore the means necessary to achieve the desired ends. Liberalism was optimistic and believed that reason (or self-interest) must prevail. Labour philosophy was pessimistic and willing to coerce for the 'greater good'. In this it reflected the influence of revolutionary ideology imported with socialist theories from Europe, or bred in resentment of the continuing power of Britain, not only in Australia, but also in Ireland. Its immediate origins, however, were in the utopian tradition represented in the Old World by William Morris and H. M. Hyndman, and in the new, by Henry George, Edward Bellamy, and Lawrence Gronlund, and imported in the baggage of working-class immigrants and intellectuals. Karl Marx's intricate arguments which in the opinion of Thomas Batho could not be assimilated, even with a wet towel wrapped round the head, were quickly and appealingly simplified into the local idiom—'socialism is being mates'.[103]

When William Lane became editor of the Brisbane *Worker* in 1890, its masthead proclaimed 'socialism in our time'. Lane, an immigrant from England to Queensland in 1885 via the USA, was in some ways typical of the men who made up the working population of Queensland at that time.[104] Many of them had recent experience of both working conditions in the Old World and the radical theories being devised to explain and assault them. Many were restless, impatient, idealistic, and not a little shocked by the kind of work and working conditions they were forced to accept in the land of promise, the workingman's paradise. The Queensland frontier had never been any kind of paradise, and even before the general economic collapse of the 1890s was far from prospering. The sporadic violence of the early 1890s, the tension between workers' camps and the local tradespeople at Longreach and Barcaldine, and the eventual confrontation between the shearers and the Queensland government in 1891 were all products of impatience among recent arrivals and distrust from those who had survived the difficult conditions for a generation, managing barely to accumulate something that looked like a future. Long-standing settlers fully supported Premier Griffith's harsh reaction to the strikers, for they saw their hard-won security threatened by itinerant outsiders.[105]

Lane with his idealism, his egoism, his personal puritanism, abhorring alcohol and anything but matrimonial sex, offered challenging leadership to workingmen. The appeal of the kind of socialism he preached can be seen in the sustained enthusiasm which equipped and despatched the *Royal Tar* to found New Australia in Paraguay in July 1893.[106] Through the columns of the Queensland *Worker* and other working-class or trade union newspapers which used his material, Lane's version of socialism with its utopianism (he admired Bellamy's *Looking Backward*), simple communism, racial paranoia, and moral puritanism reached well beyond the borders of Queensland.

Small groups of intellectual and/or theoretical socialists had been forming since 1872 when the Democratic Association of Victoria was established.[107] At this stage socialism was only one of many components of radical or progressive thought, its vague idealism appealing as a secular replacement for Christianity. By the late 1880s when the Australian Socialist League (Sydney, 1887) and the Social Democratic Federation (1889), a Melbourne group with socialist and anarchist views, were formed, socialist ideas had become coherent, more closely related to theoretical literature appearing in England, Europe and the USA. A few of the German socialists who formed the Allgemeiner Deutscher Verein in Adelaide in 1886 and the Verein Vorwärts in Melbourne in 1887 could claim to have known Karl Marx before they all emigrated, though the German socialists were not specially welcome in the still basically British Australian movement.

As outlined in 1891 by W. G. Higgs, president of the Australian Socialist League, socialism included 'the national or collective ownership and control of the means of production, distribution and exchange of wealth'. The state would provide everything, the necessaries of life and all the comforts. The whole community would be organized into one labour army under the control of the state. Hours of work would be small and leisure would be devoted to improvement of the mind.[108] 'All that any religion has been to the highest thoughts of any people Socialism is, and more, to those who conceive it aright', William Lane had written in the preface to *The Workingman's Paradise: An Australian Labour Novel*, published in Brisbane in 1892.

Without blinding us to our own weaknesses and wickednesses, without offering to us any sophistry or cajoling us with any fallacy, it enthrones Love above the universe, gives us Hope for all who are downtrodden and restores to us Faith in the eternal fitness of things. Socialism is indeed a religion—demanding deeds as well as words. Not until professing socialists understand this will the world at large see Socialism as it really is.[109]

The official membership of socialist organizations, though never large, grew when other labour and working-class organizations were losing members in the 1890s. Socialist leaders like Higgs and Ernie Lane (William's young brother) believed that a 'revolutionary situation undoubtedly existed in Australia' at the time.[110] 'We of the rebel army', wrote E. H. (Ernie) Lane, 'quite sanely thought it [the revolutionary situation] was wider spread and deeper than it actually was. As a result we thought we could sense a drastic change in the whole system of society.'[111] When no change apart from a hardening of employer attitudes and a chastening of the labour movement occurred, men like Lane moved in the direction of anarchism. Other socialists, like S. A. Rosa, a former member of the Melbourne Social Democratic League and author of *Social Democracy* (1890), believed that the capture of the state through an organized approach to democratic politics was the way forward. Only the Queensland labour movement showed an overt commitment to revolutionary socialism:

the present industrial system, commonly called the competitive system . . . must be replaced by a social system . . . which will provide for all workers . . . opportunity . . . to partake fully of the fruits of civilization and to receive the full benefit of their share of the common toil.[112]

Elsewhere, individuals and small groups of socialists gave intellectual stimulus or formed a revolutionary vanguard within the trade unions and political labour leagues, forerunners of the Australian Labor Party, schooling themselves on Bellamy's utopian novel *Looking Backwards*, the watered-down Marxism of Gronlund's *Co-operative Commonwealth* and Edward Aveling's simplified edition of volume 1 of Marx's *Capital*.[113]

Conditions in Australia in the 1890s came more to resemble the conditions in Britain and the USA upon which socialist theory was based and from which the tradition of working-class protest was growing at a powerful pace—the First International, the French Commune of 1871, the 1886 trials of the Chicago anarchists, the London dock strike of 1889. In comparison, the indigenous tradition of mateship, solidarity, and commonwealth seemed moderate, benign. The Australian workingman of the 1890s was more set on 'civilizing capitalism' than on drastic change.[114] He had opted only for what William Pember Reeves called 'a kind of socialism'[115]—some social reform, a more equitable distribution of wealth and power, as much collectivism as was compatible with individual liberty, or ensured that the tiresome tasks of government were done by someone else.

William Robertson was a struggling orchardist of Quantong, Victoria. He listed Carlyle, Ruskin, Henry George, Charles Strong, and W. K. Clifford, as well as Clodd's *Life of Christ* as the people and books that had influenced him most. He was perhaps typical of many who had read and worried their way out of orthodox Christianity towards a sense of secular responsibility. As he put it in a letter (29 August 1897) to Christopher Crisp, the stirring editor of the *Bacchus Marsh Express*, 'Sympathy for the socially downtrodden is the mainspring of my own state socialism, especially for the irresponsible women and children.' Robertson had rejected the Christian teaching that 'poverty is specially a divine decree'. Poverty could be overcome, he thought, by state provision of pensions, regulation of labour, and open acknowledgement of the brotherhood of mankind.[116]

Irish MP Michael Davitt, visiting the 'semi-communistic' labour settlements on the Murray River established in the early 1890s with the threefold aim of applying collective theories to property ownership, providing opportunities for unemployed workers, and utilizing the new technology of irrigation, noted that security of employment and housing were the real attractions at Lyrup and Pyap. Wives appreciated the absence of city distractions such as alcohol. Davitt also noted that the co-operative members tended to opt for taking their share of the property when the statutory establishment period was over. The rural isolation of these settlements imposed a social order and discipline reminiscent of William Morris's *News from Nowhere*. The only women present were wives, and they were permitted to do nothing but housework. They were not 'settlers' for collective decision-making purposes. Communal government was found to be time-consuming and coercive, producing tensions in the settlements as well as detracting from the amount of productive work actually done.[117] Even the best managed communal settlements lost their attraction as employment prospects improved towards the end of the century and the drought showed signs of lifting.

In the catalogue of late nineteenth-century ideas feared as potentially dangerous, feminism was more feared, though treated less seriously, than socialism. Both feminism and socialism rejected inequalities magnified by the industrial organization of society. Both sought restructuring to make society more equal. But socialists and feminists usually meant different things when they spoke of equality. Instead of speculating on alternative ways of organizing labour and managing society, Catherine Helen Spence's 1879 utopian novel *Handfasted* canvassed alternative ways of governing sexual relations with different forms of marriage, divorce, and child care. It remained unpublished, largely because it was thought to be immoral.[118] In her 1888 story, 'A Week in the Future', which had affinities with William Morris's *News from Nowhere* or Samuel Butler's *Erewhon*, Spence sketched a society in which 'Woman is no longer degraded as the slave or toy of man, but takes her equal place in all relations of life.'[119] This required easy access to contraception and, if that failed, abortion, also easy marriage and easy divorce until the commitment to children. Women were to vote, enter the professions, act as clergy, sit in parliament. Their clothes were to be simplified; housework was to be mechanized and centralized or collectivized. Spence thought all adults should work, whether married or single, but only for six hours a day. The rest of their time was to be devoted to voluntary family or social activities.

William Lane had written of 'two great reforms which must come if Humanity is to progress.' One was the 'Reorganisation of Industry'. The other was 'Recognition of Woman's Equality'.[120] In Lane's ideal society all women were under forty, marriageable, in bouncing health or with plump babies on their hips. Contraception he abhorred. 'There is little you can teach a girl who has worked in Sydney', says Nellie, the heroine of *The Workingman's Paradise*,

and I know there are ideas growing all about which to me seem shameful and unwomanly, excepting that they spare the little ones. For me, I shall never marry. I will give my life to the movement, but I will give no other lives the pain of living.[121]

Lane was vague about other ways of building a better life for women. They should be married to men who were considerate, clean, healthy, and good providers. They might work after their children were grown up, like Connie Stratton, at something 'womanly' like writing, or art criticism. They would live simply, be beautiful, and love, since 'love... is the keystone of this brief span of Life of ours'. Women must be fit for love; men must be fit to love them; all must cherish the 'monogamic idea'; and all be healthy enough 'to rouse true sex-passion'.[122]

Lane's ideas explain why feminism persisted. They also show the intensity with which progressive thinkers perceived the problem of female inequality. The marriage market was uncertain for men. Perhaps they could not afford it. In the labour market too, female competition was becoming more insistent. Yet men had been taught to think protectively of women. The chivalric ideals of Victorian Britain may have bypassed the colonial entrepreneur willing to exploit cheap female labour, but the workingman on a steady wage could indulge his finer feelings (and his need to wield a little power in the world) by asserting that 'his wife' need never go out to work.

Feminism produced no compelling organizations. The potential members, busy at home with the children, had no time for political or intellectual activity. Nor was it refuelled by the flow of disgruntled immigrants and works of theory. Dissatisfaction emerged when daughters of well-to-do families encountered new possibilities for women while abroad, or when thoughtful fathers and husbands began to notice the waste of talent among their womenfolk. Most feminist energy, however, was harnessed or deflected into women's expanding moral and educational role in the new puritanism. The anti-drink feminism of the American Woman's Christian Temperance Union thus seemed more relevant to women in Australia than British theories of equal rights. Indeed, the WCTU which arrived in Australia in 1882, by 1894 claimed 7400 members.[123]

From the 1860s David Syme was advocating through the Melbourne *Age* the need for greater colonial independence from Britain. He may have been provoked by deep resentment in his Scottish bones of 'perfidious Albion', but he also understood the extent to which the colonies were dominated by British economic policy.[124] In Sydney another Scot, J. D. Lang, came to the conclusion that the only future for Australia was as a federated republic.[125] Writers of the calibre of Syme and Lang made more impression in London where the colonies were a simmering nuisance than they did at home. When prominent conservative MP and historian J. A. Froude toured Australia with his son in 1885 to investigate rumours of republicanism and to argue, if necessary, the case for continuing loyalty to the Empire, he found little to concern him. This may have been because he moved only in the best circles, staying at the Melbourne Club, the Australia Club in Sydney, and visiting various government houses as a guest. It may also have been because he arrived in the wake of the Sudan affair with its heightening effect on imperial loyalty.[126]

Genuine republicans were scattered through small, radical, and mostly workingmen's organizations, noticeably in Melbourne. As well as Syme, Justice George Higinbotham's uncompromising anti-imperialism remained a source of republican inspiration. Tasmanian lawyer Andrew Inglis Clark, an admirer of Higinbotham, brought his own pro-American pro-republican views to bear on the framing of the federal Constitution, but general republican sentiment faded rather than grew in strength towards the end of the century. There was some reluctance to use the term 'Commonwealth' to describe the federation of the colonies because of 'the flavour of republicanism and the suggestion of separation'.[127] Even the Irish who might be expected to have been keen republicans seemed generally satisfied with the degree of autonomy responsible government gave to the Australian colonies and the quality of religious freedom available to the Catholic church.

Yet as republicanism faded, anti-imperial or national feeling gained strength. By the 1880s imperial federation was being advocated to formalize relations with Britain. There were various practical reasons for administrative co-operation within the Empire—for example, for running the postal services or, less successfully, to standardize the census and the law on divorce. But there was little active enthusiasm. From 1887, Australian premiers joined other colonial leaders in London, ostensibly to observe and celebrate milestones in the longevity of the Queen, but also to discuss Empire trade, defence, communications and social problems. In Australia public ceremonies honouring Her Majesty could be relied upon for an outburst of anti-monarchist or anti-imperial sentiment. Henry Lawson recalled, 'in '87 when the Sydney crowd carried a disloyal amendment on the Queen's Jubilee, and cheered at the Town Hall for an Australian Republic, I had to write then or burst'.[128] Most of these feelings whether for or against Queen and Empire, were contained by national feeling focused first on individual colonies, but by the end of the 1880s increasingly on the idea of young Australia itself, symbolically represented in the *Bulletin* as 'the little boy from Manly'.[129] The labour parties of the 1890s still found it necessary to assert that they would agree to 'the federation of the Australian colonies upon a National as opposed to an Imperialistic basis',[130] but awareness that as colonials they came second in British eyes made positive feelings about Australianness easier, especially for the increasing number who had never known any other home.

'Temper democratic, bias offensively Australian' ran the masthead of the weekly *Bulletin* magazine which began publishing in Sydney in 1880. The *Bulletin* was not averse to scoring witty or scurrilous points at the expense of Queen or Empire, but its constant enthusiasm was for the things it admired most about Australia—egalitarianism, democracy, masculinity, the bush tradition of mateship, the emerging Australian type. It contributed more than a little to the belief so clearly stated in the first federal parliament's White Australia Policy that the Australian type and Australian society had evolved into something unique and worth protecting. Unlike the Irish and other European nationalist revivals from which it took much of its inspiration, Australian nationalism had no slumbering tradition on which to draw, no folk memories or cast of mythic heroes to reinstate. Though two histories, white and Aboriginal, were linked in their dependence on the land, their experience was in conflict. White settlers looked to the sky rather than the land for the oldest sign of their separation from Europe, and found in the Southern Cross their common symbol. Ingenious use was made of the recent Australian past, the translation of the

Eureka rebellion into a nationalist, working-class revolution, or of convicts and bushrangers into folk heroes, and hard times into 'roaring days'. Yet the most flourishing sentiment was intercolonial rivalry. The Australian colonies were six separate entities with six different histories. The national spirit really only existed through their competition and in relation to Britain and the Empire.[131] So, it was 'the crimson thread of kinship', the common British heritage, that Henry Parkes invoked in Tenterfield School of Arts when he urged federation in 1889.

3
SOCIETY

THE POPULATION of Australia was officially 1 145 484 (668 560 males, 477 025 females) at the end of 1860. By the end of 1900 it had risen to 3 765 339 (1 976 992 males, 1 788 347 females). These figures do not include any full-blood Aborigines, though they do include some people of mixed race who might now think of themselves as Aboriginal. The Aboriginal population has been estimated at 179 402 in 1861, declining to 93 536 by the beginning of 1901. These figures might also include some of the people counted in the official total, though the numbers are probably not great.[1]

The study of statistics was one of the most exciting and challenging areas of human expertise in the second half of the nineteenth century. In Britain, the first modern census was held in 1851 under the supervision of Dr William Farr, and a decade later, the Victorian registrar-general, William Henry Archer, who was an admirer of Farr's work, succeeded in getting the governments of Victoria, New South Wales, Queensland and South Australia to hold a census simultaneously with that of Britain. Thereafter, arrangements broke down and it was not until 1881 that the first simultaneous census of the Australian colonies took place, as part of a grand simultaneous census, in fact, of the British Empire. Even so, the value of this census was less than it might have been since all the enumerators' sheets for New

South Wales, stored in the Exhibition Building, were destroyed in the great 1882 fire before a detailed analysis was completed.

Archer, and his successor in Victoria, Henry Hayter, along with their New South Wales counterpart, Timothy Augustus Coghlan, were ambitious and able exponents of the contemporary art or science of statistics. Hayter transformed the plain statistics of Victoria into the impressive, almost compulsive *Statistical Register of Victoria* and was the enthusiast behind the regular conferences of colonial statisticians working towards comparability and paving the way for a consolidated approach to Australian statistics in 1901. Coghlan's annual volumes, after 1886 called grandly *The Wealth and Progress of New South Wales*, made plain the rationale for his work. Statistics were the modern balance sheet of progress. They showed better than anything else the value that had been added to society since the last accounting and at what cost.[2]

In these statistical year-books and in the extensive volumes of statistics presented annually to each of the colonial parliaments, population and vital statistics occupied a prominent position, since an expanding, healthy and youthful population was the first requisite of expanding production, a strong labour market, and vigorous local consumption.

The Aborigines took what was then a logical place in these vital statistics. In most colonies (South Australia was the exception) they were not counted at all because they were considered irrelevant to wealth and progress. In some places, in parts of Western Australia, for example, they were counted where they were a part of the workforce. Elsewhere, they were seen as a charge on profits—as were lunatics, and criminals, for example—and were counted in this capacity. Obviously if they were living on missions or reserves, it was a fairly simple matter to count them and their cost. Thus in 1861, 6985 Aborigines were officially recognized in the statistics, but clearly there were large numbers living a traditional or nomadic existence in those parts of the continent which had yet to be explored.[3] By the end of the century this situation had changed considerably. Most of the continent had been traversed and most Aborigines could have been accounted for, had this been desired. It was not desired, and for very complex reasons. Statistically, as well as in other ways, the Aborigines seemed to be disappearing. The census takers tried to include as many Aboriginal people as possible among the very large group of people who gave Australia as their birthplace. There were at least two reasons for this. One was the long-standing desire to have as few liabilities on the books as possible. The other was an appreciation, on both sides, that to be classified as an Aborigine in late nineteenth-century Australia was to be a disadvantaged person, legally, politically, economically, socially. It was not impossible to 'overlook' an Aboriginal mother or grandmother in a society which traced descent through the male line, though this was how most mixing of the races occurred. It seems that people of Aboriginal parentage were passing into European society when they could in order to avoid the stigmatic treatment or the disabilities they would have suffered by claiming their Aboriginal inheritance. In this they were officially encouraged, for example by the Victorian Aborigines Protection Act of 1886 which laid down assimilation policies to be followed where it was obviously too late to try to prevent miscegenation.[4]

The question of how the Aborigines were to be counted in the Australian population was ferociously resolved for the future during discussions in the 1890s to draft the federal Constitution. 'Aboriginal natives' were specifically excluded by the Commonwealth Constitution Act 'in reckoning the numbers of the people', though for administrative reasons attempts to keep a record were still made. Those colonies with large Aboriginal populations were immediately disadvantaged by the decision that Aborigines would not count for the purpose of deciding the number of representatives for each colony in federal parliament proportionate to population. (Neither women nor children had voting rights at this stage, yet they were counted.) Part-Aborigines who were already enrolled as voters could not be excluded, however, just as the women of South Australia and Western Australia could not be deprived of their existing voting rights.[5] How many of these part-Aborigines there were will probably never be known.

Thus, despite a fondness for statistical bureaucracy, inherited, no doubt, from convict days, as well as an early enthusiasm for co-ordinated census taking, we have still only an approximate idea of the real size and composition of the population of the late nineteenth century. We can be certain, however, that it was larger than the official statistics show at mid-century. This fact has had in itself a distorting effect on all subsequent statistical calculations for the period, not only on the vital or demographic statistics, but also on all measurements of achievement in education, or health, or wages, or wealth, or the general standard of living. Australia's proud self-image towards the end of the century rests in part on a statistical self-deception. The statistics tell something about what was being done, but as a measure of achievement they are false, possibly dangerous.

Of all the colonies Victoria had the largest white population in 1860—538 234 compared with New South Wales's 348 546. This advantage was retained until 1890 when Victoria began to lose people, mainly to Western Australian goldfields, while New South Wales continued to grow steadily. At the end of the century New South Wales had 1 630 305. Victoria had 1 196 213. The colony which grew fastest in the second half of the nineteenth century, however, was Queensland where in 1900 the population was more than seventeen times what it had been in 1860—from 28 056 to 493 847. By the end of the century, Queensland had gone from being the second smallest colony (Western Australia was the smallest in 1860 with 15 346) to the largest after New South Wales and Victoria, and that was without counting the Aborigines. Much of this increase flowed from the opening of Queensland. The rest was a result of persistent (and expensive) programmes fostering immigration. Western Australia also grew rapidly, its population multiplying by twelve, but most of this growth occurred in the 1890s. New South Wales grew a little faster than South Australia, and both grew faster than Victoria. Tasmania showed the slowest rate of population growth. During the forty years from 1860 to 1900, the Tasmanian population did not quite double from 89 821 to 172 900.[6]

Apart from these differing rates of growth there were other demographically distinguishing characteristics from colony to colony, notably in the proportion of immigrants to native born, the ratio of men to women, and age structure.

For Australia as a whole, natural increase (excess of births over deaths) contributed more to the population growth rate than immigration in this period. However, the rate of immigration rose from 1860 until the beginning of the 1890s when it fell sharply. In the 1880s, 42 per cent of population growth had been from immigration. In the 1890s it was only 4 per cent. Clearly those colonies with the highest rates of population growth also had high levels of immigration. Queensland in particular maintained a series of schemes for assisting immigration which, over the last four decades of the nineteenth century, brought a total of 166 975 people to the colony. Despite official fears that Queensland was only subsidizing immigration to the more attractive and settled southern colonies, her population growth rate suggests that either migrants stayed or were replaced by internal migration to Queensland from other colonies.

Victoria abandoned assisted immigration in the 1870s. Tasmania could not afford to assist more than a few hundred, though both South Australia and New South Wales subsidized numbers varying from a handful to several thousand in most years. Advertising and administrative facilities to assist migrants in Britain also tempted immigrants able to pay their own passages. No attempts were made to regulate or restrain this flow which was steady till the 1890s.

Even in the 1880s at the height of Australia's late nineteenth-century attractiveness to immigrants, more than 50 per cent of population growth rate resulted from babies born in the colonies. In the early 1860s the birthrate was 42 per thousand of the population. This fell gradually to 34.4 per thousand in 1890–92 and to 27.1 by 1900. As the death rate also fell slightly, and the immigration rate fell dramatically during the Depression of the 1890s, the number of native-born Australians in the population began to overtake the number born elsewhere. In acknowledgement of this changing relationship, the Australian Natives Society (later Association) was formed in Melbourne in 1871 as a friendly and mutual insurance society. It grew slowly, mainly in the goldfields towns in Victoria, developing cultural and political interests as nationalism became more prominent and fashionable. Its name was not without its wry overtones, but the existence of such an organization is evidence that it meant something to be born in Australia.

By 1901, 77.06 per cent of the population (of the 'immigrant races' as the 1901 census carefully put it) had been born in Australia.[7] People with British birthplaces accounted for 18.03 per cent, Asia 1.25 per cent, 0.79 per cent of those from China, and Germany 1.02 per cent. Development patterns and immigration policies during the preceding forty years, however, ensured interesting differences from colony to colony in the composition of population according to birthplace. The most 'Australian' colony was Tasmania with 86.02 per cent of its people born here. It also had only tiny numbers of people from Asian or other non-'Australasian or European' backgrounds—a mere 0.63 per cent. Queensland was the least 'Australian' with only 64.97 per cent native-born white population. Queensland also had the largest British-born population of all the colonies, 25.34 per cent, and the highest proportions of people born in Europe, Asia, and the islands of the Pacific. The birthplace profile of Western Australia's population was similar to that of Queensland, though a little less dramatic. In Western Australia, however, 1.47 per cent of the population had been born in New Zealand and had come, probably directly, by ship to the goldfields. Of course, in Western Australia there were hardly any Pacific Islanders. New South Wales, South Australia, and Victoria most closely corresponded to the average Australian distribution for birthplaces, though South Australia's German immigrants showed up clearly, as did the tiny number (almost as small as in Tasmania) of people with Asian birthplaces in Victoria—which raises questions about the reasons for hostility to the Chinese in Victoria long after goldfields-induced anxieties should have faded.

If the distribution of the Aboriginal population is also taken into consideration, Australia as a whole was more cosmopolitan and less white the further one travelled from the south-east corner which was the real heartland of the white native born. At the outer fringes the population was immensely heterogeneous, far-from-white, and certainly less reliably counted or considered.

The two other noteworthy demographic factors during this period, relative distribution of the sexes and of young and old people, were also most marked away from the south-east corner. Table 3.1 shows the number of males for each 100 females from 1860 to 1900.

Table 3.1: Masculinity rates

1860	140.15
1865	125.38
1870	121.10
1875	118.25
1880	117.28
1885	118.33
1890	116.06
1895	113.41
1900	110.23

Source: Commonwealth Year Book, 1908, p. 155

In 1860 there were roughly 14 men for every 10 women. This evened out somewhat, so that in 1900 there were only 11 men for every 10 women. Still this was one of the highest masculinity rates in the world. Only New Zealand had a higher rate. The figure for the USA in 1900 was 104.87 men to every 100 women. In England and Wales it was 93.63, Scotland 94.58, and Ireland 97.40 men to every 100 women.[8] Variations from one Australian colony to another are worth mention. Victoria had the most 'normal' masculinity rate by 1900, 101.23 men to every 100 women. Then followed South Australia (104.04), Tasmania (107.97), and New South Wales (111.14). Queensland (125.33) and Western Australia (157.54) had the most concentrated male populations, though these figures include 'Asiatics' and Pacific Islanders who were obviously disadvantaged in the search for sexual partners. They do not include Aborigines, though Aboriginal women are known to have been included, at least for practical purposes, in the sexual equation. This still gives a poor idea of the real distribution of the sexes on a geographical basis, for women predominated in urban areas while men outnumbered women in most rural districts. Contemporary observers who noted the sex imbalances between urban and rural populations were most likely to perceive them in terms of different growth rates and to lament with Coghlan's successor, W. H. Hall, that the urban population was increasing about three times faster than the rural. This trend had been marked since the mid-1870s. Hall was content with a vague definition of the difference between urban and rural, and more concerned with what he thought was a moral danger inherent in this trend:

> it seems to be the height of ambition amongst many of the farmers' sons and daughters to procure employment in the Metropolis. This is a most regrettable tendency... our Intelligence Department should take pains to emphasise in its literature... that the man who rears a pig or a calf, or the girl who milks a cow and looks after the poultry, is a far better economic asset to the State than a clerk or policeman, or a typewriter and milliner.[9]

One of the more comforting myths about Australian society during this period was that it was easy for women to find husbands, simply because of the high masculinity rate. Marriage was thus not only a woman's right: it became her duty. Women who failed to marry were perverse. Marriage was indeed easy in the country, especially in Queensland and Western Australia, but in the larger cities and towns, this was not so. For besides the statistical distribution of men and women, other factors such as birthplace, religion, or age had their limiting effects on the availability of marriage partners. Asiatics and Islanders were virtually denied access to white women, though some marriages did occur. Scandinavian immigrants intermarried, but at the cost of their culture and church. A study of marriage and mobility in Sydney in the 1870s and 1880s concluded that 'the traditional story of the colonial man having to take what he could get in the way of a wife was possibly more real for the female Irish immigrant.'[10] Unable to marry within her own national or religious group because there were too few men, rejected as too lowly by the rest of the male population, the Irish girl may have been driven to prostitution or, if she was lucky, into the arms of a successful Chinese businessman.

The age structure of the population during this period was perhaps even more unusual than its sexual distribution, but like the sex ratio, helped create a particular image of Australian society which had important consequences in determining the structure of the workforce. Table 3.2 shows the distribution of the population by age and sex.

The population of the Australian colonies in 1860 was predominantly youthful or middle-aged, a result, obviously, of the kind of immigration stimulated by the gold-rushes and

Table 3.2: Age distribution of Australian population, 1861–1901 (percentages)

	Males				Females				Persons			
	Under 15 years	15 and under 65	65 and over	Total	Under 15 years	15 and under 65	65 and over	Total	Under 15 years	15 and under 65	65 and over	Total
1861	31.41	67.42	1.17	100	43.03	56.20	0.77	100	36.28	62.72	1.00	100
1871	38.84	59.11	2.05	100	46.02	52.60	1.38	100	42.09	56.17	1.74	100
1881	36.37	60.85	2.78	100	41.89	56.07	2.04	100	38.91	58.65	2.44	100
1891	34.77	62.02	3.21	100	39.36	58.08	2.56	100	36.90	60.20	2.90	100
1901	33.87	61.82	4.31	100	36.50	59.85	3.65	100	35.12	60.88	4.00	100

Source: Commonwealth *Year Book*, 1908, p. 156

the healthy rate of natural increase. Old people were little in evidence. By the end of the century, however, the age structure of the population had come to resemble the more 'normal' patterns of England and Wales. Just over a third (35.12 per cent) of the Australian population were under 15 years old (32.42 per cent in England and Wales), 60.88 per cent were between 15 and 65, i.e. probable producers (62.91 per cent in England and Wales), and 4 per cent were 65 and over (4.67 per cent in England and Wales).

Again there were noteworthy differences from one colony to another. Victoria had the most elderly people in 1901 (5.5 per cent of its population over 65), as a legacy from the gold-rush days. Western Australia and Queensland had the fewest old people—reflecting their recent immigration histories (1.81 per cent and 2.59 per cent respectively). New South Wales (3.44 per cent), Tasmania (4.07 per cent), and South Australia (4.11 per cent) came closest to the norm for over-65s. The greatest proportions of children were to be found in Tasmania (37.12 per cent) and Queensland (36.62 per cent) in 1901. New South Wales had 35.94 per cent of its population under 15, South Australia 35.59 per cent, and Victoria 34.08 per cent. Western Australia had the fewest children proportionately (28.93 per cent) and the greatest proportion of its population between 15 and 65—a typical age-profile for a society of recent immigrants and rather like that of Victoria in the 1860s. No obvious explanation presents itself for the large proportion of children in Tasmania, though the concentration of the population in farming and other rural activities may have been significant. So too may its relative homogeneity.

Differences in the age structure between the male and female population shaped the marriage market. The fact that the mass of the female population was constantly younger than the male led to high hopes for its marriageable and reproductive capabilities. But before 1890, 80 per cent of additions to the female population were from natural increase, whereas the immigration of young, and even middle-aged men contributed a significant proportion of each increase in the male population. There were relatively few unmarried female immigrants, most of them children or girls in their teens and twenties. So at any given time between 1860 and

1900 a substantial number of females recorded in the statistics were too young to be marriageable. Once migration ceased to dominate population growth, natural increase began to even the numbers of males and females as well as the ratio in younger age groups.

It is hard to say whether it mattered that most wives were native born whereas a proportion of husbands were born outside Australia. This may have been a source of tension, for example, in household matters or child-raising practice. It almost certainly contributed to low esteem or lack of confidence in many Australian women, and to the off-hand way in which they were treated in general—mere colonials. There was endless discussion of the quality of the Australian girl, little of it favourable.[11] Furthermore, because of this constantly youthful Australian-born female population, grandmothers were young, still likely to be bearing their own children when their eldest daughters began families.[12] Women whose age and experience might command authority were in short supply. The proud matriarch was a rare figure in this period. In a typical marriage, the husband fell naturally into patriarchy with his younger, unconfident, colonial-bred wife.

The other side of this excess of girls and very young women was the number of ageing immigrant men who did not manage to marry. For them the consequences were severe also. These men were least likely to have families of any kind, neither brothers, sisters, nor cousins. Homelessness became a problem. So did alcohol. An analysis of the earliest recipients of the old-age pension shows an excess of men over women in the applicants.[13] The price of immigration for development was paid by old and lonely men, the 'hatters', tramps and recluses, the inmates of homes for inebriates and the insane, where, by the end of the century, men dramatically outnumbered women.

Men who did marry were on average about four years older than their brides.[14] This gap narrowed from 1860 to 1900, but was always smallest in those colonies where the sex ratios were nearest to even. The smallest gap between the ages of brides and grooms was in South Australia—about 2.9 years in the 1880s. In Victoria in the 1870s and in Queensland in the 1870s and 1880s, five years difference was the average.

Perhaps this difference made marriage less intimidating to men who expected their wives to be virginal, innocent, and teachable. But for women, despite the belief that theirs was a sellers' market, the age gap made for inequality, reinforcing the socially sanctioned authority of the husband with age and experience.

The total effect of the unequal proportions of the sexes, exacerbated by their unequal distribution between country and city, and through the age groups, was to hold the marriage rate down in a period when it might have been presumed to be much higher. Even so, marriage rates for women before 1890 were much higher in Australia than in Britain during the same period. For men, however, they were much lower. Whereas only 3 per cent to 4 per cent of women in Australia had never married by the age of 50, for men the proportion was between 20 per cent and 30 per cent. The dramatic exception was in South Australia. There during the 1860s and 1870s only 8 per cent of men reached 50 without finding a wife, and although the competition became stiffer in the 1880s and 1890s, still only 15 per cent of men had failed to marry by 50.[15] Undoubtedly South Australia's slow economic growth and its settled concentration on agriculture caused the kinds of men who were least likely to marry to drift to other colonies. (In fact, a surprising proportion of the men responsible for New South Wales's crime rate late in the century had birthplaces in South Australia.[16]) On that basis, Tasmania should have shown a similar marriage pattern. That it doesn't may be because until the end of the century there were many ageing male convicts in Tasmania who had long been denied the option of marriage. In South Australia, with the difference in age between husbands and wives the lowest of all the colonies, both men and women had justifiably good expectations, not only of marrying, but of marrying someone of much the same age, background and experience. This balance was a product of the carefully planned family migration of the Wakefield generation and persisted because of the high contribution of natural increase to population growth and subsequent low rates of male migration to South Australia.

Given the considerably different rates of marriage for men and women in the other colonies, it would be surprising if sexuality was not an area of tension throughout the late nineteenth century. Women expected to marry. Men feared they might not. Women could afford to play the field, and evidently did, more or less from the time they left school. Men were frequently forced to lower their expectations. Young women were in a relatively powerful position because of their scarcity value. But once a woman married she lost that power and significance. It is not difficult to detect in the columns and cartoons of the illustrated newspapers and magazines a deep feeling of resentment against women, the power they could wield as mere girls, the constant threat of emotional blackmail as wives.

In the USA the shortage of women, especially in frontier districts, has commonly been advanced as an explanation for their early acceptance as equals and their enfranchisement. By contrast, in Britain the surplus women problem led to demands for education, entry into the professions, and enfranchisement. Australia has usually been assumed to be closest to the American model with a shortage of women. Yet South Australia, which paved the way for education and voting rights for women, had a good sex ratio, a male birthrate almost as high as the female birthrate, the highest marriage rate for both men and women and the smallest average difference between the ages of husbands and wives. (It also had the lowest insanity rate of any of the colonies.) Perhaps there was simply less sexual anxiety and tension in South Australia, and therefore no reason to fear better legal, political, and social rights for women, though it might still be asked why reforms took so long. Beyond South Australia, the idea of improving the legal condition of women may have exacerbated existing sexual insecurities. But the balance was already changing. The fall in migration in the 1890s, combined with the effects of Depression and drought, meant that a generation of women of marriageable age were suddenly without partners. By 1921, 17.6 per cent of women born in Australia between 1872 and 1876 had still never married. (Men were not affected in the same way. Many deferred marriage, but then married considerably younger women.[17]) Still the notion persisted that marriage was a woman's duty. It seems incredible now that tensions arising from passing imbalances of age, sex, and place of origin could have become so much a part of the national psyche, but this was by no means the only such legacy from these years.

Late nineteenth-century Australian society was unusually healthy by contemporary standards. The annual death rate declined from about 15 per thousand in 1880 to about 12 per thousand at the end of the century. (Comparable figures for Britain were 19 and 18.3.[18]) This was partly because of its youthful age structure and partly because the time taken by the voyage from all foreign ports to Australia served as a natural form of quarantine for severe infectious diseases. Even so, average life expectancy was only 51 for men and 54.8 for women at the end of the century.[19] Despite the declining death rate, there were at least three reasons for the low average life expectancy. One was the noticeably high rate of infant mortality. This was as high as 75 per thousand in the 1870s. Typically, the March quarter of the year was worst for infant deaths, followed by the December quarter. One child in ten in Sydney died in 1875 during particularly severe epidemics of diphtheria, scarlet fever, measles, plus normal fly- and water-borne diseases. By the 1880s, of every hundred children born, 90 at most might be expected to survive to their first birthday, 82 might see five years, and only 78, adulthood. By the turn of the century, 90 would survive the first year, 88 to five, and 85 to adulthood.[20] The improvement noted in the 1890s was attributed to better water supply, sewerage and drainage systems, as well as greater public awareness of the need for children to have high levels of nutrition including unadulterated, disease-free milk.

The second reason was the high accident rate which accounted for many male deaths, particularly in country areas where guns, horses, mining and tree felling were everyday dangers. Finally, there was an unusually high incidence of death from tuberculosis and other lung and respiratory diseases, reaching a peak of 135 per hundred thousand in 1884, although this could be explained by the number of immigrants who had already come to the colonies suffering from these diseases in the hope that the climate would aid recovery.[21] According to James Bonwick, 'Australia [was] rightly esteemed as the poor man's paradise and the invalid's sanitorium.'[22] After phthisis (the nineteenth-century term for tuberculosis), old age, enteritis, accidents, heart disease, pneumonia, cancer, then several of the complaints which carried children off at birth or shortly after were the major causes of death.[23] Apart from TB and the childhood diseases which were preventable with better sanitation, hygiene, nutrition, and medical care, the pattern was beginning to look recognizably modern. Nineteenth-century commentators had already begun to notice the increasing significance of cancer in the statistics, partly because diagnosis was becoming more precise, partly because there was already a decrease in the number of other diseases for which medicine had no effective response.

Till the end of the century, most hospitals were administered as charitable institutions and provided care only for the sick poor, those who could not afford treatment in their own homes, who had no homes, or no one to care for them.[24] In recognition of this service, especially to the homeless, governments subsidized hospitals along with other charities. Illness was treated by preference at home by the doctor who came to call. Nursing was provided by the women of the family, though if they were the patients, a professional nurse might be engaged. Nurses who were experienced in midwifery most often came in during confinements. Women's hospitals in the major cities testify to the need for accommodation for women who had nowhere else for childbirth or could not afford a private doctor or midwife. Children's hospitals were set up to isolate sick children from their overcrowded homes and insanitary environments. Country hospitals most closely resembled modern hospitals in function with their mix of accident and chronic cases. Only towards the end of the century did antiseptics and medical technology begin to give hospitals the edge over home care.[25]

A French visitor, Oscar Comettant, observed that Melbourne in 1888 was one of the best places in the world for a doctor to set up in business. If a baby had sore gums due to teething or a scratch from a fall the doctor was sent for.[26] At a time when the medical profession in Britain was building its professional standing on medical technology and apparent willingness to tackle public health problems, the Australian medical profession was capitalizing on obstetrics and paediatrics. The general practitioner who coped with rural accidents and was knowledgeable about public health and cultural and educational affairs might command respect formerly reserved for the clergyman. But with the exception of a few outstanding public health crusaders, medicine as practised in the cities and suburbs was inadequate and complacent from lack of challenge in the complaints presented.[27] Affluence produced uncritical, even self-indulgent sufferers. The insurance companies found that despite their generally better health, as they grew older, their Australian members had longer periods of illnesss than their British counterparts.[28] Doctors prospered despite the general good health, and despite the fact that their qualifications were sometimes as dubious as their remedies.

In 1861 the Australian colonies not only attempted some kind of co-ordination in the timing of the census, they also tried for a degree of uniformity in the categories of information collected. One useful category was the birthplace or nationality of residents. Though neither complete nor perfect, statistics on birthplace show that several major changes occurred in the composition of Australian society between 1860 and 1900, notably the replacement of immigrants by the native born. This weakened automatic ties with the Old World and assisted the growth of ideas and institutions based on Australian experience. In 1860 there were some families who could claim to be in their third generation in Australia; by the end of the century there were many such families, and there were some who were reaching their fourth and fifth generation. The 'Australian type' was seen as a successful mix of migration from England, Scotland and Ireland. At the same time, arriving immigrants found smaller or larger groups whose names, religion, language or colour reminded them of home. Some migrants were keen to throw off their past. Others cherished it. And their children could claim it if they wished.

Table 3.3 shows the relative proportions of English and Welsh, Scots, and Irish in the total population during the late nineteenth century. Migrants from England were always in a

Table 3.3: Origin of UK-born Australian residents (percentage distribution at census dates)

	1861	1871	1881	1891	1901
England and Wales	56.33	53.92	54.93	57.28	57.93
Scotland	15.48	14.65	14.25	15.08	14.97
Ireland	28.19	31.43	30.82	27.64	27.10

Source: Eric Richards, 'Australia and the Scottish Connection' in R. A. Cage (ed.), *The Scots Abroad*, Croom Helm, London, 1984, p. 147

majority, but with the exception of groups like those who came from Cornwall to South Australia, and middle-class women who came under the protection of special immigration societies, they have been largely taken for granted.[29]

There seems to have been no special attraction or disadvantage during this period in being English in Australia or having an English parent. An Irish background or an Irish name, however, might be something altogether different. The proportion of colonists born in Ireland was already beginning to decline in the 1860s and by 1900 was only about 4 per cent of the whole Australian population. Irish immigration to Australia peaked around 1870 with about 200 000 a year, but the proportion of people with Irish names continued to rise. The most important characteristics of the Irish migrants who arrived in the Australian colonies during the second half of the nineteenth century were that they were young and unmarried, both men and women. They were also generally better educated or had better resources than Irish migrants of the same period who went to Britain or the USA.[30] As early as 1888, the first history of the Irish in Australia was written by J. F. Hogan.[31] Interestingly, Hogan felt that it was not necessary to stress differences between Catholic and Protestant Irish. By the end of the century, however, both recent immigrants and the children of earlier Irish immigrants had to contend with a popular image of the Irish based on convict and peasant origins and assumed to be Catholic. In fact, about 30 per cent of Irish immigrants came from northern Ireland and were likely to be Presbyterian.

Among the names Hogan listed with pride were many politicians. Between 1860 and 1900, three Victorian premiers were of Irish birth, Charles Gavan Duffy, John O'Shanassy, and Bryan O'Loghlen, while other politicians like George Higinbotham exerted a lasting influence. At best Irish immigrants were possessed of talent and training for which there were inadequate outlets at home. So they entered the professions or sought security and advancement in the police force. Already politicized by their experience of English oppression or Catholic repression, they brought a heightened awareness of injustice or irrationality to colonial life. At worst Irish immigrants confirmed the music hall stereotype of comic ignorance or incompetence. One of the saddest of all was Henry James O'Farrell whose occupation is described in the *Australian Dictionary of Biography*, as 'paranoic'.[32] Family failures in Australia, his own inability to enter the priesthood, his addiction to alcohol, and the current agitation about Fenianism exploded in O'Farrell's bewildered brain, and the solution seemed to be to take a pistol and shoot the Duke of Edinburgh when he visited Clontarf Beach, Sydney, on 12 March 1868. This action set off a wave of over-indignant sectarianism and paranoia about the threat of Fenian activity in Australia. Even though, due to the skill of Lucy Osburn, Florence Nightingale's nominee as head of nursing at Sydney Hospital, the Duke recovered and made a personal appeal for mercy for O' Farrell, the unfortunate Irishman was sentenced and hanged within six weeks.[33] Fears of an international Fenian conspiracy were not groundless. The next year, John Boyle O'Reilly, transported with sixty-one other Fenians in the *Hougoumont*, the last convict ship to make the voyage to Western Australia, escaped from Bunbury with the assistance of the local Catholic priest and an Irish farmer. He was picked up eventually by an American whaler, the *Gazelle*. In America O'Reilly and his brother Fenians began planning the rescue of the others still in Fremantle Gaol. Under cover of Perth's Easter regatta in 1876, six Fenians escaped and were taken aboard the *Catalpa*, a whaler from New Bedford acquired specially for this mission. Since it could be argued that they were political prisoners, the escaped Fenians were allowed to settle in the USA.[34] These incidents, though dramatized as little as possible, brought the colonies closer than usual to the violent politics of the Old World. There was the possibility of Fenian attacks in

Australia on the symbols of Britain. Later in the century, when the relatively moderate conduct of the battle over state aid had demonstrated that there was no longer any need to fear Fenianism, the Irish in Australia became a useful source of funds for the home rule movement, with tours on behalf of the Irish parliamentary party by John and William Redmond (1883), John Dillon (1889) and Michael Davitt (1895). The generations who had learned their politics in Ireland were giving way to the Australian-born Irish who were either fairly content with their position in the scheme of things, or ineffectually romantic about their Irish heritage.

Scots migrants constituted a numerically smaller proportion of the population than the Irish, but they were more in evidence than that other group of British migrants, the Welsh, despite the prominence of names like Griffith and Hughes in our political history. Just as it is dubious to define the Irish through the Catholic church, so it is with the Scots and Presbyterianism. Edinburgh-born (Sir) Alexander Stuart, Premier of New South Wales (1883–85) and vice-president of the Highland Society of New South Wales, was all his life a devout Anglican and prominent in Anglican politics in his adopted land. The most that can be said with certainty is that the Presbyterian church saw no need to deny its Scots inheritance in the Australian colonies, even proudly acknowledging it. The church probably accounts for conforming or successful Scots.[35] Despite a tradition of emigration by the ambitious or upwardly mobile lower professional and educated classes, and despite another tradition of economic inventiveness and ingenious self-help, a proportion of Scots who came to Australia failed to fulfil their ambitions, deserted or were deserted by the church, and were fairly ordinary farmers, labourers, or fishermen. In their homeland the Scots were keen drinkers as well as fierce exponents of theology. An interesting question, though a delicate one, is whether Australia's draconian laws on drinking were the work of successful Scottish-born politicians and civil servants anxious to restrain their fellow countrymen or of English-born politicians wishing to control the activities of Irish-born brewers, hotel-keepers, and their Irish-born patrons. In Victoria in particular the Scots were seen to be disproportionately influential in economic and political life which translates to modern moral terms as arch-exponents of capitalism and exploitation.[36] In Queensland, too, they were strongly represented due in part to John Dunmore Lang's enthusiastic promotion of Queensland, continued by advocates like Arthur Macalister, a former premier who became a born-again Scot when he was appointed Queensland's agent-general in London in 1876. Queensland was the logical destination of both Scottish migrants and capital routed initially through Victoria. Subsequently Thomas McIlwraith's continuous search for new ways to finance development in Queensland kept not only investors, but lawyers and accountants anxiously aware of that colony.[37]

David Macmillan has shown that Scots migrants to Australia before the 1860s came equipped with superior educational and economic resources.[38] Like the Irish they had every reason to put British dominance behind them, but unlike the Irish they were impelled by a religious and national culture which was itself seething with dissension and uncertainty. The austerity, self-denial, and grim self-reliance which had become distinguishing characteristics of Scottish Calvinism by mid-century were perfect equipment for the rigours of pioneer life, whether in Australia or North America. The foundations thus laid ensured an encouraging environment for subsequent migrant generations. But those same qualities became uncomfortable, even unmanageable in conditions of rising prosperity and bourgeois affluence. Calvinism is conducive to the accumulation of capital, but it is no guide to consumption, either for conspicuous purposes or merely to enhance production. Among the successful Scots immigrants there were some spectacular failures in the later part of the nineteenth century, men like James Munro and A.J. Balfour who reached high places through their success in business and their subsequent involvement in politics, who had preached austerity, but who then defaulted to the tune of hundreds of thousands of pounds.[39] This was certainly a setback for other-worldliness, though it may have been a lesson that making money simply for its own sake is a narrow and narrowing activity. It is certainly not a principle on which to build a society, let alone a civilization. Yet the basic values of austerity, hard work, self-denial and canniness laid foundations of prosperity without which there could be no society, no civilization.

As the question of Irish home rule came to dominate British politics, events in Ireland began to cast the shadow of Fenianism over the Irish community in Australia. Scotland, however, developed a certain fashionable quality, thanks to the 'dear Queen's' fondness for Balmoral, the ease with which artificial silks were produced in tartan-inspired patterns, and the proliferation of mezzo-tints of highland hunting scenes, wild lochs, and romantic castles. Unlike Irishness, Scottishness carried no undesirable overtones.

The largest group of settlers of non-British origin were German—26 872 in 1861, rising to 45 000 in 1891.[40] In the 1860s the Germans were already settlers of long standing. They were scattered throughout the colonies, as skilled farmers, tradesmen, and teachers, with important concentrations in South Australia north and east of Adelaide, in the Riverina where South Australian Germans had moved in search of new farm land, and in southern Queensland. Both the South Australian and Queensland communities were based on the settlement of German pastors and their followers. Both continued to provide a strong Lutheran focus and to maintain a distinctive German identity. The church continued to attract immigrants from Germany with pastors and missionaries. As in the first generation of settlers, whether they came in search of religious freedom, were pushed by political events in Europe, were pulled to Australia by the prospect of gold, came, as did Mueller, the botanist, for the sake of his health, or Holtermann, the photographer, to escape military service, news of their success continued to encourage others.

In South Australia German schools and the German community managed to sustain at least two German newspapers in the 1870s, both weeklies, one in Adelaide, the other in Tanunda. Friedrich Basedow, owner of the *Tanunda Deutsche Zeitung*, acquired his Adelaide rival in 1875 and his father-in-law, Dr Carl Muche, became editor of the new *Australische Zeitung*.[41]

During the later part of the nineteenth century the Germans in South Australia became crucial to the development of agriculture and especially the wine industry in that colony. Names like Seppelt and Buring were attached to vintages. The Germans' approaches to farming, their industriousness and willingness to take pains, were admired, though not always emulated.[42] In the predominantly Methodist religious environment of South Australia, orderly and quietly devout Lutherans were not out of place. The Germans were accepted as modest, admirable and successful settlers.

Friedrich E. H. W. Krichauff personified this acceptance. He was elected to the first parliament of South Australia in 1857 but gave up his seat because he could not afford the time to walk the 28 miles from his farm in the Bugle Ranges near Strathalbyn to Adelaide and back for meetings of parliament. A supporter, not surprisingly, of payment for members of parliament, he spent the 1860s as a public servant, returning to parliament in 1870. He also advocated a simple land title system based on the traditional method of land registration used in Hamburg and other Hanse towns recommended to Robert Torrens by another German settler, Ulrich Hubbe, and thereafter adopted in South Australia and elsewhere as the Torrens title system. He became a driving force for reafforestation schemes, a member of the Central Agricultural Bureau, and of the council of Roseworthy Agricultural College (which was itself largely a German initiative). He was also an energetic advocate of general agricultural education, writing pamphlets and articles on such subjects as artesian water, beet sugar, and the use of fertilizers in field and garden.

In Queensland, also, the potential of German immigrants as agriculturalists was appreciated. In 1862, the Queensland government despatched J. C. Heussler, who had arrived from Frankfurt am Main in 1854 and become a successful merchant in Brisbane, as immigration agent to Hamburg. He was to recruit German settlers and offer them free passages and land orders in Queensland. It was believed that the Germans, unlike so many of the migrants recruited in Britain who did not know sand from clay, would succeed as farmers and thereby help Queensland overcome her problems in establishing agriculture. Over the next thirty years about 12 000 Germans settled in Brisbane and Ipswich. They too supported a weekly newspaper, the *Nordaustralische Zeitung*. Names like Zillmere and Marburg survive as a reminder. Farming was not their only skill. Of the eight Queensland MPs between 1860 and 1900 who were born in Germany, three were merchants (Feez, Heussler, Unmack), two were millers (Horwitz, Kates), two, newspaper proprietors (Isambert, Sturmm), and one a storekeeper (Lissner).

Nineteenth-century Germans adapted easily and quickly to life in Australia, so little of their culture survived the first generation, though a steady stream of new arrivals through the later decades of the century, as well as the church and the language, helped to keep a sense of community alive. Emil Hansel, unemployed in Brisbane in 1890, doing the rounds of the German community in search of work, though 'not very religiously inclined', sometimes went to the German church on Wickham Terrace where he could be sure of finding 'many old acquaintances in the courtyard'.[43] Another Sunday he went 'to Hermann's in Fairfield, where already there were several Germans. Social contact is as essential for life as bread and fruit.'[44]

In 1863 an article on the spiritual needs of the Aborigines in the interior by Pastor Meichel who had formerly been a missionary in India, was published in the Tanunda *Kirchen und Missionzeitung*. Following a mission rally at Blumberg, SA, a call went out to Director L. Harms of the Missionary Seminary at Hermannsburg, Hanover for assistance. Lutheran involvement in Aboriginal missions dated back to 1838 in South Australia. Competition between Lutheran congregations multiplied the number of missions, though most were also desperately poor.[45] Georg Adam Heidenreich was one of forty-five ordained men trained at Hermannsburg who migrated to the Australasian colonies between 1875 and 1877. Heidenreich established Hermannsburg Mission to the Aborigines in central Australia and for many years was a leading force behind Lutheran involvement in Aboriginal missions in central and South Australia. By the end of the century, anthropology at Hermannsburg was identified with Pastor Carl Strehlow. In 1863 Friedrich August Hagenauer established Ramahyuck and spent the rest of his life— he died in 1909—working among the Aborigines, mainly in Gippsland.

Scandinavian immigrants, who by 1891 numbered about 16 500 (Danes, 6400; Swedes, 6300; Norwegians, 3800), were similar to the Germans in many respects. Like the Germans they assimilated easily, and gained a reputation as hard workers. Many had come originally in search of gold, or as sailors, though during the 1870s large numbers of Danes especially arrived via Hamburg and the Queensland government's immigration scheme aimed originally at Germans. These Danes were 'poor and reckless. Had they been in a position to pay their own passage money, the bulk of them would have gone to the United States.'[46] Many found the Queensland climate trying on their northern European complexions and temperaments and drifted south where they ended up in dairying. Those who stayed in Queensland turned to sugar-growing, notably near Mackay, and on the Burnett, at Eidsvold. Perhaps they were encouraged by the example of Edward Knox, Scots-born, but Danish by education and upbringing, founder and managing director of the Colonial Sugar Refining Company, or the Archer family, also Scots based in Norway, whose interests ranged from a pioneering sheep station at Gracemere near Rockhampton, and Queensland politics to shipping and timber businesses in Norway and promoting Ibsen in London.[47]

Unlike the Germans, Scandinavian settlers were unable to sustain strong communities based on church, school, or language. There were too few of them. They were scattered. And they had too few women among them. Twenty-five per cent of the Danes, the main group to benefit from assisted immigration, were women in 1891, but only about 9 per cent of the Swedes and Norwegians could hope to marry a countrywoman. Intermarriage with Australian-born girls was frequent. Henry Lawson's father, a Norwegian sailor turned miner turned selector, married Australian-born Louisa Albury in 1866. Their son was born the next year. The Lawson marriage is not considered a great success, but there were many less fraught. Scandinavians merged with the 98 per cent British population rather more quickly than the Germans, and after 1891 their numbers began to decline. During the 1890s the Scandinavian contribution to Australian society was mostly imported in the guise of certain products in daily use—the Alva-Laval separator in the dairy, and in the pantry two early and popular convenience foods, Hansen's junket tablets and tinned Norwegian sardines.

There were said to be 5486 Jews in 1861, 3379 males and 2107 females, roughly 0.48 per cent of the entire population. They were either descendants of early Anglo-Jewish settlers or German Jews who had been attracted to the goldfields. By 1901 the Jewish numbers had crept up to 15 239 (8137 males, 7102 females) but they had not grown as quickly as the rest of the population and comprised only 0.4 per cent of the total. Some changes had occurred within the community which may help to explain its slow growth.

During the 1860s and 1870s there were groups of Jews in most mining towns including Newcastle, Maitland, and Lithgow, as well as the gold towns and some country towns as storekeepers, publicans, and watchmakers. During the 1870s and 1880s, however, there was a definite move from the country towns and mining centres towards the cities, especially Sydney and Melbourne. Whether this was for religious, cultural, or economic reasons differed with individuals; however the Jews became metropolitan or urbanized during the later decades of the nineteenth century. Even so, neither Adelaide, Brisbane, nor Hobart was able to attract a community above the size of a large country town. Only the twin congregations of Perth and Fremantle, drawing on renewed Jewish immigration to the West Australian goldfields in the 1890s, and isolated from the attractions of Sydney and Melbourne, were of sustainable size.

In Sydney and Melbourne the Jewish communities of the late nineteenth century developed characteristics not dissimilar from those of London. Poorer families and recent arrivals congregated in the 'down-town' areas, south of the Town Hall in Sydney and in Surry Hills and Darlinghurst around Crown and Oxford Streets, and in Drummond Street, Carlton, and Fitzroy in Melbourne. Wealthy and successful families moved 'up town' to Waverley and Woollahra in Sydney, St Kilda and Prahran in Melbourne. Small drapery shops grew into large retail stores; small carpentering businesses became substantial furniture factories; dealing and agency businesses became established and respected merchant houses. Later differences between the Sydney and Melbourne communities could be discerned in outline by 1900, for example, the concentration of German Jews in Melbourne and the tendency for arrivals from Eastern Europe to prefer Sydney. The impact of the German-inspired Liberal or Reform movement on the old established Anglo-Jewish congregations also had begun to show by the end of the century.

It is difficult to establish at what rate or in what proportions Australian Jews 'married out' or ceased to think of themselves as belonging, but both of these factors influenced the slow growth of the Jewish population between 1860 and 1900. A high level of tolerance, easy access to most trades and professions, and the widespread availability of economic opportunity worked against the coherence of the Jewish community except for religious purposes. On the other hand, the rise of well-to-do, urban, self-consciously cosmopolitan groups in both Melbourne and Sydney laid the foundations for the appearance of twentieth-century Jewish communities revitalized by immigration. Nineteenth-century Australian Jews were proud of their international traditions and connections, of their learning and civilization. Australia needed their experience and intellectual enthusiasm, though often dismissed Jews themselves with typically British superiority.[48]

Anglo–Jewish settlers could be, and were, classified amongst the '98 per cent British' population of the Australian colonies in the second half of the nineteenth century. Only the Chinese and Aborigines were seen to be different. Between the gold-rushes when they began arriving in large numbers, and the end of the century, the Chinese were a source of inscrutability and civic anxiety, especially in the cities of the east. They were approved when for special occasions, as a community, they provided displays and demonstrations, like the Chinese welcome to the Duke of Edinburgh in Ballarat in 1867, or the Chinese contribution to the celebrations in Melbourne in 1901 for the opening of the first federal parliament. At a more mundane level they ran market gardens—at Brighton near Melbourne, and Rushcutters Bay near Sydney, and grew tobacco as share-farmers in southern Queensland, near Tamworth, Bathurst, and in the Tumut, Ovens and King valleys.[49] They were regularly seen in suburban

VELLY GOOD LETTUCEE

streets hawking their much apreciated vegetables from door to door. Though by no means dominant in green grocery, they did concentrate there. For many a suburban housewife, 'John Chinaman' was a reliable and welcome tradesman. In country towns and on farms and stations, Chinese skills as gardeners made the difference between a bearable diet and an unbearable one. Housewives who struggled with their own kitchen gardens to supplement the monotony of meat, bread, jam and tea had more than their lowly status in common with the Chinaman.

There were 37 720 Chinese in Australia in 1861 (ten of them women), about two-thirds in Victoria.[50] The numbers reached their nineteenth-century peak in the late-1870s (38 533).[51] By then they were distributed between the three eastern colonies, with about 4000 in the Northern Territory. The Chinese population of Victoria fell by about half in the twenty years to 1880 and almost by half again by the turn of the century. In 1901 there were only 7349 Chinese in Victoria, both full- and mixed-blood, including 111 full-blood females and 498 of mixed blood. In other colonies the Chinese population was dwindling as well.[52] Nonetheless, major firms shipping bananas from north Queensland and Fiji to the markets in Sydney, Melbourne and Adelaide were Chinese-owned. In some country towns, especially old mining towns, the local Chinese store was a major supplier of fabrics, furniture, crockery, tin and enamel ware, often imported from Chinese sources. Despite their small numbers and fearsome reputation, drummed up by politicians for their own unscrupulous purposes, fostered by the gutter press, and reinforced by seemingly respectable government inquiries into gambling and drugs, the Chinese contribution to the quality of life of the ordinary citizen was quite important. British palates were educated to the range of subtropical and tropical produce which grew easily under Australian conditions. There were alternatives to shoddy and already declining British manufactures for mass consumption. Despite rabble-rousing, many appreciated glimpses of a different culture and allowed their curiosity to be stimulated by what they saw. Unhappily few specimens of Chinese manufacture survive—the tin billy and enamel pannikin which decorated every decently rolled swag, the 'china' cups, plates and bowls on every kitchen dresser, or the 'tin' dish in which they were washed. An export trade with China was already being fostered by British capitalists impatient with declining returns from Birmingham and Sheffield. At least some of the antipathy shown by Australian businessmen and their employees to the Chinese was based on a fundamental appreciation of China's capacity to overhaul Britain as a supplier of cheap manufactures in this part of the world.

Direct expression of these fears was seen in legislation in Victoria which imposed more stringent controls on Chinese employees than on others, and which after 1896 required that all furniture made by Chinese workmen be clearly stamped so.[53] In Queensland after 1877 Chinese miners were excluded from working on any new goldfield for the first three years of its operation. By 1888 popular agitation against competition from Chinese labour had reached such proportions that all mainland colonies from South Australia to Queensland adopted a form of exclusion legislation, showing an unusual degree of co-operation between them.[54]

Anti-Chinese feeling based on fears of economic competition translated too easily for political purposes into frenzied accounts of corrupting religious, social and sexual practices. With the Chinese, as with the Aborigines and the Pacific Islanders, there was fear of physical contact, and especially of sexual contact. Though there was no difference between the wiles used by the Chinese to trap and ensnare their female victims and those used by white counterparts, the horror of white female slavery made the Chinese seem entirely vicious. Because of their preference for opium instead of alcohol or tobacco, they were condemned as a race of drug peddlers.

In tropical Australia the Chinese seemed less out of place. Chinese miners outnumbered white on the far-away Palmer River in the 1870s. Chinese banana and sugar-growers toiled on the river flats from Cardwell to Cairns. In Palmerston (Darwin) for the last decades of the nineteenth century, the Chinese population set the tone. 'High-bowed Chinese sampans, with their square lateen sails, ply about the harbour . . . Umbrella-like hats, pigtails, wide sleeves, sandalled feet, bare legs, and indeed the sights, sounds and odours of China abound.'[55] Throughout the Northern Territory, the Chinese were growing rice and cultivating vegetables along river banks where they had come initially to search for gold. They were thought to outnumber the white population in the late 1880s by two or three to one.[56] Without them the Palmerston to Pine Creek railway could not have been built. We may never know how many perished in unrecorded battles with Aborigines on the northern frontiers.[57]

Anti-Chinese paranoia, fuelled by intellectual fashions in social Darwinism and eugenics as well as by more straightforward anxiety about economic competition and the grim accumulation of legislation restricting the rights of Chinese settlers to work or to own property, to vote, to sit as members of parliament or serve on juries, culminated in the White Australia Policy as formulated in legislation of 1901,[58] though the restrictions placed on the Chinese were not different from those applied to other groups whose colour or sex made them easy targets for exclusion. Australia's treatment of the Chinese however, did provoke international attention. Their exclusion under Queensland's first Goldfields Amendment Bill in 1876 was disallowed by the Colonial Office as offensive to other British subjects who were also Chinese and as contravening British treaty arrangements with China. In 1887 a small commission of inquiry from China toured Australia to investigate reports of adverse treatment of Chinese nationals.[59] An official Chinese protest against the 1888 exclusion legislation was lodged in London. Iniquitous as it was, the treatment meted out to the Chinese was civilized compared with that received by those not fortunate enough to have representations made on their behalf in London. The Pacific Island labour trade was somewhat restrained by the existence of the Western Pacific High Commission from 1874. But in Queensland, the Northern Territory, and to a lesser extent, Western Australia, the Aboriginal people were simply at the mercy of elected governments and local values.

Of the distinct social groups in this period, the Aborigines were the most widespread, elusive, and diverse. They ranged from those living on the fringes of the cities and in the dubious protection of missions in the south, to those who provided the bulk of casual labour in the north and west, to the tribal groups of the north, west and centre. Efforts were made to tidy the Aborigines on to reserves though as late as 1880, Aborigines and part-Aborigines begging or prostituting themselves were a common sight in the wharfside districts of Sydney or camping on vacant land in Brisbane and Adelaide. In Perth, as in most country towns, they were constantly in evidence. Aboriginal society where it could be observed most clearly was demoralized or corrupted by its unequal contact with European society. The pathos this produced was the chief justification for pity and inaction. Where it resisted subversion, Aboriginal society felt the full force of European destructive technology and was quickly altered, to the point, almost, of extermination. Traditional tribal society survived only in the harshest or most inaccessible parts of the continent. So it is probable that much of our modern knowledge derives from impoverished versions of Aboriginal life.

Traditional Aboriginal society seems to have been organized by small self-governing groups like clans within larger tribes. The clans were further sub-divided in ways which determined marriage patterns and inheritance. With these small sub-groups which consisted of a number of extended families interrelated by complex biological and structural kinship ties, the most significant division was that of sex. There was little indication of hierarchy or status except according to age. As a group, the old men wielded power over the rest—the younger men, all the women, and the children.

Anthroplogists have stressed the egalitarianism and mutual dependency of the Aboriginal clans and of individuals within Aboriginal society. Kinship determined the nature of relations between individuals and groups, their obligations to each other, their rights, and their position within society. Since all members of the society were within some degree of kinship with all others, social relations were highly complex, fiercely observed according to convention, and often very ceremonious. In such a system, what change there was occurred in accordance with the natural environment or came through reproduction. Nature in its seasons dominated all activity and could be controlled only by careful acknowledgement of its superior mysteries. Authority over women and rights in relation to their off-spring were therefore

A CAMP OF ABORIGINES

among the most significant forms of power. The clearest illustration of the importance of such power can be seen in the rules governing a man's right to bestow various of his female relatives in marriage and his expectation of reciprocity. Even though there were many constraints on the kind of marriage he might arrange for a particular female relative, in the end, his right to do so elicited respect, self-respect, and the confident expectation that he in his turn would be the recipient of a suitable wife. Such was the value of these rights that in some tribes the rules extended even to the bestowal of mothers-in-law, for a mother-in-law bestowed upon a kinsman might eventually produce a daughter suitable for a return marriage.

Although they seem to have had no concept of property in the material sense, Aboriginal societies did treat women and children as a kind of property. Women were important as objects of exchange.[60] In this respect, Aboriginal ways were little different from the British aristocratic tradition of seeking and bestowing virginal daughters to continue a dynasty or safeguard a fortune. The sexual division of Aboriginal society was observed as effectively as in contemporary British or white Australian society, though it might seem that roles were reversed. Aboriginal women were mainly responsible for the material survival of society; men took control of its spiritual and moral well-being.[61] Women were entirely responsible for the children; they also provided most of the food, estimated at between 65 per cent and 90 per cent.[62] Men devoted themselves to religion and culture and hunting big game. Where missions or white settlement had made incursions into traditional society, the old men tended to lose their control over the religious and cultural ceremonials which justified and governed the strict ordering of society. Young men glimpsed alternatives to the fear of initiation or marrying according to custom. But while women lost some of their authority as mothers and food gatherers, their instinct for survival sometimes enhanced their position in the community.

At places like Coranderrk in Victoria, Maloga on the Murray and Warangesda near Brewarrina, the men were encouraged to become farmers and the women to learn sewing and develop their crafts for sale, but a heavy overlay of Christianity made it likely that traditional aspects of society were destroyed rather than transformed. For a people who set great store on the certainty with which prescribed actions produced prescribed results, change was the hardest thing to accommodate. The stimulus of change and novelty which meant growth and life in white society was overwhelming evidence of a world out of control to most Aboriginal people.[63]

The 98 per cent British population announced by statisticians at the turn of the century may have been a rationale for federation, but it did include the diversity of Irish settlers who were emotionally anything but British, Scots who were indifferent to the existence of Albion, Jews who were only technically 'Anglo'. It also included a growing proportion of persons who, whatever their parentage, like the despised Aborigines, identified entirely with Australia. Some of them were the second, third, or fourth generation of their family to do so. And while the south-east corner of the continent appeared to be Australian-British, the northern coasts were polyglot. Not only were there Chinese, Aborigines, Pacific Islanders, and the mixed white nationalities. There were small groups of carefully supervised Japanese contract labourers in, for example, Queensland sugar mills, and appallingly exploited Japanese prostitutes who were seen as an enterprising answer to a complex multicultural problem.[64] There were the Malays, Macassans and others of unspecified South East Asian origin engaged in pearl fisheries from Geraldton, WA to Thursday Island. The experience of the majority may have been of a society which was basically British, but the experience that sent waves and ripples out through society came from the north and west. In the complex, competitive, comparative encounters on the fringes of civilization, subtle questions of status were replaced by blatant ones about individual worth and national esteem.

Both old and new patterns of family formation were visible in late nineteenth-century Australian society. In some parts of the country, large and self-contained families were common, usually in areas of recent settlement or rural districts which were in their first generation of white settlement. They consisted typically of parents who had come as immigrants before marriage or early in their marriages, and a growing number of children, who would marry to form new families. In well-established or urban areas, smaller families were becoming common. There were contemporary fears that life in Australia had a debilitating effect on the reproductive capacities of the locally born population, but in fact family limitation first became popular in this group. The average size of a family in the 1860s was about seven children. By 1900 it was about five, but both these figures conceal the number of families which had eight children or more, as well as those with only one child or none at all.[65]

The relative shortage of grandparents, and the probable youthfulness of grandmothers was felt differentially by large and small families. Where both parents were the first members of their families to come to Australia, it was unlikely that much of an extended family network existed, although brothers, sisters, cousins, and therefore uncles and aunts were not uncommon. The absence of grandparents, however, threw all responsibility for their family on a young couple. Since these were also the people most likely to have the largest families, that family was an increasing burden for the mother, with frequent pregnancies and the care of a growing number of people, at least until the eldest daughter was able to help with housekeeping and childminding. This family structure—two parents and a large number of children of whom sons assisted in working the land and daughters in the chores of the household—has been usual in rural or agricultural societies, especially where there has been no extreme population pressure on the available land or its ability to provide food and livelihood. The average size of late nineteenth-century Australian rural familes was large, almost as large as has yet been uncovered by comparable studies in other rural and colonizing societies.[66] (The median figure for births to each woman in Western Australia between 1850 and 1880 was 8.9. In rural America earlier in the century it was estimated at 8.4. In nineteenth-century Italy it was about 7.5.) This level of fertility declined steeply however, once the initial phase of settlement was over and the next generation of Australian-born parents came on the scene.

Such large families in the first generation of rural settlement have given rise to the image of the typical late-Victorian family responding vigorously to the challenge of populating the wide open spaces. As well, there were large, prosperous families in the sprawling homes of successful men of business on the outskirts of the cities. Both should be seen, perhaps, as typical responses to immigration and opportunity. In large families the first birth usually occurred within the first year of marriage. Subsequent births were usually spaced out every 24 to 28 months thereafter. This meant that the mother was usually pregnant or feeding a baby. It also meant that the eldest children were nearing adulthood when the youngest were being born. It was not uncommon for a woman and her eldest daughters to be bearing children at the same time, nor for many children to have relatively old parents or to lose one or both parents before they themselves reached adulthood. These women's lives were completely occupied with child-bearing, and many did not live to see their youngest child reach an age of independence.

Large families with their mishaps and miscarriages, meant that all members were subject to a wide experience of birth, death and child-rearing. Here was fertile ground for the cultivation and transmission of tradition in practice and attitude. Older girls learnt on their younger brothers and sisters; young ones learnt on their nieces and nephews, unless of course, these were born before their aunts. Such practical experience was seen as a source of 'old wives tales' in contrast to the scientific child-rearing promoted towards the end of the century. In smaller, more modern families, however, this cycle of experience and knowledge was broken. The woman who had only one baby, or two, had no experience or authority in comparison with the woman who had ten. As well, the time between her last baby and her daughter's first was probably about twenty years. She could hardly be expected to be as confident and knowledgeable as the woman who had given birth to her tenth only the year before. That generation of women whose first baby was born in the 1890s and who were themselves already the products of small families, responded increasingly eagerly to the advice of 'experts', the doctors especially who at least had plenty of experience of childbirth and childhood illness.

Large families gave sons the opportunity of learning and gradually moving into the family business. One of the prerequisites for sustained prosperity in pastoralism and other branches of rural industry seems to have been sons who could succeed their fathers and co-operation between brothers as in the classic ales of the Henty, Durack and Wright families.[67] The same principle worked in the world of business and manufacturing where experience and tradition were more important than formal education or professional training, but the balance here was shifting to a preference for the latter. Perhaps the greatest advantage of large families was that responsibilities could be shared. It was not incumbent on every child to satisfy parental dreams of fame or immortality or to support their old age. Thus large families could seem happy.

In small families the first birth still usually occurred within the first year of marriage. There was no dramatic change in the consequences of premarital sex or its frequency during these four decades. Women who were using contraception in the late nineteenth century were usually seeking to limit a family which was already too large for their resources. The unromantically efficient notion of family planning belongs properly to the early decades of the twentieth century. Trust in, or reliance on, contraceptives was still a long way off. Abstinence or mutual self-restraint within a companionate style of marriage were idealized as the most civilized way of approaching these problems. Freud's message about the psychological dangers of sexual repression had not yet fallen upon the world. Advanced thought, especially among the women of the 1880s and 1890s, looked forward to a world in which technology would overcome the still dangerous and often unwelcome business of reproduction altogether, eliminating not only pregnancy and childbirth, but the need for sexual intercourse and the threat of venereal disease. At worst, it was assumed, child-bearing might be left to the lower orders, much as breast-feeding had been left to wet nurses in previous centuries.

Theoretical discussion of the morality or otherwise of family limitation was one thing. Exactly how to do it or even to obtain the equipment recommended was another. Mail order overcame embarrassment. Mrs Bessye Smythe sold both literature and accessories by post, and advertisements like hers, some of them mere confidence tricks, proliferated in the more lively periodicals.[68] Still, the methods available were far from reliable and instruction in their use was often ineffective because women lacked simple physiological knowledge.

Modern writers who assume that whatever else has changed in human society, the level of sexual activity has remained constant, are apt to posit a high level of undetected abortion and infanticide to make up for failed or non-existent contraception during this period of declining birth rates. Certainly abortions, cases of infanticide and of baby-farming were prosecuted. There were also orphanages, and children were adopted or taken in by relatives, not only because their parents were dead. The Victorian census of 1891 recorded about 3000 orphans under fifteen. Only 500 of these were living in orphanages.[69] It is difficult in a society now saturated by sexuality to imagine being ignorant of the meaning of these 'natural' urges, even cultivating and admiring that ignorance as we cultivate and admire its opposite. In the last decades of the nineteenth century however, knowledge of sexuality was being transformed very slowly and untidily from its vague and haphazard traditional state to its modern, systematic and proto-scientific version. We can see questions which were entirely private matters in 1860 coming slowly to the attention of social reformers, moral guardians, statisticians, and ultimately, governments. The small family went unremarked in 1860. It was God's will, a blessing or a misfortune, depending on your circumstances. By the turn of the century it had become the subject of a government inquiry in New South Wales.[70] Though officially it was deplored or regretted, in fact the trend to smaller families was an essential component of the high standard of living of which the Australian colonies were so proud. Good wages and steady employment were not the full explanation. The careful workingman had more control over the size of his family than over the regularity of his employment or level of his wages. By the 1880s, among the working classes, a large family was becoming a sign of improvidence.

These two contrasting images of the old large happy family and the new limited family co-existed through the last decades of the nineteenth century. They had troublesome consequences for women. On the one hand there was the slightly exaggerated version of the capable bush mother of ten, the epitome of the traditional housewife, the fount of womanly knowledge and skill. On the other was the also exaggerated querulous mother of two in a narrow suburban terrace or frowsy cottage, afraid of further pregnancies, puzzled already by the problems of coping, lacking space to be creative, denied the right to earn. The second image increasingly became reality for the majority of women, yet the first persisted as the ideal. Like so many of the images by which the standards of the time were set, it belonged to a past age, to a lost world, temporarily and accidentally re-created during the pioneering phase of Australia's history, and destined to become part of our unexamined mythology.

Since the appearance of the first generation of Australian-born children, observers had been watching as if they were a new species. By the 1870s and 1880s one of the set pieces of journalism was 'The Coming Australian Race' (Marcus Clarke, 1877) or 'Young Australia' (R. E. N. Twopeny, 1883).[71] Almost invariably the younger generation of Australians was described as less deferential and respectful to their parents than their English counterpart. The boys were said to be preoccupied with sport, uninterested in intellectual or cultural matters. The girls were good fun but lacked the equipment, mental, social, or physical, to carry them into admirable womanhood.

In a comparison of records of the heights of boys transported to New South Wales in the first half of the nineteenth century with boys of a similar age born in Australia, Bryan Gandevia found that the Australian-born boys at 176.5 cm were on average about 20 cm taller. By the end of the nineteenth century the taller boys had become predominant, while the average height of Australian-born boys remained greater than that of their English and Scottish contemporaries. Gandevia attributes this taller stature to environmental influences, such as food, sunshine, and outdoor exercise, since there was no noticeable difference between boys whose parents were Australian-born themselves and boys whose parents were immigrants. This extra height, he suggests, was in itself almost enough to explain the brashness, independence, self-reliance and awkwardness of colonial youths, as well as their capacity for heavy work at an early age.[72] They would also be likely to tower over their parents, especially immigrant parents whose own growth had been retarded by malnutrition, rickets, and other environmental factors in their youth. These differences were accentuated by the rural or suburban conditions (compared with urban industrial Britain) in which most Australian children were raised. Even in Britain it has been found that rural life tended to produce taller children than urban, except where there was chronic undernourishment, as in parts of Ireland. Was Ned Kelly too tall and well-built for his own good?[73]

The same evidence suggests that colonial-born girls were about 164 cm or 11.5 cm shorter than their brothers—which made them taller than many a male immigrant. There is no reason to assume that girls did not share the benefits of sunshine, plentiful food, and exercise. Indeed, casual observers noted energy, which was deemed vulgarity and a lack of delicacy both in figure and complexion which suggests sunshine and plenty of food. There was much discussion about whether the Australian girl was really pretty—or would you want to marry one? She learned to cook, to sew, to ride, all practical skills, none of them essential to a well-bred lady of the English variety. Like her brother who was built for hard work, but who lowered his status by doing too much, the Australian girl lost caste in the drawing-room if she admitted to her skills as cook, ladies' maid or rouseabout. But unlike her brother, who might do whatever he pleased if he was making money, her success depended on her skill at insinuation. What was natural for her was almost certainly not nice. Environmental factors made her into a modern woman. Society valued her more if she was old fashioned.

Contemporary opinion seems to have been that parents in Australia lacked authority over their children. This may have been simply because they were a little intimidated by their large and vigorous off-spring. The climate also made it easier for children to lead independent lives. Picnics, excursions, sporting activities were too easily arranged always to be supervised. Except for very formal occasions, the chaperone seems to have been more theoretical than real. In 1860 Blanche Mitchell, aged seventeen, roamed all day over Sydney's eastern suburbs on foot and by omnibus. At night she went to dances at private parties and on ships in the habour, with only the vaguest adult oversight. Perhaps the youngest daughter of the widow of a senior public official was different from other girls, though Blanche seemed to have plenty of friends with freedom to join her.[74] The Australian climate did not, as the wife of the Bishop of Adelaide remarked, 'tend to attract the family round the fireside'.[75] Adolescent freedom, which in colder, wetter climates had to await the invention of the motor-car or the installation of central heating, came early in the land of the great outdoors. In 1869 Edmund Barton was twenty. A diary he began then reveals a charming young man, earnest, though not excessively, respectful towards his parents, fond of his brothers and sisters, more interested in cricket and rowing than in his studies, and somewhat romantic. He met Jeannie Ross when he was in Newcastle playing cricket. Later he recorded sentimentally 'the last walk we will have together' at the end of a visit to Sydney. It was 1877 before he was in a financial position to marry her.[76] For most young people, meeting and marrying depended on the availability of opportunities and resources.

> Take me down the Harbour
> On Sunday afternoon—
> To Manly Beach or Watson's Bay
> Or round to Coogee for the day,

ran the refrain of a later popular song.[77] During the 1890s Depression, marriage bureaux sprang up in Sydney and Melbourne, presumably to fill a gap left by lack of means to join organizations or pay for outings.[78]

Some thought parents had too easily surrendered their rights over their children's lives. To J. F. Hogan, historian of the Irish in Australia, the education system was too lenient, the state having usurped parental rights.[79] R. E. N. Twopeny thought parents were too easy-going with their children. He advocated more frequent use of the rod or the strap. The trouble in his view went all the way back to babyhood.

> I have a holy horror of babies, to whatever nationality they may belong; but for general objectionableness I believe there are none to compare with the Australian baby... the little brute is omnipresent, and I might almost add omnipotent.[80]

Nurseries were not common in Australian homes. Nurses were expensive and hard to get. So the mother herself was obliged to act as nurse and the baby became part of the family circle almost from birth. In small families a baby disrupting the household was both a temporary and infrequent phenomenon, and could be tolerated as such. Oppressed by his own children because of the shortage of good servants and adequate nursemaids, poor Twopeny was confronted by babies wherever he went.

> Wherever his mother goes, baby is also taken. He fills railway carriages and omnibuses, obstructs the pavement in perambulators, and is suckled *coram populo* in the Exhibition. There is no getting away from him unless you shut yourself up altogether. He squalls at concerts; you have to hold him while his mother gets out of the onmibus, and to kiss him if you are visiting her house.[81]

In Twopeny's ideal English world, babies were either consigned to nurseries until they were old enough to emerge as little ladies and gentlemen, or they were bundled into corners with something to keep them quiet while their harassed mothers got on with their lowly paid work. Some of the conditions in Australia, like good wages and improving housing, should have been conducive to that formality in family life for which Twopeny yearned. Other circumstances, notably the shortage of servants, led to greater informality. How could a mother who had to set her own dinner-table present a dignified image to her children? How could a father required to play with the baby till tea-time be remote and autocratic?

Fashionable separation of child and adult worlds could not be maintained either in a small family or an almost servantless house. It was silly to have separate meal-times for children and adults, difficult to provide separate living areas, nurseries, servants quarters, drawing- and dining-rooms, in a four-roomed villa. Relaxation of the distinctions between child and adult worlds, however, weakened the socialization process and broke down the sense of graduation. The right to eat adult food or participate in grown-up conversations was rarely withheld. Forms of dress remained as the only obvious distinction between childhood and adulthood. When a boy graduated to long pants and a girl put up her hair and lengthened her skirt, each was symbolically 'grown up'.

Some of the formalities of family life were retained and even exaggerated according to half-remembered ideas of how things were done 'at home'. Or they were imposed as recommended in the now-appearing guides to etiquette, household management and child-care.[82] Parents still referred to each other as Mr and Mrs in the presence of a third person, or as there was increasing justification for it, 'Father' and 'Mother'. Sons were often required to address their fathers as 'sir', whatever they might think of them. Rules of punctuality were developed to replace those formerly imposed by space and the presence of a serving class. The family was organized round meal-times which were in themselves often dictated by the precision of the railway or tram timetable. 'Manners' were impressed at meal-times. So was Christianity, in the form of grace, or more extensively in family prayers. Without servants to wait on them or serve the food, these functions became a source of power for parents. Father was in command of the carving knife, wielding it on the joint or the loaf of bread. Mother allocated portions of vegetables and poured the tea.

Except among well-to-do families, few meals were eaten away from home, and then mainly of necessity or on festive occasions. Workers could not always go home for lunch, which was the meal most likely away from home. Entertaining friends to meals at home was a practice ordinary people had scarcely begun to copy from the well-to-do, though affluence permitted generosity in casually shared meals. Upwardly mobile young marrieds in the suburbs plied their friends with afternoon tea or supper following an evening of music or cards. Suburban life itself tended to discourage dinner parties. As Oscar Comettant discovered when he was bidden to dine in St Kilda during his stay in Melbourne in 1888, by the time he got there and caught the last tram home, there was not a lot of the evening left for the pleasures of the meal.[83]

Only men ate in public, at official dinners, in restaurants, in hotels and in their clubs. During the prosperous years hotels frequently provided substantial 'counter lunches' to attract the drinking patrons. Women rarely ate in public, only in hotel dining-rooms if they were travelling, or in tea-shops, delicately, when they were in the city. Sydney restaurateur, Quong Tart, made an easy success of his discreet upstairs tea-room designed especially to accommodate ladies[84] and 'retiring rooms' within the large department stores were introduced as an attraction to modest women for whom a day in town was otherwise a considerable trial of self-restraint.

Even in households where the most tyrannical control of meal-times was maintained, parental authority ended when permission to leave the table had been granted. In churchgoing families, attendance as a family at the Sunday service was usually obligatory.[85] Beyond that, the cohesion of the family depended variously on simple affection and the ability to provide social and economic advantages.

F. D. BEACH & SON'S
Commercial Dining Rooms,
HINDLEY ST., & GILBERT PLACE, ADELAIDE.

THESE centrally situated and well-known Dining Rooms have recently been considerably enlarged and remoddled to meet the requirements of our daily increasing trade.

THE HINDLEY STREET DINING ROOMS
Both for Ladies and Gentlemen, have been tastefully re-decorated, and will be found equal to anything of the kind in the colonies.

THE GILBERT PLACE DINING ROOMS
Are now replete with every convenience. The New Upper Room, 54 ft. x 28 ft., is both lofty and well ventilated, and in addition to the usual facilities, F. D. B. & Son have added a Lavatory and other conveniences.

THREE COURSES ONE SHILLING.
Soup, Meat, Pastry.

SPECIAL ROOM FOR LADIES.

ALL COOKERY UNDER THE SUPERVISION OF THE WELL-KNOWN CHEF F. D. BEACH.

A Large Table contains the ture of the day, and also a Envelopes. These are reserved

The present agitation in favor Industries, brings to the fore

CANDIED PEEL &
has long been successfully & SON. These goods are superior to the imported price.

Newspapers and current literature—supply of Pens, Ink, Paper, and for the use of customers.

of the fostering of Native the fact that the Manufacture of

LEMON SYRUP
carried on by F. D. BEACH pronounced by *connisseurs* to be article and at a much lower

Candied Peel, 1s. per lb., or 6 lbs. for 5s. The trade liberally dealt with.

For centuries the Christian church had been the main source of rules and ideals for the family. During the nineteenth century, however, there were important transfers from church to state in authority over family life. State intervention could be seen in the codification of the marriage law, also in registration of births—an activity formerly carried out church by church and parish by parish. At mid-century, the churches still retained most of their traditional influence on the family as an idea. Visiting parishioners in their homes to monitor church attendance, family well-being, and the nature of the children's religious education was considered one of the most important duties carried out by clergy of all denominations. Clergy wives and daughters, and other religiously suitable women assisted in this work. The right of the church to direct the education of children was widely accepted, though to intervene in an unsatisfactory relationship between husband and wife was less acceptable. While gratuitous interference in such matters as housekeeping practice was most resented, the clergyman or priest remained the first resource in time of trouble, and the church a link to advice and a range of support services.

Some traditional ideas about the family were more easily sustained than others. While the churches appeared to retain some kind of attraction to those wishing to marry (only 6 per cent of marriages were conducted in the new civil registry offices[86]), civil divorce was legalized in all colonies before the end of the century. In the 1901 census, there were 1228 men and 1147 women who admitted to being divorced. More than half the divorced men and three-quarters of the women lived in New South Wales.[87] Ultimately it was the state's ability to lay down a set of rules about the family as a unit of income and to command universal adherence to those rules which made the churches seem voluntary, if not bumbling and old-fashioned repositories of traditional values about the conduct of family life. Legislation which assumed the family as an economic or social unit at first tended to reinforce the teaching of the churches. The role of husband and father as head of the household was already important in the Protestant churches. His legal and economic authority were strengthened by legislation in such areas as property, contracts, child custody. But so was his responsibility as worker and wage-earner. His income became family income; his wage, the family wage, especially as paid employment for wives and children was restricted. In Catholic families where a higher degree of personal answerability for one's spiritual condition was expected, the imposition of Protestant concepts of the legal and economic headship of husband and father may have been subtly undermining, though there was nothing subtle about the attack on 'the Christian family' denounced by Revd G. F. Dillon in this sermon in 1873:

Already, almost everywhere, infidels have succeeded in breaking the legal unity of the matrimonial bond, and have decreed, in defiance of Christ, that 'what God has joined' it should be lawful for 'man to put asunder.' Callous alike to the interests of children and to the helpless weakness of woman, they have sought but the gratification of passion, and in the confessed principles of Socialism, aim at the reduction of men and women, made 'into the image and likeness of God,' to the level of the cattle of the field.[88]

At the same time, the subservient role required of wives in both the Protestant family and the family as it was being defined in nineteenth-century law was at variance with the dominating and resourceful behaviour thought natural for the mothers of Irish Catholic families. Whether their strength came from their greater economic contribution to families close to their peasant origins as small farmers or selectors in rural Australia, or from the high degree of respect accorded to the mothers of sons in true biblical fashion does not matter. In important ways which cut at the heart of traditional or instinctive behaviour, Catholic families were more threatened than Protestant ones by evolving legal definitions of their role. The Catholic church could outlaw divorce, forbid mixed marriages, insist on educating the children in church schools. It could do nothing to assist the man required to assume unequal responsibilities for himself and his family, or the woman denied the authority and support tradition taught her to expect. There is no way of showing the effect of such tensions in a generation, even two, though both the mid-twentieth-century conservatism of the labour movement and the prominence of convent-educated girls modern in Australian feminism may owe something to those tensions.

Late nineteenth-century legal and economic definition of the family was hastened by the fragmentary support given by church-based institutions to those who lacked a framework of family or kin. Traditionally the church had supported those who, through no fault of their own, were the victims of society—the widows, orphans, those born with afflictions, and those stricken by accident or incurable illness. The facts of Australian history were already making this difficult for the churches. Because there was no dominant or established church, it was possible to evade responsibility, to discriminate between those who deserved support because they supported their church, and those who did not. As well, the circumstances of Australian settlement made the traditional criteria of blamelessness difficult to apply. The convict heritage cast grave doubts on the innocence of many who seemed in need of support. Yet immigration produced new categories of need. Often in dire straits, dislocated immigrants were also eminently recoverable. Orphans too seemed innocent, unable to be charged with neglecting their religious duties. Most people who could not be cared for within the confines of their families were seen until fairly late in the century as dubiously deserving of charity. The deserted wife could blame no one but herself. The unmarried mother and the chronic drunk lacked self-discipline or a sense of purpose. Even lunacy was touched with deliberate defiance, flouting the rules of society, and treated as such.[89]

The resources of the churches were stretched providing pastoral care and ordinary charity in their hospices, refuges, asylums, orphanages and homes. By the end of the century these institutions fell into two main categories. Some were supported by the Catholic church and sustained by the work of religious. Others were semi-secular institutions which depended on the administrative skills and fund-raising abilities of a small number of dedicated men and women drawn from across the Protestant denominations. As in education, there was sectarian competitiveness in the establishment and management of charities. Winning souls from the other side sometimes got the better of humane good sense. This did little to counter the charge of church hypocrisy. The inadequacy of the churches and the uncertain status of the family therefore made it inevitable that governments became involved in providing for those who were unable to provide for themselves.

Well before 1860 the state's responsibility for minimal support of the outcasts of the convict system had been recognized. Hence the provision in New South Wales and Tasmania of hospitals for the sick poor, lunatic asylums, orphanages, and some outdoor relief for widows and deserted wives. In the later part of the nineteenth century it was assumed that the need in these areas should diminish, and also that as the economy provided plenty of opportunity, self-reliance should be encouraged. Not a few of the committees and institutions providing relief for the homeless or destitute simply refused to consider assisting able-bodied men.[90] A similar rationale lay behind the provision of tools and equipment for work. The Queen's Fund in Melbourne arranged the supply of sewing machines on time payment to deserving women who would in theory become self-supporting and pay them off.[91] Work experience was offered in labour colonies in the 1890s.[92] There was assistance in kind rather than cash through soup kitchens and the distribution of rations, clothes, blankets. In this, the impoverished members of white society were not treated differently from the Aborigines. The unemployed could cut firewood for sale or distribution to needy families and thus earn food or shelter. Meanwhile, unemployed women of the middle classes met in sewing circles to stitch garments for distribution to the needy.

A Sydney directory for 1861 listed the following charitable and benevolent institutions: the Sydney Female Refuge, the Sydney Female Home, the House of the Good Shepherd, the Benevolent Asylum, all devoted to the walfare of homeless women or unmarried mothers; two orphan schools, one Protestant, one Catholic, each with separate sections for boys and girls; St Vincent's Free Hospital, the Sydney Infirmary, and the New South Wales Alliance for the Suppression of Intemperance and for the Social, Moral and Intellectual Elevation of the People.[93] Two lunatic asylums, one for ex-convict lunatics, one for free lunatics, were heavily subsidized by the government, as was a home for destitute children. Eleven country towns had hospitals which also took in homeless, destitute old people. There were benevolent

BENEVOLENT ASYLUM, GEORGE STREET, PARRAMATTA—WARD NO. 1

asylums for unmarried mothers at Liverpool, Parramatta, and Penrith. By 1870 New South Wales had 31 hospitals, 7 benevolent asylums, 11 orphan schools, and 5 lunatic asylums (4 government, 1 private), all sustained jointly by state and voluntary contributions. Voluntary contributions alone kept the Home Institution, the Sydney Female Refuge, and the House of the Good Shepherd, as well as new charities such as the Sydney Sailor's Home and the City Night Refuge and Soup Kitchen.[94] By the turn of the century the government had its own Department of Charitable Institutions and an Aborigines Protection Board.[95] There were thirty-four voluntary charitable institutions now listed, including the Society for the Prevention of Cruelty to Children, the Animals' Protection Society of New South Wales (founded 1873), the Aborigines Missionary Association, and the Carrington Centennial Hospital Home for Convalescents.[96]

Changing attitudes to needs, rights, and the responsibility for care can be seen most easily in the funding and use of hospitals during this period. In 1860 most hospitals were funded and administered by boards of governors on behalf of groups of subscribers, Subscribers were usually well-off citizens who made an annual contribution to the hospital as a kind of insurance, though not in order to be eligible for admission themselves. They and their families received any necessary treatment in their own homes and beds, which were cleaner, safer, and more comfortable. The hospital was a place for sick or injured servants and employees whose homes were unsuitable or who had no one to look after them. They could be sent to hospital without additional worry or expense if they were injured or too sick to work. It was not easy to gain admission to most hospitals without referral from a subscriber, though by the end of the century, hospitals had begun to admit those who sought treatment and to charge what it seemed they could afford to pay. The better-off began to seek admission (sometimes pretending to be poor) because hospitals were now both safer and more effective. With increasing levels of funding, both from government and through widespread public appeals, hospitalization came to be seen as a right rather than a charity. Certainly access to expensive medical equipment and specialized skills such as the use of anaesthetics and antiseptic procedures would not be kept for the poor alone.[97]

The institution in New South Wales at the end of the century of a universal though fiercely means-tested old-age pension also showed the idea of assistance from the public purse overhauling an older, stigmatizing notion of charity. As old age crept on a hitherto youthful society it was managed in an *ad hoc* manner. By the 1890s Tasmania, South Australia, Queensland, and Western Australia all had machinery whereby poor though respectable old people could apply for assistance to enable them to live outside the forbidding and rightly feared old-people's asylums which were largely financed by government. In the interests of economy, and also to discourage potential residents, these institutions were run along workhouse lines, separating the sexes and providing a minimum of food and shelter.[98] It was nonetheless cheaper to provide small cash grants as a special charity to some elderly people who could thereby manage on their own. By contrast, everyone was invited to apply for the old-age pension. Whether it was granted depended on the individual's existing resources or whether family members were shirking responsibility for an aged relative. Respectability and sobriety were essential. Assumptions about moral worth and the responsibility of the family to care for its own, inherited from the Christian tradition, had not changed. What had changed was the right of those who fell outside the family to expect assistance from the public purse. As the number of elderly people in the community grew, and as the ability of the churches to finance and administer their traditional charities faltered, it became more difficult for them to establish or assess the extent of need. Some city neighbourhoods were largely untouched by traditional surveillance. The police and the law courts were just as likely to uncover cases of destitution, desertion, homelessness here as the priest or clergyman. This had long been the case, of course, in thinly populated rural districts where the law was the only institutional presence. Thus through the work of the police and their own health officers, governments found themselves turning charity into social work.

From a statistical point of view the convict remnants in the Australian population became unimportant during the gold-rushes and the period of subsequent immigration. Psychologically too, the golden age overwhelmed memories of old Botany Bay. But the statistical averages and the excitement of gold obscured the fact that some of the colonies were rather closer to their convict origins in the second half of the nineteenth century than others—Western Australia, for example, was still taking convicts from Britain till 1868. The British government continued to maintain the penal settlement at Port Arthur in Tasmania until 1871 and thereafter to support it at the annual rate of £36 19s 8d for each convict remaining. It contributed a further sum (about £6000 p.a.) for the upkeep of the Tasmanian police force and to subsidize institutions caring for 'imperial paupers'.[99] Signs of the convict past were visible in the populations of both Western Australia and Tasmania in the later stages of the century.

In addition to convict-built roads and buildings which were obviously remnants of the prison system recycled for other purposes, often of a welfare nature, Anthony Trollope thought he saw 'the Bill Sykes physiognomy' all about him during his visit in the early 1870s.[100] Staff and officials trained in the convict system did not quickly change their ways or habits of mind, nor did the convicts simply go away. Both Western Australia and Tasmania had higher masculinity ratios in the 1860s than other colonies. Both subsequently had the problem of ageing, often unmarried, or homeless men. Tom Stannage has shown how the crime rate in Western Australia as late as the 1880s reflected the effects of this criminal or criminalized group. In 1878, Western Australia's crime rate was seven times higher than South Australia's; 40 per cent of cases in the Western Australian supreme court between 1880 and 1890 involved ex-convicts.[101]

The Tasmanian convicts were more pathetic and hapless than those arriving in Western Australia. The main Tasmanian legacy from the convict system was poverty, both individual and social, for no society could easily support such a burden of repressed and wasted human resources. Fear of convictism was evident too in Tasmanian institutions and in the attitudes adopted by those in authority, since former convicts could not be excluded from economic life or from the social and political rewards of wealth. So, in the interests of self-respect, the past was obscured and the present repressed. The many thousands of convicts who left Tasmania or Western Australia to begin new lives in other colonies often under assumed names, had no desire to recall the past. Though it was not often used against them, the possibility was always there. When it did come up, it was quickly suppressed, as in the case of W. H. Groom, victim of the most effective of several cases of ex-convict mud-slinging in Queensland politics.[102] As early as the 1870s stories of the convict days were becoming a part of colonial mythology, thereby keeping the threat alive for those who had been closer to it all than they wished to remember. Novelists Marcus Clarke, Rosa Praed, Rolf Boldrewood, and Price Warung made use of the dramatic and violent possibilities of convictism.[103] Undoubtedly they reinforced latent fears of exposure.

Both Tasmania and Western Australia were remote from

160 THE OXFORD HISTORY OF AUSTRALIA

NATURAL PAVEMENT, PORT ARTHUR

Romanticized picture of Port Arthur c. 1888

the more bustling colonies and their real problems of uninspired economies and disproportionate numbers of social misfits were contained by distance. A trickle of ex-convicts remained a worry, especially in South Australia where the stigma of convictism was resisted and resented. Until the turn of the century most passenger shipping called only at Albany in Western Australia, thus effectively isolating Perth and Fremantle. All male passengers departing from Albany were still required to present a certificate signed by a magistrate to show that they were not and never had been a 'prisoner of the crown' and to pay 1s for it.[104] In the end the threat of criminalized societies in these two colonies and the fear of discrimination in the labour market could not be avoided. Similar anxieties about cheap controlled labour were also provoked by the plantation system in Queensland, and to a lesser extent, the use of Aboriginal labour.

By 1868 nearly 10 000 male convicts had arrived in Western Australia. Under regulations laid down by the British government they worked for a period on government projects, then were granted tickets-of-leave and sent to work for private employers. They were still subject to summary jurisdiction, still a charge on the imperial government, though their employers could be required to feed, clothe, and house them or pay minimum wages in lieu. Once granted conditional pardons they became virtually free, no longer subject to summary punishment, and permitted, if they wished, to leave Western Australia as long as they did not return to Britain. Thus they ceased to be tractable labour; nor did the British government accept any financial responsibility for them. It was tacit policy to encourage migration of former convicts from Western Australia after conditional pardon, for they were seen as damaged human beings, prone to ills and misfortunes, liable to be a charge on the economy instead of an asset. The House of Commons Select Committee on Transportation in 1861 discovered that Western Australia was taking 300 convicts a year and sending 300 conditionally pardoned men to the eastern colonies—a situation which the Tasmanians might envy.[105]

Mounting complaints from the other colonies—the fact that Britain refused to take these men back was very insulting to the non-convict colonies—led to the decision in Britain that henceforth conditional pardons must be served out in Western Australia. This somewhat lessened the attractions of transportation in the west. There was not much disappointment when Britain decided to discontinue the system after 1868, though concern about the effect of withdrawing so much imperial funding was justified by Western Australia's feeble economy during the following twenty-five years. Western Australia itself was large enough to accommodate, even to lose its former convicts, but no society could easily accommodate the accompanying habits of easy exploitation and the failure of human imagination from which they arose. Aboriginal people paid dearly for their inability and unwillingness to step in where the convicts left off.

The 'taste for slavery which has not yet lost its relish' lingered most obviously in the west.[106] In Queensland it was a more insidious thing. There, in theory, the convict system had come to an end in 1840, but because Queensland had been a place for retransportation from New South Wales, most convicts were transported for life and serious or desperate crimes. Ray Evans has shown how Queensland institutions to cope with the old, the diseased, the insane, the unemployable, and the unfit grew initially from the need to deal with the human remains of the convict system.[107] In this respect, Queensland was like Tasmania. The important difference was Queensland's size and unexamined potential. It was easy to forget the significance of a few solid stone buildings in the south-east corner as enthusiastic settlers rode north and west. What was not forgotten were brutal skills, useful for the rough work of development, made more effective by draconian legislation once self-government was instituted. Elsewhere in Australia, those skills and attitudes had come close to exterminating the Aborigines who resisted their role as a substitute for convict labour. In Queensland impatient and enterprising settlers sought and found a substitute supply of cheap, enforceable labour barely twenty years after the last convicts were transported to Moreton Bay. Queensland was not subject to economic constraints brought about by Tasmania's size and proximity to Victoria's protective tariffs, nor to official control by the British Colonial Office like Western Australia, where, despite the questionable behaviour of some of Her Majesty's representatives, and their tendency to collude with local values, moderation sometimes prevailed.[108] Queensland landowners overcame their shortage of labour for the gruelling work of early settlement by importation from the nearby islands of the Pacific. There were some significant similarities between the Islanders and convicts. The Islanders were young men, without wives or families, and subject to discipline. The differences, however, made the Queensland experience both more horrific for those who suffered it, and in the end, more momentous for Australia as a whole.

The crucial factors were these: the size and scope of the operation—over 62 000 Islanders were brought in over forty-three years;[109] its management—by private individuals under regulations devised and administered by a democratically elected parliament. This was not something that could be explained or excused by responsibility located elsewhere: it was the responsibility of Queensland society as a whole. But it was also easily, even callously, brought to an end. Whereas the convict past lived in the Australian psyche for generations as a reminder, a resented debt, and whereas the USA never freed herself of her obligation to her African slaves, a generation after their repatriation, it was almost as if the Pacific Islanders had never been. The few who stayed faded into grey-haired Tommy Tannas in hessian sheds under houses, gardening and doing odd jobs for their keep.

The ease with which young men—70 per cent were between the ages of sixteen and thirty, though there were boys as young as nine, and about 6 per cent in the end were women—were recruited, initially from the New Hebrides and Fiji, later from the Solomons and islands closer to New Guinea, was one of the attractions to those engaged in the trade and a cause for dismay among its opponents. There was no bribing these 'temporary immigrants' with 'meat three times a day' or 'farms of their own'. It was not intended that they should stay. Nor is it likely that many of them understood where they were going or what it would be like. Under pressure from bookish or religious types who understood the possibility of analogy with the Atlantic slave trade, regulations limiting methods of recruitment and the period of indenture were introduced and regularly rewritten, beginning with the landing of the first cargo of New Hebrideans at Robert Towns's Logan River plantation.[110] But the coast of Queensland is very long. It was hardly explored in 1863, and scarcely settled forty years later. There were enough islands to lose or confuse mere bureaucrats, just as there were plenty of men willing to take risks or bribes, bend rules or interpret them generously.

The Islanders were lured with the promise of guns and knives or money to buy them, cajoled or coerced on board small island trading ships for the relatively short voyage to Bundaberg, Mackay, or one of the many other Queensland ports. None of the hard-won laws which stipulated a minimum of decency, health or safety on other immigrant vessels was seriously applied, though a system of inspection did exist. In most cases both ships and their captains had seen better days.

The earliest arrivals went to work on cotton plantations in southern Queensland. Cotton was seen as Queensland's crop of the future in the 1860s, replacing American supplies disrupted by the Civil War. But forced coloured labour proved no better than the labour of wives and children advocated by earlier enthusiasts for cotton, and the industry did not flourish. Sugar-growing was still in its experimental phase, but the Islanders found immediate acceptance on inland sheep stations, ever hungry for unambitious labour, and even in the towns where they were welcomed as cheap and believable casual labourers, especially in the house and garden. The belief that white people were incapable of physical labour in tropical climates had yet to be tested and proven more or less wrong. In the meantime, 'dark-skins' were thought better adapted to digging gardens, chopping wood, and carrying water. Once the technical mysteries of milling sugar were resolved, the Islanders became invaluable for clearing thick scrub to plant cane, then in the intensive work of weeding, cutting, crushing and milling. Early sugar plantations were large, self-contained in that they did their own milling, and labour intensive at crushing time.

The Islanders were housed in huts or sheds, sometimes of their own construction in the traditional island style which they preferred to boxes of corrugated iron with no ventilation. Overcrowding was invariably a problem and so was disease. Sanitary arrangements were frequently left to nature. Sugar was planted right up to the huts and the practice was simply to duck outside, except that for several months of the year the ground was nothing but mud and slush. Dr Thomas Bancroft, government medical officer, reported from the Johnstone River in 1885 that 'fecal accumulation' at some places was so great that 'stepping stones were placed to prevent one going shoe tops' deep'.[111] The consequences were 'mild typhoid, severe dysentery and diarrhoea', all diseases easily transmitted across racial boundaries without any physical contact.

It was possibly an underlying concern about health questions which tipped the balance of public opinion in Queensland against the plantation system in the end. The death rate amongst the Islanders had reached scandalous proportions by 1880—62.89 per thousand rising to 147.74 per thousand in 1884, and dropping to 58.2 per thousand in 1886, the year after Griffith first announced that the system would come to an end by 1891. When it is remembered that these deaths occurred among men in the prime of life labouring out of doors, the figures are the more shocking (though not when compared with a death rate of 160 per thousand among the Aboriginal population in 1848–50).[112] Queensland was not a healthy place for anyone. Of 265 cases of leprosy recorded in Australia between 1860 and 1901, 111 were in Queensland 'and in addition, many kanakas' comments the Commonwealth Year Book for 1908.[113] Tropical conditions, remoteness, and an imperfect understanding of the connection between sanitation and disease combined to produce a comparatively high mortality rate among European settlers also. Kay Saunders has put it at 17.49 per thousand in 1884, the year when Islander deaths reached their peak (compare 15.91 in 1883).[114] But the difference between European and Islander mortality is still enormous, the consequence of a totally inadequate diet, excessively hard work, and debilitating living conditions. In their weakened condition the Islanders succumbed not only to the range of tropical diseases and fevers, but to scurvy, tuberculosis, influenza, measles, and whooping cough. Regulations which laid down a minimun dietary scale for plantation workers were flouted openly. The fallacy that black skins protected their owners from sunstroke or heat exhaustion was stoutly maintained.

There was never any question of a revolt or an uprising. There was not enough energy or freedom for that. Kind-hearted attempts to teach a little of Christianity were discouraged for fear the 'darkies' would learn a message of hope. Instead, it was made easy for them to spend their money on alcohol or occasionally visit prostitutes.[115]

With such attitudes towards the health and welfare of the Islanders who were at least productive labourers, it is not surprising that the Aborigines fared worse in Queensland. The Aborigines took their chance in the open labour market or fought a surreptitious war for their lands and livelihoods against advancing settlers and native police. In settled districts they became objects of private charity, but concepts of health, welfare, and decency which were applied even to the Islanders were not seen as relevant to them. Where the diet laid down for the Islanders was merely inadequate for the kind of work they were required to do, the Aborigines were starving. When the scientific expedition on HMS *Challenger* called at Somerset, Cape York in 1874, botanist H. N. Moseley reported, 'The natives were in a lower condition than I had expected.'[116] 'Food is their greatest desire' he went on, and described how the two Aborigines he was using as guides made a quick fire as soon as he had shot some parrots for them, and ate the birds, entrails and all, half-cooked. He was surprised to find that they knew the difference between a shilling and a florin and spent what money they were given immediately on biscuit at the store.[117] 'Square gin bottles, of which there were plenty lying about the camp, brought from the settlement' had replaced traditional containers.[118] Noel Loos has described a large-scale experiment in food relief undertaken in North Queensland in the 1890s. Rations were provided to the Aborigines at various centres, like Atherton. This stopped raids and the Aborigines camped quite peacefully, even working occasionally for the settlers. It was found that food relief cost less than maintaining the native police.[119] In 1892, 400 Aborigines were being 'controlled' by the expenditure of £20 per month on rations for distribution. In 1895 the residents of the Cardwell Range petitioned Horace

Tozer, Queensland's Colonial Secretary, for protection against Aborigines who were setting fire to their properties and threatening life. Tozer remarked, 'If a distribution of food could be arranged by some competent person near this place all outrages would cease. I prefer this if it can be arranged to native police.'[120] With the Queensland Aboriginal Protection Act of 1897, the position of the Aboriginal people was spelled out. Tozer again: 'This Bill endeavours to do as a charity organisation does: focus the assistance in some definite channel.'[121] But unlike the ordinary recipients of charity, the Aborigines were excluded from institutions which might help them, from hospitals and schools, from the system of wage regulation, from trade union protection.

Between 1860 and 1900 there were 111 executions by hanging in New South Wales, 109 men, 2 women, 11 of the men for rape, the rest, men and women for murder or attempted murder.[122] The number of executions in Victoria (where neither rape nor attempted murder was a capital offence) was about 73.[123] Figures for the other colonies seem less certain. Those for Western Australia and Queensland may be nearly as high as those for New South Wales and Victoria, while those for South Australia and Tasmania are somewhat lower. There was a trend, of which all colonies were proud, for the frequency of capital convictions to decline towards the end of the century, a sign that society was becoming more settled and civilized. The fact that per head of population the rates in Western Australia remained high till the end of the century seemed to confirm this belief.

It was difficult, and as colonial statisticians pointed out regularly, unwise, to compare crimes rates in the different colonies, but comparisons were still made. Factors such as population structure or the size and efficiency of the police force made a difference. South Australia and Victoria with their 'normal' age and sex ratios had lower crime figures 'per 100 000' of the population simply because the most violent and lawless element—unmarried males between the ages of twenty and thirty—was statistically balanced by greater numbers of women and children. Western Australia had an excess of rootless young men, first ex-convicts, then miners. South Australia, the least policed colony, had 950 inhabitants to each policeman in 1900 compared with Western Australia's 354.[124] New South Wales seemed heavily policed with the number of police per head rising towards the end of the century. Everywhere police fulfilled a variety of public service functions such as collecting statistics, supervision of vaccination programmes, truancy prosecutions, and old-age pension applications. They took initial responsibility for lunacy or cases of mental breakdown, and in Victoria were charged with keeping walls free of obscene grafitti and footpaths clear of orange peel. In New South Wales, it was said, the police and the magistracy were more strict in their treatment of drunk and disorderly persons than they were elsewhere.[125] Such factors undoubtedly affected the comparability of crime statistics.

Similar trends, however, did seem to be emerging in all colonies. The combination of more generally available education, work, and income, especially among the young to middle-aged male population, and improving opportunities to marry and set up a household went along with a decline in crimes against property and persons. Greater efficiency and consistency in making arrests and in interpretation of the law flowed from a better educated police force and the professional magistracy which gradually superseded the old system administered by honorary justices of the peace. There were comparatively fewer arrests towards the end of the century than during the 1860s and 1870s, and higher conviction rates in the lower courts with fewer cases finding their way to higher courts. This suggests not only greater confidence and efficiency in the operation of the system, but also a sharpening perception of 'the criminal element' in society. Indeed there was a tendency to accept and identify a small habitual criminal class. By the end of the century, the hope or belief that crime could be eliminated by deterrence, transformation of individuals or reform of the social system itself had faded. At the same time the fear that criminality was inheritable was laid to rest.

Such changes in attitudes among those who had become professional administrators of law and order encouraged liberalization in the treatment of some forms of crime. By

THE ABORIGINALS' CAMP

1900 T. A. Coghlan, for example, was describing drunkenness as a 'pseudo-crime'[126] and wondering what significance should be attached to the fact that the first conviction of many habitual criminals was for drunkenness. There was some understanding as early as the mid-1880s that the disproportionate appearance of certain national groups in the courts, especially the Irish, or Scandinavian and American sailors in Sydney, was due to their social and economic position, to structural factors such as their lack of education or their temporary residence, rather than to inherent criminality. In adopting the abolition of capital punishment as part of its programme, the Labor Party followed contemporary informed opinion on the relationship between social and economic disadvantage and inability to evade the full weight of the law. Working-class offenders were least likely to have their crimes viewed in a sympathetic light or to be able to buy justice or exoneration. Such disability was most graphically illustrated in the Aboriginal population. In Western Australia towards the end of the century, crime statistics for Aborigines were registered separately because they so magnified the overall figures when included.[127] Yet most Aboriginal crime was slight—sheep and cattle stealing, or drunk and disorderly behaviour—offences which would ordinarily receive only a fine or some sharp advice from the magistrate. Aborigines were rarely in a position to pay fines. They stole stock because they needed food. The numbers of Aborigines in gaol were grossly inflated, both because of their hapless economic position and the tendency of the law to have a cumulative criminalizing effect.[128]

It was sometimes said that natural selection had intensified a natural Australian propensity for lawlessness and idleness as exemplified by the larrikin pushes and the troubled labour conditions of the 1890s.[129] The most striking theme in the administration of law and order, especially in late nineteenth-century New South Wales, however, was an élite preoccupation with sex.[130] There in 1883 both forcible rape and carnal knowledge of a female under ten years became punishable by death. Other offences such as attempted rape, indecent assault and wilful exposure could be punished by both whipping and imprisonment. The 'severity of these penalties seems to have had no parallel elsewhere in the British empire at the time'.[131] By the end of the century all colonies dealt severely with 'crimes of lust', and rape was everywhere a capital crime. Both Queensland (1868) and Tasmania (1879) introduced contagious diseases legislation modelled on British legislation of 1866. In both cases this legislation was conceived as protecting society against prostitutes, whereas the legislation on rape was to protect good women against male violence.[132]

The reasons for such severe legislation on sexual matters are not easily unravelled. Peter Grabosky suggested the Anglican establishment in New South Wales as the strength behind both the legislation on moral questions and the temperance movement.[133] The severity of the law on sexual behaviour in general, and especially the addition of whipping as a punishment for rape, was probably a direct consequence of the absence of ecclesiastical courts. Whipping was deemed the most appropriate punishment in cases of incest (which were presumed to involve rape).[134] In Tasmania an early shortage of women led to repressive sexual legislation, not only against rape and carnal knowledge, but also against sodomy, which was a capital crime until 1887 though no executions occurred after 1863.[135] The Tasmanian legislation against contagious diseases was in response to a threat that the British navy would withdraw its base unless steps were taken to ensure clean, safe prostitution, though there was also a lingering anxiety that feeble-mindedness, a legacy from convict days, would be transmitted to the next generation through prostitutes. Fears about the rapid spread of veneral disease certainly influenced contagious diseases legislation elsewhere. Queensland knew itself to be a frontier society under siege, with a desperate need to differentiate safe and respectable sex from the dangerous and disreputable variety. Contagious diseases legislation, however, may have made things worse, creating a false sense of security in prostitutes who were officially free of disease. Of course they were not. There was no medical treatment known at the time. Later in the century intercourse between white settlers and the Aborigines was also outlawed in Queensland in a futile attempt to halt the spread of the disease.[136]

Everywhere the legislation on sexual matters meant both more and less than it said. Its effect as a deterrent was most visible on the police who became unwilling to lay charges, and on juries who became reluctant to convict when the death penalty was involved. Much as he was reviled for his judgement that nine youths must hang for rape in the Mt Rennie case and his impatient handling of the Dean murder trial, Mr Justice Windeyer was abreast of the times, as his doubts about Christianity and his acceptance of birth control show.[137] The traditional authority on sexual behaviour, the church, was losing its effectiveness. Community sanctions seemed weak. In some cases, for example, with Aboriginal men who were accused of rape, and Aboriginal women who served as prostitutes, neither church nor community ever had any authority. At the same time, the wild extremes of public response to the rapes at Mt Rennie or George Dean's blatant attempts to murder his wife and frame his mother-in-law indicated something of the level of latent sex hostility in the society.[138] Legislative attempts to control sexuality imply a strong current of repression, magnified in the male population by the extreme vulnerability of the female population. Sexual activity was an obvious outlet, as was sport or war, for the energy and aggressiveness of healthy and vigorous people, but in this period sex was rather more dangerous than sport, at least for the female participants, and more probable than war.

Legislation failed to contain sexual violence or to curb drinking. The average Australian male of the late nineteenth century was more likely than either his father or his grandfather to douse his aggression or unhappiness with alcohol. There were 'no fewer than twenty-three liquor Acts' passed in Victoria in the second-half of the nineteenth century.[139] As well, the proportion of deaths caused by violent accident among young men was unnecessarily high. Suicide, the 'collective sadness' described by Emile Durkheim in 1897, seemed to G. H. Knibbs worthy of special examination. His study of suicide in Australia was published in 1912. He found there had been a slight tendency for the suicide rate to rise in 1887–88, and again in 1893 and 1897, then to fall in 1900, the rises coinciding with economic crises and drought, the fall with the Boer War. The suicide rate in Australia was approximately half way between the highest rate (Switzerland) and the lowest (Ireland), which may have been an interesting comment on the loss of authority of the church in Australia, but it was also higher than that in England and Wales. With men, the rate of suicide seemed to rise, reaching a peak at age sixty-two. With women there was little suicide at all, and that mostly among young women before they were married and had responsibility for children.[140] To a certain extent organized sport damped down larrikinism or youthful hooliganism. But Henry Lawson, with his unsatisfactory marriage, his uncongenial employment prospects, his aggressive militarism, his drinking problem, his melancholic lack of direction in life, spoke easily from his own experience to many of his contemporaries of things they knew.

Towards the end of the nineteenth century social theorists interested in such questions as the meaning of sexuality, the significance of heredity, the causes of aggression or the origins of criminality—Havelock Ellis, Sigmund Freud, Francis

Galton, Emile Durkheim, Ceasare Lombroso—began publishing their observations. Except for the Aborigines, Australia did not interest them as it did a large troupe of visiting social reformers, like Beatrice and Sidney Webb[141] or Henry Demarest Lloyd,[142] though S. A. Rosa kept Lombroso informed about developments in criminology in Australia,[143] and William Chidley sent his sexual memoirs to Havelock Ellis who put them to good use.[144] There was much discussion of Australia as a 'social laboratory', though this was mainly in political and economic terms, of the development of democratic practices, or the management of relations between labour and capital.[145] Interest in the new 'Australian type' was more anxious than scientific, concerned to maintain purity of colour and quality of strength and character. Other aspects of the society which made it a working model of comparative development were overlooked: the effects for example of the many variables which had gone to produce six different but similar colonies; the impact of varying rates of urbanization and industrialization on a basically rural society; the management of sexuality in a society where the dominant ideology was excessively masculine, but where females were in short supply; the relationship between heredity and environment, or affluence, aggression, health, and leisure. It is easy now to explain late nineteenth-century Australian use of alcohol and tobacco as the product of increasing affluence and modern stress. Likewise the sporting preoccupations of the male population reflect aggressive tendencies highly prized, even cultivated, by the needs of pioneering and economic development, but for which acceptable outlets were diminishing. The successful innovations in Australian society in the second half of the nineteenth century were possible because of economic growth and the absence of traditional institutional or social restraints, yet in seeking a moral or social order to preserve those gains, unexpected difficulties were encountered. The state could not simply replace the church as the mainstay of the family. Nor was the family large enough or suitably flexible in its nature to form the sole basis of social organization. As personal happiness began to replace social conscience in the value system, new forms of measurement and regulation were required.

4

CULTURE

THE SPEECH of the Australian-born working man of the late nineteenth century was probably a little less broad than is usual today, though it may have been slower, more deliberate, more self-consciously correct than that of his English-born counterparts, with a tendency to inflate the uncomplicated and diminish the serious. His language at first was colourful, inventive, or pedantic where self-education brought out the autodidact. Later generations tended to pomposity, evasion and cliché, though these too were sometimes a cover for shyness or embarrassment at being in a position to speak out at all. 'Whilst conveying my own unobtrusive individuality into Echuca on a pleasant evening in the April of '84' was Joseph Furphy's choice for a low key beginning to *Rigby's Romance*. Among educated Australians, especially those whose teachers were immigrants, a recognizable form of educated English was heard. Only its idioms were sometimes obviously Australian. This was more noticeable with men than with women, for whom 'correct' English was most important as an indicator of social status. A lady did not use 'slang'. Since the population still included a fair number of recent immigrants, it was not difficult to stay in touch with the many variations of current English speech. Australian-born working-class women, if their occasional portrayal in print is any guide, practised a slightly whining or alternately brawling version of the laconic style of their menfolk. Neither their education nor their experience was likely to have provided them with the embellishments and circumlocutions which appeared in male speech patterns. Indeed any pretensions to self-improvement were quickly observed and ridiculed. As guardians of tradition and taste, women could not be permitted freedom to innovate.[1]

Two themes stand out in the history of Australian culture in the second half of the nineteenth century. One is the way in which ordinary people, given time and means, adapted the classical forms of British and European culture for their own purposes. The other is the gradual acceptance of environment and ordinary experience as legitimate expressions of a national culture.

Most culture in the Australian colonies was based on British forms, also being democratized in the second half of the nineteenth century. The factory system, by defining hours of work, defined hours of leisure. As operatives were trained for precision and efficiency in numeracy and literacy, they learned also to think and read. Urbanization made 'nature' and the countryside a striking alternative to 'culture' and the city. By entering into the mass production of wool, minerals, meat and flour for Britain (and other commodities on a small scale for her own people), Australia experienced something of the factory system without the iron mills. By accepting eagerly the uses of new technology to compensate for a sparse population, Australians experienced the efficiency of modernity. By clinging to their few cities, they kept the bush at bay until they were ready to accept it on their own terms. Culture had been what superior and leisured people did to amuse themselves and fill in their time. In Australia it became what ordinary people did with their leisure to amuse themselves and make them feel superior. For culture, leisure was essential, whether statutory holidays or evenings no longer cramped by tiredness and lack of means. Most Australian culture of this period was of and for people released into time and affluence, able to think of creative leisure instead of sheer rest and recreation. But because of that they were dependent on known or classical forms. Indigenous forms had not had time to develop. So the landscape, the environment, came to be of immense significance, initially because it provided the setting, then because it suggested challenges to traditional forms and styles, and eventually because it was a barrier to the wider world, a badge of provincialism.

The forms and knowledge of the Old World were immediately evident in buildings, the basic expression of culture. By the second half of the nineteenth century the whole range of classical styles was being deployed on official buildings as well as on the homes of wealthier citizens, though availability of local materials produced significant regional variations, especially in domestic architecture. Both Melbourne bluestone and Sydney clay provided distinctive colour and texture. Something approaching an Australian style began to appear in buildings constructed for particular purposes, as in wool-stores and shearing-sheds. But the conjunction between means, needs, and taste was most clearly expressed in the modest homes of ordinary people, in the cities and in the country. W. S. Jevons observed in 1858 that Australians cared more about the ownership of a house than the quality of accommodation it provided.[2] Houses were often constructed of poor or flimsy materials, quite mean in size and appearance. The simple single-storey cottage which grew by the addition of a lean-to out the back was the height of aspiration. A paling fence, whitewashed or not, was almost mandatory. Individual taste was expressed in the garden which was the most quickly and inexpensively established feature of owner-built houses. Gardeners tried to re-create flower and vegetable gardens from cooler, moister climates.[3] Visitors to the colonies were surprised by the 'clusters of wooden houses . . . painted white to keep off the sun. Gardens and flowers were, as usual, universal.'[4] Gardening seems to have been a popular, even serious practical pastime. Most weekly and monthly journals carried a gardening column of some kind. The seedsman's advertisements vied with those of the draper and the grocer, while flower shows and horticultural exhibitions were eagerly followed. This practical botany was perhaps the nearest many people got to science. As the Australian environment's capacity to sustain a wide range of the world's plants in ordinary outdoor conditions became appreciated, so opportunities to learn and enjoy more about gardening were multiplied. Especially interesting in this regard were the botanic gardens, the oldest and most easily appreciated government contributions to learning, culture, and leisure. They epitomized contrasts between the local and the exotic, scientific inquiry and aesthetic pleasure, education and entertainment.

By the 1870s botanical gardens in both Melbourne and Adelaide had achieved international reputations for their beauty, their extensive collections, and their contributions to the study of botany. In Sydney and Brisbane, warmer climates and choice locations, in Sydney on the harbour foreshores, in Brisbane on a sharp bend in the river set off by the pink and lilac cliffs of Kangaroo Point, the gardens were more lush and fantastic than scientific and orderly. The gardens in Sydney brought pleasure and wonder to visitors already overwhelmed by the harbour itself. Anthony Trollope thought them 'perfect'. (He stayed opposite the gardens in Melbourne at 'Fairlie' but found them 'pretentious'.[5]) Sydney residents, however, were becoming accustomed to riotous intermingling of tropical and subtropical plants from all over the world in their own gardens. Frangipani, hibiscus and magnolia competed with Moreton Bay figs and local and imported palms in gardens, both public and private, singly and in stately avenues.[6] The Brisbane gardens created a feeling of order and design for those who knew the bush. The wilder hillsides and mountain gullies had yielded some of their more precious and spectacular specimens, fragrant hoyas and gigantic staghorns, the brilliant stenocarpus, the glistening macadamia, promising to match gifts like the delicate bauhinia, the brilliant poinciana, the dreamy jacaranda from India, South East Asia, and South America.

Acclimatization societies struggled still to discover the useful and tameable among the millions of species inhabiting the east coast ranges, but their more determined work lay in adding to this bounty innumerable varieties from the known world's store-house of favourite and favoured plants. So the almond and the chestnut were planted in groves while candlenuts and coconuts slumbered by northern beaches. Lemon trees proliferated in suburban gardens while native limes fell uselessly to the ground. Where acclimatization societies left off, individuals took up. It was difficult to say which was the greater challenge, coaxing a delicate English plant to flower in the upside-down Australian climate, or wresting the secrets of propagation and cultivation from the native flora. The former easily gained over the latter however, as a sign of affluence, civilization, and superiority. A creeping addiction to neat lawns, formal gardens, borders and floral displays was dramatized early in disagreements between Baron von Mueller, botanist in charge of the Melbourne Botanic Gardens, and his employers, the Victorian government. Mueller's passions veered between a comprehensive listing of the Australian flora and its adaptation for use elsewhere, and the introduction of commercially viable plants to Australian conditions for cultivation. He was willing that the public should enjoy themselves among the trees and plants, but regarded his scientific work as paramount in the arrangement, management, and display of the gardens. Not for him the 'useless' floral display, the avenue planted for impact, the sweeping lawn which only wasted water, the contrived setting for secluded encounters.[7] His successor, William Guilfoyle, had grown up in northern New South Wales and learnt his gardening in Sydney. He was fascinated by the subtropical potential of Australia, and experienced in the flora of the Pacific region. It was easy for him to combine his own interests with the demand for dramatic and elegant public gardens. Under his management, the Melbourne gardens became more showy and subtropical, less obviously Australian and improving.[8] Such developments were in harmony with the confidence of the 1880s, and moves towards formality and elegant effect, indoors and out, in private and in public. Native plants with their fantastic shapes and tendency to straggle picturesquely or to resist pruning, lost favour to the more disciplined and well-bred species of the Old World.

Indoors, necessity was being smothered by comfort. In the bush, household essentials might still be improvised from kerosene cases, cut-down kerosene tins, and old bottles, but in town, elaborate furnishings and fittings were expected in even a modest home if the extensive advertising of hardware and furniture stores is any guide.⁹ At 'Nasturtium Villa', home of the Wapshots, on 'the Saint Kilderkin Road', Marcus Clarke found 'evidence of wealth without taste'.

> The drawing-room furniture, most expensive, and therefore most excellent, was green picked out with crimson. The curtains were yellow damask (Heaven only knows how much a yard at Dungaree Brown's!), while the carpet represented daffidowndillies, roses, and sun-flowers on a pink ground. The pictures on the walls were either chromo-lithographs, or—more abominable still—oleographs of the most glaring, hideous, yellow staring nature... The dinner at Nasturtium Villa was an infliction under which all have suffered. Soup (*bad*), fish (*indifferent*), sherry (*very bad*), mutton (*good*), vegetables, own growing (*most excellent*), *entrees* of fowl and some other nastiness (*both infernally bad*), champagne (*that is to say moselle*), cabinet pudding, tarts, custards (*all good*), cheese (*colonial and so so*), dessert (*good*), wine (*tolerable*), cigars (*very shy, W. not being a smoker*), and brandy (*the most admirable which could be bought in the city*) ... In a jolly bachelor camp, in a pleasant manly meeting of friends frying a lamb chop, baking a damper, standing a drink, or uncorking a bottle, Joe Wapshot would have been excellent, charming, beneficent. But in a badly furnished drawing-room, blocked up with a grand piano, a bird-cage, an indifferent plaster cast of the *Venus aux belles fesses* (he doesn't know it under that name!), and five spoiled chromo-lithographs of Der Günstumper's *Windelkind Gurken-salat*, he is much out of place. Mrs Joe, who would be delightful on a desert island, is simply an unobtrusive nuisance in her own house...¹⁰

It is kinder not to think of Marcus Clarke's actress-wife resented and neglected in a suburban villa, struggling to feed their children.

The archetype of gracious living, Australian style, was the station homestead, low, long, cool with wide verandahs and comfortable furniture, surprisingly elegant, but relaxed.¹¹ Tension between good taste and pretentiousness was a recurring theme. Visitors (or immigrants like Clarke) could see the potential for a more comfortable version (with Mediterranean overtones) of the English style they still expected. Native-born Australians were handicapped by their ignorance of Mediterranean sophistication and still inclined to suspect that colonial meant inferior. At Nasturtium Villa, the local produce was good, though methods of cooking continued to be English and unimaginative. Australian wines

Advertisements. xci

F. H. FAULDING & CO.,

IMPORTERS,

Wholesale Druggists and Drysalters,

42 AND 44 KING WILLIAM STREET, ADELAIDE,

AND

NORTH PARADE, PORT ADELAIDE.

CHAMPAGNE.

A Continental Wine Circular, under date 1st September, 1880, reports—"That in the Rhine scarcely any Wine is expected to be made. In Champagne the yield will be again very small, the vines having suffered severely from the cold winter and from the subsequent dropping of the leaves." Some years ago this report would have been of much more importance to us than now, since the

CHAMPAGNE

of Australia is coming into the first rank, and with the Cricketers, the Oarsmen, the Wool and the Wheat, claims for itself a place and a name. It is found by a French winegrower of great experience now in South Australia, that the soil and climate of the colony are admirably adapted for the production of the light and sparkling wines so necessary to give zest and enjoyment to life in our adopted country.

F. H. FAULDING AND Co. beg to announce that under the advice and management of Monsieur Bourband they have established the manufacture of South Australian Champagne from judicious blends of the lightest and choicest vintages of the country, and they confidently invite consumers to try, with a certainty that the result will be mutually advantageous and satisfactory. The alcoholic strength of this wine being below most of the imported, and the price less than half for any approved brand, will, they trust, bring it into general family consumption.

IN QUARTS, ONE DOZEN CASES; IN PINTS, TWO DOZEN CASES.

WHOLESALE FROM F. H. FAULDING & CO., ADELAIDE,

AND RETAIL OF NUMEROUS AGENTS THROUGHOUT THE COLONY.

PLEASE NOTICE THE LABEL.

Champagne

Cuvée speciale

TRADE MARK—GRAPES.

Pour les Colonies Australiennes.

were already recommended in preference to spirits (which were for hardened drinkers) and beer (which was of uncertain quality). Australia's champagne was said to have first-rate potential like her cricketers, oarsmen, wool and wheat, and in Adelaide, J. A. Froude enjoyed an 'Australian hock, light and pleasantly flavoured, with some figs and apricots'.[12] The lifestyle of the suburbanite was already being deplored:

By far the greater number of people dawdle in bed till the last possible moment, when all at once they jump into their bath—that is, if they take a bath—swallow a hasty breakfast [probably consisting of chops, steak, or sausages], and make a frantic rush for their steamer, train, or tram, in order to begin their daily work.[13]

Gradual acceptance of Saturday as a half-holiday for increasing proportions of the workforce engaged in regular office and factory work brought a new perception to the meaning of leisure. Long holidays were another matter. For the wealthy there might be a sea voyage to Europe, travel on the continent, an extended stay in Britain visiting relatives and old haunts. For wealthy landowners a version of the 'Season' prevailed as well. The slack months of the winter permitted long periods in town while husbands attended to parliamentary and other business and wives dealt with matters of succession, social standing, and shopping. For those who could afford to live more or less permanently in town, the country seemed to have fewer attractions than it did for their British counterparts. The well-off, whether rural gentry or urban business or professional men, liked to send their sons 'home', ostensibly for pleasure and education, but really in search of a wife—a distant cousin, or the daughter of a family vouched for by relatives, friends or business associates. The shy, mysterious, financially attractive but no longer entirely youthful bachelor from the colonies in search of a wife was a favourite for romantic tales of the time. Likewise, the beautiful débutante of uncertain colonial origins and perhaps even more uncertain parentage could be found. She was, however, rarely a match 'at home' for her wealthier and more aggressively 'ladylike' American competitors. She would as likely as not end up as a respectable and reliable if somewhat drearily cultured spinster aunt to the children of her British-born sister-in-law.[14]

Such 'holidays' were undertaken once, twice, perhaps three times in a lifetime. For shorter, more frequent escapes from home and business, cool climates were preferred, though the seaside was not without attraction. Several Sandgates and Brightons promised, along with resorts like Lorne, Queenscliff, Watson's Bay, Manly, and Victor Harbour, invigorating air, sea bathing, and the restorative effects of long walks on cliff tops within sound of the waves. A combined visit to the city and seaside at Christmas time was a welcome treat for many country families.[15] Because all major cities were on the coast, seaside picnics and excursions were easy and attractive, especially by public transport. 'On holidays' wrote Edmond Marin la Meslée of Sydney in 1883,

the entire population pours out of the city. Some go to Botany Bay, others go to 'picnic' at the many beauty spots in the harbour, which is covered with pleasure-boats and steamers packed with all classes of people. They set out after breakfast in the morning and return about six in the evening after a day in the open air. Quite a few are not so fresh and sprightly on the morrow and have to rest for a few days more to recover from the fatigues of their holiday, perhaps because they had not enough wisdom to drink moderately. The *larrikin* element is particularly prominent on these occasions, and sometimes it creates disorders which the police cannot handle. But nothing can give the foreigner a better idea of the easy-going ways of all classes, than to land in Sydney in the middle of a public holiday. Omnibuses, ferries and trams overflow with people in holiday attire, all making their way to favourite resorts. There are no rags and no beggars: everyone has the means to dress decently and have his share of the fun. Well-being is universal. Orange-sellers reap a great harvest, and these days make the fortunes of the publicans at Manly beach and Botany.[16]

On Saturdays and Sundays too, Bondi and Coogee, St Kilda and Brighton, Henley Beach and Glenelg were crowded with trippers.

Some chose the sea, others the mountains, according to personality or identification. Among colonial politicians, Deakin, Robertson, Dalley, Macalister all opted for the sea. Parkes developed Faulconbridge in the Blue Mountains. Among certain groups there was a marked preference for mountain resorts, Medlow Bath and Mt Victoria, the Dandenongs, or following vice-regal precedent, Sutton Forest and

BRIGHTON BEACH ON A PUBLIC HOLIDAY

Moss Vale, Mt Macedon and the Adelaide Hills. Even small towns in the ranges behind Brisbane were thought of as hill stations. In the 'highlands', deciduous trees ensured fresh, green, shady summers—when there had been enough rain—sharp evenings, the possibility, even, of log fires, lawns, and 'English' gardens. If a spa could be established, so much the better. Gentle games of tennis and croquet, picnics, walking and riding filled the days. For those who required something more vigorous, Tasmania was considered the Australian equivalent of the English Lake District, the Welsh mountains or the Scottish highlands. The 'inducements which Tasmania offers as a place of residence to those who have been enervated or invalided by a lengthy sojourn in tropical climates' were not less attractive on a short visit.[17] Tasmania was also a good excuse for a short sea voyage, a fashionable cruise on a private yacht as Mr Anthony Churchill, hero of Ada Cambridge's *A Humble Enterprise* (1896), discovered in his need to evade the romantic complexities of Melbourne. Tasmania became the 'social rallying point' of the summer season, and Hobart 'the place where Australian society may be seen under its most charming aspect'.[18]

K. S. Inglis has described the patterns which celebrations of traditional holidays, the monarch's birthday, the sabbath, Christmas, assumed in the Australian colonies.[19] Because it was officially a Protestant society, there were none of those saint's days which brightened the routine of working people in Catholic Europe, and perhaps for this reason, the English 'Boxing Day' was retained, especially as it fell at the height of summer in the antipodes. Irish Australians were keen to retain and revitalize their traditional St Patrick's day celebration. From the 1860s, depending on the state of Anglo-Irish relations and the levels of Irish national consciousness in Australia, it was celebrated with greater or less sectarian vigour and response. Occasionally when it fell on a Saturday, St Patrick's day was proclaimed a public holiday, but mostly, those who wished it a holiday made it one. Other national days were added to the calendar, though not as public holidays, rather as occasions for ceremonial dinners or concerts. There were also the Australian national days. January 26, Anniversary Day in New South Wales, was accepted in 1888 under pressure from the Australian Natives Association, and in a centenary gesture, as a national holiday also in Victoria. Separation Day, formerly a Victorian public service holiday, became a mere bank holiday.[20] By the end of the century only South Australia and Western Australia persisted with public holidays to mark their foundations.

Where rural or agricultural activities dominated the economy, traditional work patterns—periods of concentrated effort interspersed with periods of light and intermittent labour—made specific holidays unnecessary. Following British experience, the need of city-based clerical workers for a break in the second half of the year (Easter made a break in the first half) brought about the bank holiday and the public service holiday. For the rest of the urban workforce who were hired casually by the day or the week, holidays seemed unnecessary or unwelcome. But the notion of a paid holiday as a reward for constant service or skill was gaining ground. Various skilled trades—butchers, for example—fought for and won the right to a specific trade holiday or picnic day during the late nineteenth century. Such holidays were in essence a link with the pre-industrial past, an urban equivalent of harvest home or a celebration of the trade or craft, an occasion for professional unity and solidarity. The continuing tradition of friendly society activities and demonstrations also maintained the link. Towards the end of the nineteenth century, the number and extent of friendly society and trade society holidays, half-holidays, processions, picnics, and sporting competitions was such that in Melbourne and some of the larger Victorian towns pleasure grounds were set aside for these occasions and known as friendly society parks. These holidays were usually granted on the understanding that employees attended whatever activities had been arranged. Most working people probably attended no more than one or two of these gatherings a year, yet the frequent processions, often with floats, banners, and a band, made a colourful addition to the life of the city. The eight-hour day movement with its demand for a universal workers' or labour day was a culmination of these variegated practices.[21] Certainly the institution of 'Labour Day' or 'Eight (or Six) Hour Day'—it went under various names—led to the decline of many individual trade and friendly society outings, though some were written into and carefully safeguarded in early awards. The tradition of a picnic, sporting activities and an evening entertainment at a concert or theatrical performance lingered in the annual school or Sunday school treat for children.

Regulation of work and leisure proceeded gradually on the premise that there was a division of labour. It was assumed, for example, that married women would shop for their families while their husbands were at work. Thus the early closing movement became viable. By the end of the century only shops dealing in certain perishable goods or providing special services could trade after dark or at weekends. Wives were advised to finish their housework while husbands were away so they could be decorative and amusing in the evenings. Nonetheless they might find themselves at home minding the children while husbands were at lodge, union, or other meetings and at football matches or the pub at weekends.

Access to leisure was rationed, as was paid work, according to sex and marital status. There were no statutory half-holidays for housewives. Organization and commercialization of leisure made it more difficult for mothers with children to participate.

The Australian workingman's high earning capacity produced its equivalent in time off and money to spend in recreation. The drinking sprees of bush-workers were a notorious illustration of this conjunction. By contrast, the drinking culture of the urban worker during this period was significant mainly because it competed with married life.[22] Gambling on a large scale as witnessed by two-up schools and betting shops also required a certain amount of cash to sustain it.[23] Tobacco was the cheapest escape, used by itinerant labourers, the unemployed, the Aborigines. The 'smoko' was enshrined in the Australian vocabulary as a symbol of superior working conditions, but for the poor, tobacco usually came as a hand-out. Coarse, dark plug tobacco for chewing rather than smoking was consumed as part of their rations by rural labourers, especially the Aborigines and Pacific Islanders. The plaintive request 'Gib 'em bacca, boss' echoes through these years.[24] 'There is something in the climate', thought Harold Finch-Hatton, 'that brings out the flavour of tobacco, and a good deal in the way of living that encourages smoking.' He also thought that

> about ten years is as long as a man can go on smoking [fig tobacco] without finding that it is knocking his nerves to pieces. A fig a day, or just short of an ounce, is a common allowance, but a Bushman's pipe is never out of his mouth. He is always lighting it to have a few whiffs, which is a most poisonous form of smoking. The last thing he puts away at night, and the first thing he looks for in the morning, is his pipe, and if he wakes in the night, he has a smoke then.[25]

Those who hoped that the provision of leisure and opportunities for self-improvement would lead to an elevation of morality and cultural standards were probably disappointed. But rising concern about the easy, simple pleasures of drinking, smoking and gambling was probably alarmist. Activities

GIB 'EM BACCA, BOSS

which took him out of doors and used up physical energy began to appeal to the black-coated worker, clerk or salesman. Leisure activities, invented by the wealthier members of society for amusement and exercise, began to attract recruits lower down the economic scale. This hastened organization so that entry was controlled and behaviour ordered by the codification of rules.

Sailing, horse-racing, and shooting were among the oldest and most aristocratic of 'pastimes'. They were also obvious, spontaneous activities for Australian conditions, and among the earliest to develop organizations and controls. Yacht clubs had been formed in New South Wales, Victoria, and South Australia before 1860, and shortly after in Queensland, Tasmania and Western Australia though their survival was in some places uncertain and not continuous. A shortage of members with the wealth as well as time and enthusiasm for management was usually the problem. Royal patronage in Sydney and Melbourne encouraged exclusivity and relegated common owners of working boats to the massive regattas, staged on anniversary days and other special occasions demanding public spectacle and entertainment.

Horse-racing also became organized in the 1860s. The high rate of horse-ownership made necessary by distance and fostered by affluence had a strong impact on our culture. The horse became a major theme for writers of both prose and verse. Hoofbeats drummed their way out of Adam Lindsay Gordon's poems into Rolf Boldrewood's novels and through the late nineteenth-century balladists, 'Banjo' Paterson's *Man from Snowy River*, Will Ogilvie's *Fair Girls and Grey Horses*, to Barcroft Boake, and 'Breaker' Morant with guns and violent death as optional extras.[26] 'I soon discovered', wrote English literary man Douglas Sladen, who spent many years in Australia and married an Australian,

that nothing was of any importance in Australia except sport and money. If Tennyson or Walter Scott had gone to a bush township, he would have been judged merely by his proficiency or absence of proficiency as a groom. Horsemanship is the one test of the inhabitants of a bush township.[27]

Between 1860 and 1900 the number of horses increased at a slightly faster rate than the number of humans. The value of these horses was estimated in 1900 at about £5 million.[28] There was, needless to say, widespread interest in the quality of horse-flesh and in horsemanship, and this gave rise to all manner of competitions involving horses and riders, not only races, regular and spontaneous, but games like polo, and ploughing and pulling competitions for draughthorses. Perhaps working with horses produced a kind of gentleness and sympathy not seen in men who worked with machines. The horse had aesthetic qualities and aristocratic or rural associations.[29] As city dwellers became more numerous, a knowledge of horses elevated its owner beyond those who were mere passengers on trams and trains. Even so, Michael Davitt thought horses were 'cruelly neglected all over Australia', that far too little care and consideration was given to ordinary working horses, who could be perhaps too easily and cheaply replaced.[30] Australia sent 16 175 men to the Boer War, and 16 314 horses. Few of the horses returned.

A wide and free market in horses created problems for breeding and quality control. The Victoria Racing Club's Melbourne Cup, first run in 1861, the Australian Jockey Club's Derby (1861) and other famous races, most of them established during the 1860s, set a goal and a standard. In 1866 the *Australian Turf Register* consolidated and standardized racing season dates, thus regulating all the small country meetings whence owners of winners came to enter for the more prestigious races.[31] Intercolonial competition was very strong. The racing calendar began to exert its influence over the social calendar. 'Whatever may be the Governor's tastes, he has to regard the 'races' as *the* event of importance in the year, and the racing set as the best people of the colony.'[32] Audrey, Lady Tennyson, complained to her mother that the boredom of always having to attend the races was one of the least desirable aspects of being the wife of a governor in the Australian colonies.[33] Unfortunately for Beatrice and Sidney Webb, the schedule of their visit in 1898 coincided with the spring racing season. Mrs Webb was most critical of 'dressy, snobbish and idle' well-to-do Australian wives who thought it 'unwomanly' to take any interest in public affairs, but were only too willing to be seen parading their latest finery at a public racecourse.[34] The rituals of the Melbourne Cup were well-established by the 1880s.

Some parties take their lunch in their carriages, the ladies seated and the gentlemen helping them to the various good things spread out on a tablecloth on the grass... Others of the less wealthy, enjoy their fowl, or ham-sandwich; and they laugh, and joke, and quaff their bottled beer, and sip their whiskey, and get up their 'sweeps', with the same utter disregard of everything but the enjoyment of the passing hour.[35]

Like horse-owning and racing, gun-owning, shooting and hunting were enjoyed by much of the male population—an ironic comment, perhaps, on the English game laws which brought so many early settlers to this country. By the 1860s the gentry were introducing and breeding deer, rabbits, and foxes in order to sustain their fondness for riding about in brightly coloured coats. The impact of introduced animals on the native fauna and flora, and the indiscriminate shooting and trapping of native animals for their fur, for food, for fun, or because they were regarded as 'vermin', began to worry the more thoughtful hunting enthusiasts. Expeditions such as those arranged for the Duke of Edinburgh in 1867 could not be held too often. The Duke was taken possum shooting on the Government Farm in the Adelaide Hills. He shot fifty-two possums one moonlit night. He also went kangaroo shooting several times during his stay. The most satisfactory killing though was provided by rabbits on Thomas Austin's Barwon Park property near Geelong. Austin's reputation for his part in acclimatizing rabbits for hunting in Australia has veered between hero and villian, but he was certainly generous towards the Duke's party. Eight-six were killed by the Duke himself the first day, and of a thousand which fell to a dozen guns the next day, the duke shot 416. Sixty-eight of these were accounted for in ten minutes while they were trapped in a corner. Two guns were used, one being loaded while the other was fired. They became so hot they blistered the hands and could only be held by the stock.[36]

Legislation to protect native game during the breeding season was introduced in Victoria in 1861, and in south Australian in 1864, though even in quite knowledgeable circles, native Australian fauna was thought to be dull, scarcely worth preserving. Acclimatization and zoological societies had as one of their aims stocking the bush with a more lively class of game and a more tuneful set of birds, though this may have been partly because even in the 1870s, native wildlife was becoming sparse. Trollope noted his disappointment in 1872 that the only emus he had seen were those kept as pets in

CULTURE 191

THE MELBOURNE RACE-COURSE ON CUP DAY
Note the carriages lined up by the rail

Adelaide.[37]

Perhaps this was a reason also why coursing was made legal in Victoria in 1873. During the 1870s and 1880s the pursuit of kangaroos or introduced hares by greyhounds became a fashionable winter pastime for the Victorian and South Australian social set. A leading centre for coursing was the Hon. W.J. (later Sir William) Clarke's Rupertswood near Sunbury. The Hon W.J. not only arranged excursion trains to Sunbury for followers of blood sports,[38] he was also the first president (from 1877) of the Victorian Football Association, the organizing body for Australian Rules.[39]

In Ada Cambridge's story, *Across the Grain*, published late in 1882, one of the characters reminisces about the beauties of life on the Murray 'in the old days', the blackfellow, fishing for cod and trout, but above all, the game he could shoot 'kangaroos and emus, native companions and wild turkeys, black swans and black ducks, quails and bronze-wings, not to speak of cockatoos and parrots of all sorts and kinds'.[40] Alas no more. The Aborigines and poor white farmers to whom possum, wallaby, kangaroo and rabbits meant food from their carcasses and cash for their skins were the ones who really missed the game. Those who wished to hunt looked to the northern parts of the continent. In southern Queensland till late in the century it was possible to make a living from it. 'Marsupialists' were paid bounties for the ears of each kangaroo or wallaby they brought in. Lady Tennyson might refuse the gift of a possum skin rug and suggest that she would prefer an Australia-wide campaign to protect native wildlife,[41] but Albert Searcy, a South Australian public servant, would still include in his memoirs of life in the Northern Territory an enthusiastic chapter on the possibilities of hunting crocodile, buffalo, and numerous birds. He noted, without much interest, that birds once plentiful in some places were no longer there. The local Aborigines attributed this to 'bad spirits'.[42] Perhaps they were right.

The passion for hunting was close to the surface of the Australian male psyche. Some of the more memorable verse of the 1860s and 1870s was about hunting—Charles Harpur's 'The Kangaroo Hunt', and Henry Kendall's 'Wild Kangaroo', 'Oppossum Hunting by Moonlight' and his best-known 'Song of the Cattle Hunters'. 'Banjo' Paterson's 'Man from Snowy River' is another poem about hunting wild stock. It was not only hunting wild animals that appealed. Not a little of the interest in the running battle between police and bushrangers, and especially in the long drawn-out search for the Kelly gang, lay in the hunt. Adam Lindsay Gordon caught the sentiments precisely in his poem 'The Sick Stockrider' (1870). 'We led the hunt throughout, Ned, on the chestnut and the grey'. The hunt he is remembering is for Starlight and his gang—'we emptied our six-shooters on the bushrangers at bay'.[43] The Aborigines were sometimes referred to as 'black game', or 'black vermin'. Arthur Bicknell expected 'a brace or two of black game before morning', when he was out camping.[44] Former Western Pacific high commissioner, Arthur Hamilton Gordon, reported privately to British Prime Minister Gladstone in 1883:

I have heard men of culture and refinement, of the greatest humanity and kindness to their fellow whites, and who when you meet them here at home you would pronounce to be incapable of such deeds, talk, not only of the *wholesale* butchery (for the iniquity of *that* may sometimes be disguised from themselves) but of the *individual* murder of natives, exactly as they would talk of a day's sport, or of having to kill some troublesome animal. This is not the spirit in which to undertake the government of native races.[45]

'Banjo' Paterson undoubtedly knew his audience when he wrote in one of his dispatches to the *Sydney Morning Herald* from South Africa during the Boer War: 'This is something like sport, this shooting at human game with cannon over three thousand yards of country.'[46]

Alongside hunting with its riding and shooting, cricket was a slow and cerebral game. The way cricket, and its winter alternative, football, could harness youthful masculine aggression and burgeoning sexuality into regulated and exhausting activity had been demonstrated in the English public schools. Both these games became popular in the colonies through the education system. During the 1860s rules were codified. The 1870s saw the beginning of competitions, intercolonial and 'international' matches, thus establishing a hierarchy for achievement. The year 1876 is generally seen as crucial in the history of cricket, for that was the year in which a combined Victorian and New South Wales XI met and defeated a touring all-England XI.[47] In 1877 the Victorian Football Association was formed, and by then there were sixteen clubs playing in the rugby competition in New South Wales. Unhappily, for those who believe that there must have been a time when the game was the thing, played for its own sake, there were never intercolonial or international competitions without sponsorship or an eye to the gate. Most of the rule changes made after the basic structures had been worked out in the 1860s were to add spectacle or entertainment. Melbourne entrepreneurs, Spiers and Pond, not only arranged trial shipments of tinned Australian meat to England: when Charles Dickens was unable to undertake a reading tour of Australia they planned for 1862, a touring English cricket team was substituted. They also backed young T.W.S. Wills when he took a team of Aboriginal cricketers touring in

England in 1867. (The Aborigines combined cricket matches with demonstrations of boomerang throwing and corrobboree dancing.[48])

John A. Daly shows in his study of sport, class and community in colonial South Australia how extensively the Adelaide gentry sponsored, organized and contributed to the formation of early sporting clubs which were based on school, church and workplace as well as local associations.[49] At first there was considerable participation from the gentry, who played in teams, provided hospitality and refreshments, and maintained facilities such as courts and pitches. This fostered a belief in the equalizing virtue of sport itself.

It is one of the greatest charms connected with manly exercises that . . . they level all social distinctions. When we see a fine strapping fellow throwing a cricket ball . . . or in the tough game of football we do not stop to enquire whether he is a blacksmith or a professional man, we applaud because we recognize him as a man amongst men.[50]

Within a decade or so, the names of the gentry appeared only as patrons, financial supporters and occasional committee men.

In all three South Australian country towns he studied (Gawler, Kapunda, and Strathalbyn), Daly found that the team sports, cricket and football especially, had a unifying effect on the community and that inter-town competitions, with only a little encouragement from local newspapers, were almost warlike in their rivalry. The same kind of local identification and rivalry became evident in the suburbs of the larger cities. It was easily translated into national spirit when international teams were being fielded. In country towns the local squirarchy was expected to provide major support—land, facilities, sponsorship—while the town professionals and civic leaders contributed organization, management, and perhaps some players, cricketers more probably than footballers. Most of the competitors were of the lower orders.

A similar pattern is revealed in Brian Stoddardt's account of sport and society in Western Australia, only there, the value of organized sport in maintaining social order during the otherwise disruptive gold-rushes of the 1890s was underlined.⁵¹ It is possible that both South Australia and Western Australia were untypical in that their societies were small and amenable to this almost eighteenth-century style of squirarchical influence and deference. The eastern colonies, New South Wales and Victoria especially, were more urbanized and industrialized, though the activities of the squire of Sunbury show that some opportunities for influence were available. Evidence from toward the end of the century, especially of the growing links between sport and politics, suggests that men of wealth, status, and power still supplied leadership, patronage, and financial assistance.⁵²

The latent nationalism and repressed martial spirit exhibited during England–Australia cricket tours has been extensively discussed.⁵³ So has the way in which both cricket and football, by the turn of the century, had become desirable alternative forms of employment and upward mobility for working-class youths, and even the occasional Aborigine, like cricketer Jack Marsh.⁵⁴ An even more old-fashioned working-class activity—the working-class equivalent of hunting perhaps—fighting with bare fists—was also tamed and transformed into a less respectable but more efficacious route to fame and material success. During the 1870s boxing gloves and Queensberry Rules began to oust the older, more primitive trials of strength, endurance, and cunning. By the end of the century, exceptional speed as a runner, tenacity as a fighter, or skill in handling a ball might carry a boy's name as far as the headlines of the daily press and reward him with a season or two as a local hero. (The financial and other rewards were more exacting, and just as likely to feed back into ancilliary worlds of drinking or gambling.⁵⁵)

Organization and competition transformed most outdoor activity into sport. Ploughing, timber-cutting, bicycle riding, walking—elevated to 'pedestrianism'—all acquired their clubs, their events, their champions.⁵⁶ Healthy activity with a purpose was deemed a good thing. But the creation of 'champions' and the manipulation of crowds began to acquire their own skills. This process is most dramatically illustrated by the history of rowing.

In the middle of the century rowing was, like horse-riding and shooting, something that large numbers of people, women and children as well as men, did out of necessity.

Rivers provided quicker and easier transport and communication in many places than roads. Getting around Sydney, for instance, was often simply a matter of rowing oneself. Two of the early sculling champions. Ned Trickett and Harry Searle, rowed as part of their daily labour, Trickett from his home at Greenwich to his work as a quarryman, Searle to school from his father's farm on the Clarence River. In 1876 Trickett was taken to England by James Punch, an innkeeper and former sculler. He beat the Englishman James H. Sadler on the Thames and was acclaimed 'sculling champion of the world'.[57] Thereafter Australians continued to do well in 'world sculling championships', though there was as yet no international organization. All championships were decided on the basis of heavily promoted challenge matches. Like boxing, rowing demanded both the strength and stamina fostered by a labouring life and a hearty diet. As well as Trickett the quarryman, and Searle who worked on his father's farm, the other notable rower of the period, Bill Beach, was a blacksmith from Dapto. Unlike boxing, rowing was a 'clean' sport, suitable for public schools, and increasingly popular with young men from the public service, banks and insurance companies. 'Banjo' Paterson rowed as a young man, of necessity perhaps, while he lived in his grandmother's home on the Sydney foreshore at Gladesville.[58] Steele Rudd's pen-name began as 'Steel Rudder' beneath his humorous reports of the activities of his rowing club on the Brisbane River while he was still a junior clerk in the Queensland Justice Department.[59] Like boxing and horse-racing, and unlike the many team sports, the mass attraction of rowing lay not in the prize money to be won by a single champion oarsman, but in the much greater sums that would be wagered on the outcome of a single race.

Ned Trickett was credited with placing young Australia on the map. He reassured us that the Australian type was 'of world class', or so it was said. Even more remarkably, federation became feasible when Harry Searle was accepted and acclaimed as a representative not of New South Wales, but the whole of Australia.[60] Maybe. Certainly his death from typhoid at twenty-three distracted attention from unpleasant revelations about his personality and behaviour. What were the motives of those who arranged the incredible sequence of public demonstrations as his body was carried home from Melbourne to Maclean via Sydney in December 1889? Why was it thought necessary to stage funeral processions and memorial services in both Melbourne and Sydney as well as at his last resting place by the Clarence? A few brave souls voiced their uneasiness at this distasteful manipulation of public emotion and questioned the disproportionate claims made for Searle's significance. There was a disturbing shallowness in such hero worship—and this was by no means an isolated example.

Admiration was lavished on all manner of dead heroes, not only on sportsmen like Harry Searle. The funeral of explorers Burke and Wills in Melbourne in 1863,[61] the death of General Gordon in Khartoum in 1885,[62] the death in office in 1898 of the thirty-seven-year-old premier of Queensland, T. J. Byrnes,[63] all brought forth incredible public displays of grief. Adam Lindsay Gordon's reputation as a poet was enhanced by his suicide at the age of thirty-six. Then there was the sturdy growth of bushranger legends, culminating in the apotheosis of Ned Kelly after his trial and hanging in 1880.[64] Searle's biographer, Scott Bennett, linked 'exquisite agony in the contemplation of [that] young man's fate' with K. S. Inglis' discussion of the response to Gallipoli: 'Australians [have] cherished the heroes struck down suddenly in the days of their strength.'[65] Gordon, it was said, 'glorified the manliness of a man, making it lean a little too much perhaps to the physical side'.[66] One senses now, both in the hero-worship and the response to death, relief granted by legitimate opportunities for the display of emotion and enthusiasm. In death, at last, it was possible to pay the tribute which had seemed dangerous, even unmanly, in life, and to mourn much more than loss of promise. Already the level of material comfort, of security, or dull conformity had so altered the meaning of life that purposes had become confused. Sporting prowess invoked tribal memories of endurance and survival, while the tragic reality of death confirmed the meaning of life for a dull egalitarian society. The size of the crowd undoubtedly said something about its fondness for sport or a particular hero. It might also say more about popular taste for spectacle, the paucity of alternate outlets for emotion, or the immense skill of those who promoted and organized sporting events and public demonstrations. Thus Charles Tait, who for a decade had organized and supervised crowds for concerts, theatrical tours, and other large-scale public entertainments in Melbourne, was called upon to stage the opening of the first federal parliament in 1901.[67]

Australian historians have generally placed an optimistic interpretation on the growth of sport in the later part of the nineteenth century. They stress egalitarianism, national maturity, cohesiveness, pride. Other meanings are less inspiring. Organized sport clearly fostered Australian conformity and discouraged intellectual effort. Imperial ties were probably strengthened at the expense of national self-confidence. Republicanism was dealt a mortal blow. Organized sport provided a safe outlet for aggressive and militaristic tendencies. It also fostered them, not only among the players. Spectators were free to scream 'kill the umpire', to shout 'put in the boot', to riot at cricket matches.[68] Organized sport provided incentives and outlets for a handful of creative entrepreneurs. It also soaked up economic and emotional resources which might have produced different outcomes if directed elsewhere. Consider simply the money, skill, and imagination that was poured into innumerable, and sometimes amazingly innovative forms of gambling. But the saddest effect of the pursuit of organized sport in the late nineteenth century appears in the way sport mirrored work by becoming sexually segregated. A problem already marked —the uneasiness and lack of communication between the sexes—was exacerbated by the segregation of leisure. Certainly women and girls, sometimes in large numbers, followed most outdoor sports as spectators. Many engaged in and enjoyed the physically liberating effects of riding, swimming, tennis, bicycling. Girls no less than their brothers were affected by the equable climate, relative affluence, high levels of nutrition, opportunities for freedom on horseback or rowing on the river. No doubt those girls who took up cricket or tried hockey were responding to the same instincts and challenges as their brothers. As well girls whose families could afford to keep servants had a great deal of enforced leisure though others had no leisure at all between earning a meagre wage and helping in the house. In the organization of sport or statistics, the easiest classification was by sex. This was further justified by the emphasis on competition. Only the well-to-do who could afford private tennis courts, croquet lawns, and stables had the choice of sporting activity which was either competitive or involved co-operation and mingling of the sexes. Social activity which led to easy mixing for ordinary people became more dependent on contrived market situations. The basis for many a marriage was thus a little less secure.

Sports like rugby were thought to counteract sexual repression or the homoerotic fantasies characteristic of English public school life at this time. We can only speculate whether the Australian cult of mateship was altered or strengthened by the sporting life. Organized sport and leisure certainly restored superiority to the primitive physical characteristics of male strength and aggression, though, or perhaps because, technology was easing the differences between the sexes. Sport or the life outdoors became a young man's preferred way of spending his leisure time, of developing and expressing his personality. Leisure indoors was for women or weaklings. And this dichotomy flourished so that sport, identified with the climate or landscape, was seen as Australian, an aspect of national character, whereas indoor leisure or cultural activity became suspect as inherited from Britain. It was a deceptive distinction, as a moment's reflection on the origin of most sports will show, but powerful, in retarding the development of a more broadly based national culture.

Education was viewed quite favourably when it led to greater opportunities for cricket and football. It was also resented as the imposition of dead, irrelevant, or subversive culture. The task of equipping the coming generation with learning and knowledge and skills was a challenge to the post-gold-rush educationists, most of whom were themselves immigrants, optimistic about the links between education, efficiency and progress. The Australian colonies in the 1860s were an educational theorist's dream—such scope, such potential, such freedom from vested interests and handicapping attitudes. Distance was the main problem which could surely be overcome by organization and sound logistics. William Wilkins aimed to be able to say what every child in New South Wales was studying at any minute on any day.[69]

While bustling modernity and efficient organization appealed to some, there were others who sought to preserve what little there was of tradition in the society. These two views appear in our history as the battle between church and state for control of the education system. Traditional values were identified as repressive, mass regimentation as progress. In the end there was a thin and bitter taste to education too, which gave us an on-going subject for argument but did little to deepen or enliven cultural life.

The parents of children in the Australian colonies in 1860 had wide choice in the education of their children. Not bothering to educate them at all was a distinct possibility. Until education became compulsory in Victoria in 1872 (1880 in New South Wales)—and indeed, not even then, for it took a long time to instil the necessity and enforce attendance at school—many children of poorer parents began working as soon as they were able. As Margaret Barbalet has remarked, 'Work in itself was not thought to be degrading or a hardship to children.'[70] Girls of four and five were set to 'watch' a baby or 'help' some not-quite-so-poor or young or harassed housewife. For a lad there were plenty of opportunities for street-work, fetching and carrying, scavenging, selling small items as a disguised form of begging. Farmers' children invariably 'helped'. It has been estimated that more than half the children aged between five and fifteen in 1861 could not read and write.[71] By the end of the century the figure was reduced to one in five. Tasmania had the highest illiteracy rate, followed by New South Wales. Victoria had the highest proportion of people who could both read and write.[72]

Schools varied enormously in 1860. The intention of the national system introduced in New South Wales in the 1840s was to provide an elementary education at as little cost as possible wherever church or other private schools were not available. But the problem of reaching the less densely settled areas remained an intractable one in all colonies. Wealthy settlers hired tutors or governesses. Havelock Ellis began to think about the significance of sexuality while tutoring the Platt children near Carcoar in 1876.[73] James Brunton

Stephens, tutor to the Barkers at Tamrookum station on the Logan River in 1866, was bold enough to decline young Rosa Murray Prior's offer of publication for his 'Convict Once' in the magazine she wrote and edited for the amusement of her family on neighbouring Marroon.[74] Neither Ellis nor Stephens enjoyed life as a tutor, finding the work unremitting and the situation open to exploitation as bookkeeper, accountant, and general factotum. Signs of bookishness were despised as effeminate.

Governesses also found their position difficult. 'They are Governess, Nurse and Sempstress all in one. They are expected to be highly accomplished and to make first rate scholars out of wild, petted, vulgar, passionate, ungovernable children.' Catherine Helen Spence's Clara Morison had the sharp sound of personal experience and caught well the pretensions of families who liked the sound of French and music in a governess's qualifications but really wanted a mother's help and maid-of-all-work.[75]

Some children were sent to stay with relatives in town or board at one of many schools established under church auspices or by private citizens. Until the end of the century the churches put most of their effort into boys' schools. For their daughters, parents tended to favour small schools in which a family atmosphere prevailed, like the school in *Picnic at Hanging Rock*, though the tradition of a convent education as the best training for a girl lingered, even among Protestants.[76] Most country children were taught what their mothers could manage or spent long hours walking or riding to and from the little bush schools which were among the more modest inventions of the state education system. Some of these schools were so small and remote that they could only command the services of an itinerant teacher.

In the towns, the choice of elementary schools varied immensely. There were state and/or public schools with discretionary policies about the level of their fees. (Education did not become free in most colonies before the end of the century.) A notice headed 'Government Schools' in an Adelaide directory for 1870 went on: 'Children of the Public can be taught in these schools at a fee not exceeding One Shilling per week, and Destitute Children free of cost on receipt of a certificate from the Relieving Officer.'[77] These schools provided a primary education in reading, writing and arithmetic. As well there were elementary schools run by the churches and private individuals, contracting in number and variety as state-provided education became cheaper and more widely available. By the 1880s only the Catholic church maintained a network of elementary schools.

Moves to improve and expand the education system, tied as they were to political ambitions discussed later, were also influenced by developments in educational theory and practice beyond Australia. William Wilkins, responsible from 1863 into the 1880s for New South Wales education policy and administration, had been influenced by Swiss educationist Pestalozzi, though he found the organizational innovations of James Kay-Shuttleworth more applicable to the colonial situation.[78] A. B. Weigall, headmaster of Sydney Grammar School, and E. E. Morris of Melbourne C of E Grammar School encouraged the development of their schools (and by example, many other private secondary schools) according to the views of Thomas Arnold of Rugby.[79] Edwin Bean, classics master at Sydney Grammar from 1875, a product himself of Clifton College, and later headmaster of All Saints, Bathurst, was another exponent of the college ideal. (His son, C. E. W. Bean, carried this ideal into his interpretation of Australia's part in the First World War.[80]) Arnold was an influence on J. A. Hartley, headmaster of Prince Alfred College in Adelaide 1871–75, and later as Inspector-General of Schools in South Australia, the man responsible for that colony's public education system.[81]

Administrators tended to come from overseas. Some were more adaptable than others. F. J. Gladman, principal from 1877 of the Melbourne Training Institution for teachers, had begun in a Lancastrian school in England and worked his way as a pupil-teacher through the system, gaining additional qualifications through external examinations.[82] (James Bonwick, one-time proprietor of his own school and later a prolific writer of texbooks for Australian schools, also came out of Lancaster's monitorial system.) Despite his background and preference for regimentation, Gladman could see dangers in training teachers without educating them. In England culture was omnipresent and pervasive. In Australia the teacher might be the only representative of culture and knowledge in a community.

Rank-and-file teachers, both Australian-born and immigrant, were a mixed group in their education and training. An outstanding problem was finding enough teachers sufficiently trained or educated to meet requirements. The Lancastrian practice of employing older children as monitors, some paid, some unpaid, lingered till the end of the century, mainly for reasons of economy and because of staff shortages in some areas, though it was perceived to be educationally undesirable. Likewise, the apprentice system of training teachers by accepting both boys and girls of thirteen and fourteen as pupil-teachers survived well beyond the establishment of more professionally oriented training colleges. Again, the reasons were connected with economy and staff shortages. In 1862 in England, Robert Lowe's Revised Code introduced payment to teachers by their results as demonstrated by inspections and public examinations. In the interests of economy and efficiency this system was adopted in the colonies, most enthusiastically in Victoria. The use of monitors or pupil-teachers alongside payment by results produced an instrumental approach to teaching which was at odds with Pestalozzian ideas of cultivating the talent in every child. Nor did real education flow from an extensively detailed syllabus rigidly examined and inspected. There was a great deal of rote learning of acceptable texts and answers. Children learned to read, write and do practical arithmetic. The rising literacy figures testify to that. A great many must have gained something useful from their object lessons, nature study, geography and training in manual and domestic arts, singing and drawing. For the evidence points to an alert, intelligent population, skilful at improvisation, both for survival and entertainment.

Such an education, though lacking depth and intellectual stimulus, was not necessarily as limited and stultifying as later historians have seen it. In South Australia, for example, in the 1880s, one of Hartley's teachers, William Gatton Grasby, who later gained a reputation not only for his criticisms of Hartley, but also for his advocacy of Froebel and the part he played in the development of agricultural education, was developing school libraries and nature study museums.[83] Grasby was one of the first of thousands of Australian-born teachers to undertake that cultural rite of passage, the trip overseas. In Europe in 1881 at the age of twenty-one, at last the link between the world he learned from books as a boy in the Adelaide Hills and his experience as a pupil-teacher became clear. Grasby was able to see how his remote colonial experience related to great traditions of which, in theory, he was an inheritor. His views on the nature and purposes of education found ready sympathy among the generation of native-born teachers coming into the system. In 1891, his 'very suggestive and forcible treatise', *Our Public Schools*, appeared, as did his *Teaching in Three Continents*.[84] He wrote for various educational publications and used *Garden and Field*, a monthly agricultural journal which he owned and edited, as 'the means of communication between all progressive people in Southern Australia'.[85] He was not alone in condemning the pupil-teacher system—'the attempt to educate children by means of boys and girls themselves uneducated' as he described it.[86] Other progressive teachers were coming to see the incompatibility between real teaching and the annual examinations required for payment by results. Grasby's perception of the importance of a scientific education even for mere farmers was a little ahead of its time, but it illustrates the importance of education in a society without traditional ways for people to follow. Roseworthy College, of which Grasby was an early, if short-term principal, was soon replicated in the other colonies (for example, Hawkesbury Agricultural College, 1891). Grasby moved eventually to Guilford in Western Australia to continue, through the *Western Mail*, his mission of producing better informed and more versatile, innovative farmers.

Other teachers began to read the same message as Grasby in the educational literature. Children needed to understand more of what they were learning, and why. There was no point in a child reciting the letters of meaningless words. Elementary science was of no use if there were accidents and errors in workshops and factories through want of application. The kind of technical education advocated by men as different in background as Norman Selfe, the successful Sydney engineer and designer,[87] and C. H. Pearson, historian, educationist, farmer, and Melbourne's 'professor of democracy',[88] appealed to teachers who had become enthusiastic about the New Education movement in the 1890s, the 'training of hand and eye' and allied ideas.

By the end of the century, education seemed ready for the considerable reforms proposed by the Knibbs–Turner Report in New South Wales and the Fink Commission in Victoria. Since the 1880s, in fact, practical reforms had been proceeding quietly with the generation of Australian-born teachers and administrators like Frank Tate and C. R. Long in Victoria, Peter Board in New South Wales, and those who formed the nucleus of new professionally self-aware teachers unions.[89] Though problems of costs, distance, and regimentation remained, small things like the quality of school equipment and the design of school furniture were tackled. For instance, backless forms on which children sat rigidly for hours were deemed undesirable (except in Queensland where they remained standard issue till well into the middle of this century).

The kindergarten movement, established in the 1890s, was quickly and easily corrupted by inexperienced teachers and inadequate resources.[90] An anxious desire to demonstrate results for money spent focused attention on the neatness with which children cut out their paper mats and shaped their clay. Froebel's message about the value of the activity in developing ability and character required a sophistication not often found in infant teachers, usually only girls themselves. The need for visible and quantifiable results permeated all education as an admirably responsible attitude to the expenditure of public money. Such accountancy is not infallible about values. There was a tendency for costliness to become the value of worth instilled by education.

Despite the changes in educational theory and practice of the last two decades of the nineteenth century, the experience of education probably changed little for the children. The sound of multiplication tables chanted both forwards and backwards, the sing-song of spelling lessons, the false intonation of the perfectly memorized but meaningless answer still echoed along corridors and verandahs.[91] William Wilkins tried to provide reading books more suited to Australian needs than the imported materials generally used, and Grasby to teach geography, geology, and botany from the examples at hand instead of from British textbooks. There was cause for pride in the fact that throughout this period a higher proportion of Australian children was receiving a basic education than their British counterparts. The need in Australia, it might also be said, seemed greater.[92]

The values espoused and the methods adopted to achieve widespread elementary education acted as a deterrent to higher education. Secondary schooling and university education were luxuries when educational resources were already so thinly stretched. Most of the need for professional skills and higher education was supplied by continuing immigration.[93] Education at a British university was still the making of an Australian gentleman. Even so, expensive universities were established (and maintained at a meagre level) mainly as a matter of prestige. Given the size of the country's population even at the end of the century, one university would have been adequate to its nineteenth-century needs. There were four. They had few students, a minimum of staff. The admission of women to the universities from the 1870s contained an element of pragmatism: women added to the number of students, though prudishness imposed segregated classes in some subjects. Little thought was given to the necessary or possible connection between the universities and the rest of the education system, known by the end of the century in most colonies, significantly, as 'public instruction'. Through their matriculation examinations, however, the universities provided an acceptable service in certifying educational quality and achievement in the otherwise variegated secondary market.

Several of the church or independent schools destined for recognition as great institutions were already in existence in 1860. Others switched to secondary education as the state system began to satisfy the need for primary schooling. Some 'high', 'grammar' or 'advanced' schools were set up within the state system, with courses leading to university matriculation, especially in New South Wales, Queensland, and South Australia. In Victoria and Tasmania, non-sectarian secondary schools were late arrivals. An important feature of these state secondary schools was that they offered an academic curriculum to girls as well as boys. Though a few private girls' schools began to prepare their pupils for matriculation examinations in the 1880s, real educational opportunities for girls were limited. Girls received less education than boys. In 1861, 18.5 per cent of bridegrooms signed the marriage register with a cross. Compare this with 30.69 per cent of brides who were unable to write their name. The position had improved somewhat a decade later when the figures were 10.58 per cent of bridegrooms and 16.4 per cent of brides, but even in 1881, the difference was still noticeable: 4.34 per cent of bridegrooms and 6.78 per cent of brides. By the end of the century, no doubt because of the move towards compulsory education which required parents to send their daughters to school instead of keeping them home to 'help' or sending them into service, only 1.29 per cent of brides were incapable of writing their names, slightly fewer than the 1.35 per cent of grooms recorded.[94]

Generally inferior education with an emphasis on 'accomplishments' left girls unprepared even for their adult responsibilities, namely the care and education of young children, and the creation of a civilized and cultured environment for husbands and families. As a consequence, Australian girls were noted for their empty-headedness, their frivolity. The quality of recruits to teaching, a majority of whom were girls, became a cause for concern. In Adelaide, The Advanced School for Girls, for many years the only state secondary school in South Australia, was established especially to educate girls to become adequate teachers, though matriculation was still considered an unnecessary extravagance.[95]

Universities existed in Sydney (1850) and Melbourne (1853). Charles Lilley's agitation in the 1870s for a university in Queensland came to nothing for nearly forty years,[96] but South Australia and Tasmania set up their own universities in 1874 and 1890. Among the educated immigrants of the nineteenth century, graduates of Irish and Scottish universities were much in evidence, not surprisingly given the depressed economic opportunities of those parts of the British Isles and the stranglehold of Oxbridge on politics and the professions in England. The views of these graduates counted for a great deal in shaping the new Australian universities. So did the fact that it was easier to recruit professors sympathetic to the modest requirements of the colonies through the old-boy network in the provinces. There was a preference for a classical education in staff recruitment, but little use for classical scholarship. As a first-year student (aged fifteen) at the University of Sydney in 1860, Samuel

Griffith studied classics, mathematics, chemistry and experimental physics. In his third year he won medals for his translation of part of Milton's *Comus* into Greek trimeter iambics and his composition in Latin hexameters, both of which he was required to read at the graduation ceremony.[97] The use of the university as a centre for research was barely understood. As well, universities were virtually excluded from debate on the great question of the day, the future of Christianity, by their official non-sectarianism. Nor were political or social matters much discussed. At the University of Sydney, G. A. Wood encouraged his students to examine large moral questions concerning the use of power through his teaching of history, but when he came out against participation in the Boer War, an attempt was made to dismiss him from his chair.[98] The very lively professor of music at the University of Melbourne in the 1890s, G. W. Marshall-Hall lost his job in the course of a controversy which raged round his 'libidinous' poetry.[99]

Education and culture were discontinuous. Universal elementary education created the capacity for intellectual growth, but those who succeeded usually did so by persevering in libraries and bookshops. At Tom Collins's camp on the Murray, one of the drovers had drawn up a list of the 'ten masterpieces of poetry' he should read:

Paradise Lost and *Regained*, counting the two as one; Goethe's *Faust*, especially the second part; Dante's *Divine Comedy*; Spenser's *Faerie Queen*; Thompson's *Seasons*; Young's *Night Thoughts*; Cowper's *Task*; Tennyson's *In Memoriam*; Edwin Arnold's *Light of Asia*; and lastly, any poem of Walt Whitman's . . . I wrote to Cole for the prices of the best rough-and-ready editions delivered at the Melbourne Railway Station. The end of the matter was that the parcel was waiting for me at Echuca when I crossed the river on my way to this bend to spell for the next wool.[100]

After trying without much success to read them, he said, he gave them to the public library at Hay.

A yearning for more education or culture is evident in the formation of schools of arts, mechanics' institutes, and workingmen's colleges. By 1900 there were well over a thousand of these institutes in Australia. At offical levels there was support for practical education, schools of mines, evening classes teaching commercial skills, but finding teachers was a problem. In most communities, those with more than elementary education were usually clergymen, trained in the classics, without modern or scientific subjects. Public lectures and debates were popular and inexpensive, though here too there was always a shortage of performers, aggravated by the fact that women rarely spoke in public. Concerts, soirées, and the occasional dramatic performance were preferred as they gave a boost to institute funds and had the advantage of acceptably involving both sexes.[101]

Stories of the fortunes made in Australia by men who were virtually illiterate tended to discourage young people from persevering with education. Certainly the prospect of well-paid employment distracted young men from higher education. Not all wealthy men, however, were ill-disposed to education and culture. In 1884, Thomas Fisher, the child of convict parents, orphaned at twelve to become a bootmaker, left £33 000 to the University of Sydney for a library. Without the endowment of chairs, colleges, scholarships, and buildings, the universities could not have survived. Likewise public art galleries, libraries, and museums depended on generous bequests and benefactions. This did tend to encourage appreciation on grounds of cost rather than any other quality. Contemporaries disagreed over the level of cultural and intellectual leadership available among the professional classes and clergy. A shortage in most professions meant that it was still not difficult for mediocrity to become established and make a good living. And the pressure of work and money making may have worked against much culture. As Henry Bastow wrote in 1861 to his fellow Dorchester apprentice, Thomas Hardy, he had no time to keep up his Greek since migrating to Tasmania to practise as an architect.[102] The immigrant who brought his culture with him could choose to abandon what he deemed inappropriate for colonial life. The native born were without this choice. When Alfred Deakin set to work in the Melbourne public library to make a comprehensive study of aesthetics,

he did not succeed in building up a satisfying philosophy of Art. For a young man who had as yet seen none of the world's masterpieces of painting or sculpture or architecture, the task was merely

MELBOURNE PUBLIC LIBRARY
1. PICTURE GALLERY 2. READING ROOM 3. NEWS ROOM 4. THE PUBLIC LIBRARY AS IT IS TO BE WHEN COMPLETED

impossible: he lacked the necessary basis of experience. When he leaves the other arts and comes to poetry, this disqualification ceases to apply; few indeed were the masterpieces of poetry with which he had not made himself in some measure familiar.[103]

Cultivating taste was difficult. Crude or vulgar attempts to interpret ideas imported from London or the continent and refracted through a mirror of Australian money were an easy target for satire. How was the young Australian housewife to develop her taste, given the gaps in her education, official preference for the elevating rather than the stimulating in art, music, and literature, and range of goods available commercially and valued according to their costliness and showiness? Badham thought the universities should be able to help, but even he, the editor of Plato's *Philebus*, spent most of his time and effort marking children's public examination papers.[104]

In harmony with contemporary views of the role of architecture in elevating public taste, classical, Renaissance, or Gothic styles of architecture were adopted for educational institutions, museums, art galleries and public libraries. Educational theorists thought that schools should be as inspirational for children as churches were to worshippers. The church analogy was carried into school architecture literally. A preference for Gothic in university buildings and classic forms for museums and galleries indicated that an inspirational message was attached. Such grandeur may have been forbidding, as it was in banks and office buildings, discouraging to ordinary people who were suspicious of officially sanctioned improvement. But whereas the business of banks and insurance offices was essential, there was no need to enter the temples of culture.

Australia's first permanent art gallery was opened as the Museum of Art in 1861 in the building which housed the Melbourne Public Library. It contained mainly plaster casts, but pictures were gradually imported, mainly story pictures or pictures of dramatic scenery, with or without figures and animals. They were crowded on the walls.[105] Such galleries were most useful to the educated seeking reinforcement or extension. However the politicians, trustees, and self-appointed watch-dogs of the public purse believed that an art gallery should be morally uplifting, accessible and agreeable as a place where families could go on Sunday afternoon (once that question was resolved and visits to cultural institutions were accepted as near equivalents of church-going). 'Advanced', difficult, or suggestive paintings were eschewed. Size and price tag were considered the best indicators of value and significance. Only public libraries attracted people who were actively seeking self-improvement, or a warm place on a wet Tuesday. These readers preferred light reading, fiction, and periodicals to the more earnest volumes of classics and theology.

'Ain't got sich a thing as a swappin' book on you, I s'pose?' Dixon, the bullock driver, asked Tom Collins as they settled down for a meal at Cameron's Bend. 'One o' Nathaneal Hawthorne's here waitin' for a new owner. Can't suffer that author no road. He's a (adj.) fool; too slow to catch grubs.'[106]

'What are you reading now?' Collins asked.

'Bible,' replied Dixon, with a touch of self-righteousness, whilst indicating with a sideward glance the noblest and least understood compilation on earth, where it lay in a kerosene box, together with a supply of tobacco and matches, a large dictionary, a well-worn pack of cards, and the insufferable Hawthorne. 'Got her in a swap for one o' Ouidar's,' he continued.[107]

This enthusiasm for print led to many short-lived magazines and newspapers based largely on the contributions of their readers. The epitome of this interaction was the Sydney *Bulletin*, founded in 1880 and sustained by news and other items contributed by a wide circle of non-professional but aspiring writers.[108]

In 1861 the postal services carried about 12 letters or postcards and 8 newspapers for every 100 persons in the colonies. By the end of the century these figures had risen to 62 and 30.[109] Literacy was expected for the efficient functioning of society. From birth until death there were forms to be read and filled in. Job applications became a test of literacy and style. Running a business was difficult without clerical assistance. Imports, exports, tariffs, wages, awards, factory regulations all relied on a high level of literacy, and on the fact that one language, English, was universally spoken and understood. It is now impossible to calculate how much of the achievement, especially the social progress of this period, was due to these two factors—the high rate of literacy, and the assumption that English could be used at a quite sophisticated level for all government and administration. Though German continued to flourish, and in their own communities some groups like the Chinese spoke and wrote their own language (and Aborigines spoke theirs), no concessions were made to non-English readers.

The man with the printing press seems to have arrived in every township with the first settlers, the arch-exponent of the links between tradition and progress. 'Though our field be limited and local,' Henry Laurie wrote in his editorial in the Warrnambool *Examiner*, 1 October 1867, 'our best energies will be brought to bear on it.' He went on:

Writing for a Victorian audience, and for the most part on subjects which may be of little interest beyond the colony, we yet hope to use our pen in the spirit of the most cosmopolitan journals. We are not of those who think that journalism in this country must of necessity adopt a tone inferior to that of the best journalism at home. Here the newspaper is almost the only representative of indigenous literature, and as such it should be respectable and respected.[110]

Many newspapers did not survive longer than their editors chose to stay. Those associated with mining towns were particularly vulnerable. Often they appeared irregularly or infrequently, once or twice a week. Little is known of their circulation or influence. Often not even the name of the editor can be established. The mere fact that they existed and continued to spring up vigorously and in great numbers shows that there was a place for them, or perhaps more accurately for their publishers, editors, and writers.

Richard White has suggested that Australia in the late nineteenth century was as good a place for a freelance writer as anywhere in the world.[111] This was largely because of the number of newspapers bought by a prosperous reading population. And not only newspapers: magazines, weekly fortnightly, monthly, quarterly, and annual proliferated with similar ease. Of 449 items listed in Lurline Stuart's bibliography of nineteenth-century periodicals, 314 belong to this period, and this is only a survey of those giving space to creative literature.[112] Though her list includes a variety of sporting, theatrical, religious and political publications, there were others devoted to professional and business matters, and to special concerns like boxing, racing, betting, fashion or finance. Many organizations published bulletins and newsletters, and during the 1890s there was a notable growth of newspapers in the labour movement.[113]

The Armidale school of arts which opened in July 1859 charged an annual membership fee of £1. With this it subscribed to the *Edinburgh Review* and *Quarterly*, *Home News*, *Punch*, *Illustrated News*, *Illustrated Times*, *Household Words* and *London Journal*; also the *Sydney Morning Herald*, *Empire*, *Maitland Mercury*, *Goulburn Chronicle*, *Tamworth Examiner*, *Bathurst Free Press* and *Moreton Bay Examiner*. By 1866 it had abandoned its British journals, with the exception of *Punch*, as Australian papers became more substantial.[114] One of the earliest decisions of many a friendly society or trade union was to take out subscriptions to certain papers or periodicals. Private individuals paid their subscriptions through booksellers like Robertson and Mullens in Melbourne.[115] In 1866 there were 1750 copies a month of *Good Words*, 1500 of the *London Journal* and 690 of *Punch* being distributed in Sydney alone, with smaller numbers of the more serious reviews like the *Cornhill*, *Blackwoods*, *Edinburgh* and *Quarterly*.[116] Even the local newspaper of the late nineteenth century carried a much heavier burden of information than we expect today. There was news, local news which spread more slowly and less surely by word of mouth, news from nearby towns, from the city, political news especially, and news too, often of a fantastic kind, from the world abroad. Then there was advertising—people buying things, selling things, losing and finding horses and cattle, offering and seeking work, even advertising, as E. W. Cole did, for a wife.[117] Mail times, sailing times, produce news, financial news, births, deaths, marriages, government notices, law reports, warnings, pleas all had their places. No pictures were used, only small illustrative blocks, decorative scroll-work, a pointing hand, a stylized design of fruit or musical instruments at the head of the appropriate column or as part of an advertisement. By contrast the weekly illustrateds employed full-time artists to sketch scenes, objects, portraits, caricatures and cartoons.

The typical newspaper consisted of a large sheet folded to give four pages. Advertising and notices appeared on the first page. The inside pages carried most of the news and editorial comment. Any remaining space and the back page were given over to reading matter of various kinds, depending on what was available and what might appeal to the editor and the readers. Country papers like the *Town and Country Journal* and the *Queenslander* regularly contained a digest of articles from the British and American press, selected for amusement as much as for news value, though this changed as the telegraph gave greater immediacy to the meaning of news. (The Franco–Prussian War of 1870 was fought and the treaty signed before any news of either arrived in Australia.) Each month, to coincide with scheduled shipping services to Britain, city papers brought out a résumé of local news specifically for posting 'home'. This service was so much used that post offices sold pre-stamped newspaper wrappers.[118] Within Australia, newspapers were sent through the mail free or at special low rates, an indication itself of the importance of the service they were thought to provide to a scattered community. City newspapers were larger and more sophisticated than their country cousins. They could devote more space to both news and advertising. They could also afford to publish original local material. Indeed, they had more incentive, since many of their readers would see the British and American journals excerpted by country editors.

An immense volume of fiction, descriptive prose, and poetry was published first in newspaper format in late nineteenth-century Australia.[119] In Britain at this time, serialization of new fiction was the staple of monthly and quarterly publishers' magazines. Novels were then consolidated and produced in permanent form for the library market. In Australia the weeklies filling their columns with escapist fantasy were the best market for stories and short stories. Marcus Clarke's *For the Term of his Natural Life* was written as a serial for the *Australian Journal*. The story ran for three years, on and off, from 1870, and then appeared as a book, somewhat altered, in 1874. An earlier novel, *Long Odds*, had come out in the *Colonial Monthly*, with some of the episodes filled in by a friend, G. A. Walstab, while Clarke was laid up from a hunting accident.[120] Rolf Boldrewood's *Robbery Under Arms* began to appear in the *Sydney Mail* in 1881. It had to wait till 1888 before it appeared as a book. Boldrewood wrote to his London publisher that 'booksellers in the country complain very much that they cannot be supplied in sufficient quantities so that customers go away disppointed'.[121] Ada Cambridge's Australian writing career began when her first novel, *Up the Murray*, appeared as a serial in the *Australasian* in 1875. Many of her subsequent books were published in Melbourne weeklies before becoming available under the imprint of a British publisher. At the same time, British writers like Trollope, Hardy and Meredith appeared first in serial form in both Australian periodicals and imported literary reviews.

This method of publication reached a wide public and not one necessarily trained in the formalities of literary appreciation. Many writers earned a living as journalists or editors and were pragmatic about their work. Marcus Clarke contributed articles, sketches and short stories to papers and periodicals all over eastern Australia before he acquired the *Australian Monthly Magazine* in 1867. This he renamed the *Colonial Monthly* and although it survived only three years under his editorship, its office became a centre of literary life in Melbourne, much as in the 1890s the *Bulletin* office in Sydney, with A. G. Stephens as literary editor, attracted aspiring writers and illustrators.

The *Bulletin* is the best known and most studied of the weekly newspapers which fostered a local literature, but it is by no means the only one. Melbourne journals like the *Australasian* were a little too eclectic or too urbane to play a large part in making the legends of the 1890s. Others have simply been overlooked. Mary Hannay Foote contributed only some small, near-perfect verses in 'Where the Pelican Builds' to Australian literature, but as literary and social editor of the long-lived *Queenslander* from 1886 and 1896 she carried on the work of Gresley Lukin and W. H. Traill, publishing Brunton Stephens, Essex Evans, and Ernest Favenc.[122] Much of the style later identified as *Bulletin* style was evident in the

Queenslander. Indeed Traill came south, eventually to edit the *Bulletin*. Archibald and Stephens also brought their larger-than-life experience on the raw edge of civilization in Queensland to journalism in Sydney. They were not the first to revitalize the metropolis with the experience of the provinces, but the impact of Queensland's rough and often violent conditions on Australian culture through Sydney and, eventually, Melbourne, has been underestimated.[123]

In their willingness to publish local writers, nineteenth-century editors created the real venue for the growth of an Australian literature, and grounds for the emergence of an Australian publishing industry. Their readers liked reading about themselves or what they or their friends had written. Many a story or poem with little claim to that elusive 'literary' quality was included because of its subject matter. If the 'Answers to correspondents' columns in many of the weeklies are to be believed, it was the search for 'literary' quality which was the downfall of many an aspiring writer. Earnest young women like Catherine Helen Spence or Menie Parkes, well read in the best English authors, aware of literary traditions and style, but with nothing more thrilling than their housebound experience and imagination to draw upon, could not easily compete with plain tales of true adventure 'outback'.

Writers began to use Australian themes and settings quite consciously. This was to be expected from the rising generation of Australian-born writers like the poet Henry Kendall, who knew only the beautiful coastal scenery of New South Wales in which to realize poetic imaginings inspired by Tennyson and Matthew Arnold.[124] Novelist Rosa Campbell Praed was able to translate the exotic aspects of her girlhood experience in southern Queensland into what might otherwise have been ordinary three-volume romances.[125] Immigrants like Marcus Clarke, Henry Kingsley, and Rolf Boldrewood wrote consciously of their experience in Australia, seeking its origins in the now receding past. For *His Natural Life* Marcus Clarke went to Tasmania to research the convict system which was not even a ghost in his memory.[126] Boldrewood's tales of station life in the Riverina evoked the mythic days of the squatting age before the gold-rushes came to complicate relations between master and servant.[127] A steady flow of yarns, memoirs, and stories of the 'old days' helped to distinguish between new chums and old hands. In adapting true stories a degree of exaggeration was almost invariably employed, so it became difficult to tell where history ended and fiction began. Perhaps because they badly needed a sense of belonging, reassurance of survival and progress, readers preferred to think that the stories were true or real. To disentangle fact from fiction in Price Warung's tales of the convict days is practically impossible.[128] The number of fictional up-country stations almost equalled the number of real ones, while the 'originals' of many a fictional character vied with each other for recognition. An imagined past was better than none at all.

As with writing, it was the subject matter of art more than style or technique that became noticeably Australian. Like writers, artists depended on journalism or commercial work for a living. Some survived by teaching or portrait commissions; however, the photograph was replacing the miniature or portrait as a likeness.[129] There is scarcely a collection of letters after 1860 which does not mention the taking or transmission of photographs to or from relatives and loved ones in another country or colony. Photographs were the ideal and obvious modern way of staying in touch. The traffic in likenesses was enormous and as photographs became less expensive, views were also in demand. Surviving collections probably underrepresent the state of late nineteenth-century photography, for the most common postcard-like views were less cherished than portraits and both were more liable to incineration than a packet of letters. Oscar Comettant thought 'Photography in Melbourne has reached a level of near perfection.' The work he admired in particular was at Johnston O'Shannessy's studio—'portraits of all sizes leaving nothing to be desired, and also some remarkable collections of views of Australia and New Zealand'.[130] There was some continuity between art and photography. Working at Hewitt's studio in Collingwood when he was thirteen and then at Richard Stewart's in the city, Tom Roberts decorated photograph mounts with designs of native flowers and leaves. In 1885 the promise of a job reorganizing Barrie and Brown's photographic studio brought him back to Melbourne after years of trying to become an artist in England.[131]

Until the mid-1880s artists, especially those who were immigrants and had learned their craft elsewhere, had the same aspirations as their contemporaries in Britain and Europe, though they tried to adapt their techniques to Australian conditions. Portraits of people or their property (for example, the landscapes with houses of Conrad Martens) were in greatest demand.[132] Art was either of heroic proportions or a ladylike activity. (Of 30 students at the National Gallery painting school in Melbourne in 1888, 18 were 'young ladies'.[133]) A high degree of technical skill in oils, watercolours, flower painting and sketching is evident, though, with the exception of the heroic oils and some of the flower painters, little of this work has been taken seriously.[134] The tradition of studio painting on a grand theme remained strong for the few like John Longstaff who could afford the luxury of a studio or a grand theme. The struggle to emulate and compete with the Academy in London was intense. The major patrons of art in Australia were already public art galleries who wanted paintings to be both aesthetically improving and morally uplifting. There was also the question of value for money. It was not easy for Australian painters out of touch with the avant-garde, and especially for the rising generation of Australian-born painters whose experience of the European tradition was necessarily limited, to supply the kind of work desired by the galleries. Towards the end of the period, large and impressive paintings depicting great natural disasters—floods, bushfires, mining accidents—or which elevated Australian types—pioneers, shearers, bushrangers, and of course, horses—became more acceptable and were bought, but most painters could not afford to work on a grand scale without the certainty of a commission, and commissioned work had its constraints. To celebrate federation, the National Gallery of Victoria strangely commissioned *The Landing of Captain Cook at Botany Bay 1770* from E. Phillips Fox. A commercial firm, the Australian Art Association Ltd, paid Tom Roberts £3000 for a painting of the opening of the first federal parliament. Roberts was required to include 'no fewer than two hundred and fifty portraits from the life' and to sign 500 photogravure reproductions.[135]

From 1855, when Adelaide Ironside's mother scraped together the money so that she might show her work to John Ruskin and study further in Italy, there was a trickle of aspiring artists to London and Europe. Eventually native-born painters like E. Phillips Fox brought their overseas experience and techniques home to create a more sophisticated and cosmopolitan art world in which even immigrants like Girolamo Nerli and young Charles Conder could find interest and inspiration. The *plein air* school of the late nineteenth century adapted splendidly to Australian conditions and subjects. The climate encouraged outdoor painting with artists leaving their studios to camp at Heidelberg and Balmoral.[136] Thus the bohemianism of the European artistic tradition was adapted to an Australian fondness for the outdoors and the sporting life.

Neither Heidelberg nor Balmoral were far from civilization. The images of the Australian landscape and its inhabitants (though not its Aboriginal inhabitants, who were no longer idealized) were tied to the south-east corner of the continent. Black-and-white artists had as subjects the complete range of human types (including some excruciatingly caricatured 'outsiders', idiot Aborigines, loathesome and deceitful Chinamen, ludicrous and vicious women) and landscapes. Their technical competence assured them of a living until processes for transforming photographs into letterpress were developed. Even more utilitarian were the thousands of advertising illustrations; every conceivable item in household and rural use appeared in massive catalogues sent out to facilitate mail-order shopping. Any one of these catalogues now provides a carefully illustrated guide to consumerism in the late nineteenth century, but then it was work for a commercial artist.[137]

A living could be made if the artist was versatile. Julian Ashton worked on the Melbourne *Illustrated Australian News* and from 1881, the *Australasian Sketcher*. He covered the arrest of the Kelly gang at Beechworth in 1880 alongside the official police photographer. His sketches and the photographs of the police photographer, not only of the gruesome occasion, but of each other at work, catch perfectly that merging of one era into the next.[138] Later, Ashton moved to Sydney where he was employed to illustrate one of the many sets of commemorative volumes celebrating the centenary of white settlement in 1888, the *Picturesque Atlas of Australasia*. Subsequently he set up a school where a great many young ladies and some young men learnt to paint landscapes with figures in them. Ashton accepted a commission to decorate the walls of the marble bar in the Adams Family Hotel in 1892 and was an important influence behind the New South Wales Art Gallery's purchase of the work of Australian artists. By the 1890s, however, those of his pupils who found employment in painting and drawing found it mostly as commercial artists.

'Sardonic' is a word frequently used to describe Australian writing in the last few decades of the nineteenth century. Its visual equivalent can be seen in the many cartoons and satirical illustrations in newspapers and magazines. Two of the best known of the cartoonists from this period, Phil May and Livingston Hopkins, were transient, the one from Leeds, the other from Ohio, but there were plenty of others. Vane Lindsay lists G. R. Ashton, D. H. Souter, Frank Mahony, George Lambert, Percy Spence, A. H. Fullwood, Fred Leist, B. E. Minns, Alfred J. Fischer, Alf Vincent and Ambrose Dyson. They 'made Australia, in the second half of the nineteenth century, one of the most important centres of black-and-white art in the world.'[139] Of these, only Ashton, Mahony and Lambert reached the heights and were able to 'draw a man sitting on a horse'.[140] As well as contributing sardonic humour, cartoons and caricatures eased the chore of reading for readers who may have been literate, but not always to the point of persevering with long paragraphs. The 'popular' press achieved the same effect with 'conversations', short paragraphs, scurrilous suggestions as in the *Bulletin* and then John Norton's *Truth*.[141]

Decorative artists, craft workers and designers, men and women (of 157 pupils at a Melbourne school of design in 1888, 112 were girls[142]) easily and effectively exploited distinctive Australian symbols in their work. By the time an Australian coat of arms and flag were needed, there was an embarrassing choice of possibilities. Carved into the backs of chairs and sideboards, repeated on numerous cast-iron balconies, mounted and reproduced in silver, shaped and painted into pottery and enamelware, worked into patchwork and crochet, stamped and printed on an increasing range of commercial products, Australia's unique flora and fauna (along with the Aborigines and their weapons) were a natural source of decoration and identification.[143] Lucien Henry arrived in Sydney in 1879 and was appointed art instructor at Sydney Technical College in 1883. His discovery of the waratah resulted in some spectacular designs in an art noveau style to which it adapted perfectly, but local artists were already confidently working gumnuts and eucalyptus leaves, flannel flowers, wattle, emus and kookaburras into their patterns and had developed a sophisticated understanding of native timbers and local clays.[144] Most of this work so obviously marked as Australian was considered inferior when compared with imported objects, but utility and price were important too. Still there must have been some demand for Doultons of Burslem to produce a wattle pattern semi-porcelain dinner-service.[145] As the unashamed messages of late nineteenth-century advertisers show, Australian symbols were perceived then in a simple and straightforward manner, not yet laden with cute nationalist symbolism, nor touched with nostalgia. They were natural, and to the younger generation, more easily understood than anything from the Old World.

The quality and nature of popular entertainment relied, as it usually does, on a balance of what was available and what entrepreneurs thought they could sell, in theatre, music, circus acts, or public lecturers. As the size of the potential audience grew, it became possible to sustain interest in a greater variety of entertainments, and at differing levels of seriousness. This period saw the rise too, of serious, intellectual, and respectable theatre, more suitable for educated people (as well as their wives and daughters) than the common, vulgar, popular amusement that was usual. Increasingly serious theatre set standards, though audiences remained small. By the end of the century, however, Australian themes with their nationalist overtones were bridging the gap between the superior consumers of culture and those who were less well-informed or less demanding.

The theatre catered to all the escapist instincts which now find outlets in moving pictures, television, and the club variety shows.[146] Its vitality began with audiences, willing and able to pay for their amusement, night after night, but depended on early entrepreneurs and actor managers like George Coppin and Bland Holt, the fascinating and gruelling international touring of late nineteenth-century 'stars' and theatre companies, and eventually the establishment of a permanent company in J. C. Williamsons.[147] 'Live' theatre was far from a minority taste. It has been estimated that 93 000 people saw *Struck Oil*, the play that made J. C. Williamson's name and fortune in Melbourne during its 1874 season. The population of Melbourne was about 110 000 at the time, so many people must have seen it more than once.[148] *Struck Oil* was outstanding and is perhaps not a good example. But another play, the Garnet Walch–Alfred Dampier melodrama, *The Scout*, was seen by a total of 31 075 people at the Alexandra Theatre in Melbourne during a period of twelve nights in 1891.[149] The knowledge that audiences of this magnitude did exist kept many a less successful manager in business.

Theatre managers would use anything they thought might entertain an audience. Programmes frequently included two dramatic features, perhaps a whole play and the best-loved scenes of another, probably the source of the tradition of a main film and supporting feature in later picture theatres. Songs, dances, comic recitations fitted between the acts or around the intervals. Live animals, troupes of bell-ringers, christy minstrels, talented 'juveniles', any number of solo performers in fact, who would have been equally at home in a circus or music hall, could be and were included alongside more recognizable drama, especially where the number of venues for these various kinds of entertainments was limited.

Nineteenth-century audiences did not have the habit of sitting quietly and listening.[150] They talked and ate their way through performances, as in the 'family circle' where

paterfamilias takes his brood of little ones, when they go to see the pantomime or whenever he so far relaxes as to give them a treat to the theatre.

It is a pretty sight to see a father, and mother and five or six children sitting in the front row of seats, all as happy as mortals can be, except perhaps the mother, who is in constant fear that some of the youngsters, in their anxiety to see what is going on in the pit, will fall over the parapet on to the heads of the people below. The father is at the head of the commissariat, and is kept fully employed in serving out lollies and apples as long as they last.[151]

Audiences seem to have been as fickle and demanding as modern equivalents who cause television series to come and go with dizzying frequency. Constant change was necessary, not only in the repertoire, but in the content of plays which were written and re-written for effect, or topicality, or to suit the expressed preference of the audience, or to accommodate actors who drank or actresses who fell pregnant. Before the audience began to tire of a programme it was time to mount a new one, or to move to another town. With a small company, often the members of one family, moving was a way of life. There were always towns where a one-night stand was a welcome event, regardless of the content of the show. Audiences came to be entertained. Their tastes were diverse and their critical faculties too little used to be discriminating. Comedy, tragedy, farce, melodrama, played for utmost effect with conspiratorial theatricality, all went down well. In his study of melodrama, Eric Irvin used twenty-four different cateories, including opera and musical comedy, extravaganza and vaudeville, to classify works written in Australia during this period. In noting their first production, he listed a total of ninety-two theatres round the country. And as he points out, the Australian shows were a mere drop in the bucket of cheap imported plays and touring productions.[152]

Changing taste and fashions were shaped by the skills of the most successful managers, actors and entrepreneurs, and reflected changes in theatre technology. The amazing effects achieved with lighting in the 1860s became commonplace (as did fires caused by limelight), and gave way to spectacular waterfalls and rushing streams, exploding buildings, trick-riding and pistol-firing from the saddle. George Darrell's *Sunny South* featured the female lead administering a horse-whipping to the villain. This scene was so popular it was written again into a subsequent play. The main players, both male and female, were more promoted than the plays. So were the effects. The better players (men more than women whose careers faded as they lost their looks or became too busy with motherhood) could dominate a company and, either because of their capacity for drawing crowds or because of their stake in its ownership, were able to choose their own material and the way it was presented. Actor-managers wrote or commissioned plays to show off their skills and those of their companies. Alfred Dampier and his daughters, Lily and Rose, appeared in specially written tear-jerkers called *All for Gold* and *Helen's Babies* until the girls were too old to carry off the parts any more.[153] J. C. Williamson's *Struck Oil* was commissioned to contain roles in which both he and his first wife excelled.

Plays were often adapted or dramatized, with greater or less freedom, from a novel, story, or actual incident. Some Australian writers found work re-writing plays bought cheaply in job lots in London or the United States. Both *Robbery Under Arms* and *For the Term of his Natural Life* underwent dramatization soon after achieving literary recognition. Nat Gould was a popular author made more so through stage versions of his, mostly horsey, stories. Conversely, authors of work which did not lend itself to dramatization missed publicity and earnings. Few women novelists achieved dramatization, though the serious theatre of the late nineteenth century did begin to treat moral and social themes appearing in their work. By this time, however, Australian subjects in Australian settings were all the rage.[154] It was not easy to create a convincingly Australian drawing-room. There was no mistaking convicts, gold-mining or gambling as Australian, though none of these could compare with bushranging as a subject for drama. Endless versions of the Kelly story were seen in theatres through the 1890s. Any other bushranging story was worth a try. The themes and stock situations were similar to wild west themes popular in American theatre and sometimes exported. Adaptations worked easily on both audience and material. It is no wonder then, that it was a bushranging drama which was filmed first in 1906 and that twelve of the first twenty films made in Australia were bushranging stories.[155]

Music was an important part of theatre. It was a short step from songs which entertained the audience during and between acts to more integrated musical entertainment, from *Struck Oil* to operetta and musical comedy. Gilbert and Sullivan found an audience in the colonies more readily than Ibsen, beginning in 1879 with *HMS Pinafore*. Only months after its first performance in London, Margaret Lyons of Balmain had seen the *Mikado* in Sydney with her father, and named her boat after Nellie Stewart's 'Yum Yum'.[156]

Grand opera was the least accessible form of musical theatre during this period. Tours were expensive and so was the cost of a permanent company in Australia. Melbourne was the main centre for opera with the remarkable W. S. Lyster drawing on local and imported singers and musicians for his regular seasons at the Royal Victoria Theatre during the 1860s and 1870s. Occasional seasons in other cities, and country tours, were arranged, though there were few suitable theatres and the cost of moving the large cast and orchestra was always a problem. Lyster's death in 1880, not long after a fire which destroyed the Royal Victoria Theatre, put an end to Italian Opera for nearly a decade, though J. C. Williamsons continued to stage new operettas. Towards the end of the 1880s, the Martin Simonsen Opera Company gave occasional seasons in Sydney and Melbourne. In the early 1890s they were joined by the Montague–Turner Company, and JCWs, which brought out a touring Italian company in 1893. This seems to have been a successful venture, but it was a long time before it was tried again.[157]

Keeping a large group of suitable singers and instrumentalists together seems to have been one of the major difficulties of staging opera. Nor could a permanent orchestra be established. Yet there was an immense amount of amateur singing and music-making, much, though not all, occurring under church auspices. Katie Hume wrote to her mother from Drayton, Queensland ('that is sufficient address') on 16 April 1867:

We have been to Church this morning, when I performed at the Harmonium for the first time i.e. I played the Hymns. We had no chanting... It is a poor little instrument. No stops. One cannot make much of it. I played 'Agnus Dei' from Mozart's 1st. I long to hear yr beautiful Easter Music, & to join in yr hearty Services.[158]

Four years later she could report the opening of the new harmonium, purchased after a fund-raising campaign.

I had a singing class at our house for a few nights previously to practise the Hymns which were 'Jerusalem the golden', 'Hark, hark, my soul' and 'An endless Alleluia'. The old instrument has been sold to a Washer woman for £5, so we are quite out of debt.[159]

Church music was less stylized than it has become.[160] As well, liedertafels, philharmonic societies, glee clubs, musical unions all flourished, some surviving from the 1860s through to the twentieth century. In Adelaide in 1859 Christmas was marked by a performance of *The Messiah* by the Adelaide Leidertafel with an associated ladies choir, both under the direction of Carl Linger. A German immigrant, Linger was a special source of musical inspiration in South Australia, and through his tune for the unofficial anthem 'Song of Australia', a lingering presence in Australian history. Within a few years there was a similarly thriving liedertafel in Brisbane as well as a philharmonic society inspired by the less well-remembered but equally enthusiastic Sylvester Diggles rehearsing a Christmas oratorio. A liedertafel was formed to complement the well-established Melbourne Philharmonic Society in 1868. A Christmas performance of *The Messiah* was mounted in Sydney in 1860, but Sydney and New South Wales seem to have had the least persistent singers.[161] None of the country towns except Newcastle, where Welsh miners organized an eisteddfod in 1875, could show anything to rival the Ballarat eisteddfods and South Street competition in Victoria[162] or the musical life of country towns in Queensland and South Australia. In 1879, however, a chorus and orchestra of about 700 altogether was raised for a series of concerts celebrating the International Exhibition in Sydney. The chief attraction was an 'Exhibition Ode' with words by Henry Kendall set to music by Signor Giorza, a composer, pianist and conductor resident in Sydney since 1874. Another Kendall ode, 'Victoria', was set by Leon Caron for the opening of the Melbourne Exhibition in 1880 and sung by a mixed choir of 900 voices.[163]

Choral singing was immensely popular in Melbourne, and in Adelaide, where the British tradition of choral music was maintained and derived strength from its religious, especially Nonconformist connections. As a young man, G. L. Allan, founder of a Melbourne music publishing house, taught music in the schools, conducted church choirs, and held classes for aspiring opera singers as well as his choir members.[164] Musical societies followed settlement along the coast of Queensland almost as quickly as printing presses. There the eisteddfod movement soon gained enthusiastic followers. Music was undoubtedly inexpensive fun and fine sociable activity. By remote camp-fires, the man who could sing or play one of the portable instruments, a mouth organ, 'squeeze box' or fiddle, even a gumleaf or Jew's harp, was an asset. Snatches of half-remembered traditional tunes and hymns mingled with more recent popular songs from opera and operetta as they made their way 'up country' and into 'folksongs'.

Ease of access, cheapness, portability—none of these quite explains the popularity of the brass band in Australia during this period. Perhaps its effectiveness out of doors was the secret. An afternoon stroll in the park was often accompanied by the strains of a brass band playing selections from opera, the light classics, and the standard repertoire of marching music. Rotundas, some quite elaborate, were built to house the bandsmen and their instruments. Military bands often accompanied parades and played on official and ceremonial occasions. Before amplification and loud hailers, the commanding noise of a band was an asset to a trade union demonstration, a circus procession or a Salvation Army meeting. For an organization, whether it was a union, a factory, a mine, or even a small country town, a band provided a focus for social activity and a sense of purpose—raising money for instruments, practising, marching and playing in public, even in competition. Furthermore, band music could never be considered effeminate. Perhaps that and the opportunity to wear uniforms were among its greatest attractions.

An English musician, Frederick Cowan, was invited to organize orchestral concerts in Melbourne in conjunction with the centenary celebrations. He brought a group of musicians with him, added to them players recruited in Melbourne and Sydney, and gave a series of concerts in both cities.[165] Attempts were made to keep the orchestra together after 1888 without success. Yet, when Marshall Hall began organizing another orchestra in Melbourne in 1891 he found a nucleus of players. The combination of Hall's energetic personality and the breadth of the repertoire of orchestral music, most of which had never been heard in anything other than arrangements for piano or brass band, made these concerts very memorable for those lucky enough to hear them.[166]

At the same time in Sydney the Amateur Orchestral Society under Roberto Hazon, who arrived in 1887 to conduct Italian opera and who stayed as a singing teacher, came into being. Along with the Philharmonic Society, also conducted by Hazon, they provided Sydney with performances of music otherwise unheard before the days of radio and gramophone recordings.[167]

It was easier to organize and perform chamber music. Amateur groups of friends made music together with a piano, a couple of violins, a flute or clarinet, some singers. Sales of instruments and sheet music became the basis of fortunes accumulated by large music houses like Allens, Palings and Beales. At a professional level, some of the most effective chamber music groups were also family affairs—John Deane and his brother in Sydney in the 1860s, R. T. Jeffries and his daughters in Brisbane after 1871.[168]

Australian tours by noted individual performers kept alive a notion of prevailing standards elsewhere. In the main, however, music was still more a participant activity than a spectator sport. For every appearance by a visiting artist there were a dozen concerts of the semi- or non-professional kind, in which people displayed their talents and amused each other, not only singing and playing instruments, but with recitations, dance routines, and comic turns.

The musical culture of Australia before 1900 has been dismissed as imitative and of little interest.[169] Perhaps it was if one is searching only for the beginnings of a truly Australian musical style. The most original and strangely nationalist of early Australian composers, Percy Grainger, born in Melbourne in 1882, left Australia in 1895 and was not fully appreciated until well after his death in America in 1961.[170] Regardless of its creative achievement or intellectual value, music-making produced a lively social and cultural atmosphere. Its imitativeness resulted from enthusiasm for the latest, perhaps even the best developments in the Old World. Derivative in form, ephemeral in practice, but a popular and pervasive link with tradition, music was also least affected by the environmental influences which produced an early sense of Australian style in literature, art, and drama. Isolation from trends and developments elsewhere was itself a hardship and a challenge. The orchestras and conductors who struggled with Mendelssohn and then Wagner, and the audiences who were lucky to hear compositions more than once, were demonstrating a form of heroism not at all unlike the pioneer spirit.

Music was also important in a culture where there were otherwise demarcated spheres for women and men. It was one of the few organized social activities in which both sexes participated, for how could *The Messiah* be staged without sopranos? Women had a 'natural' place in music, although the liedertafels specialized in doing without them. Theatre also offered undeniable parts for women, and it is not surprising that so many of the women's names in the public domain were those of singers and actresses, nor that Australia's reputation for fine singing voices was so high. This was one area where we drew on the whole of our talent and fostered it. Singing, playing an instrument, charades and amateur theatricals were encouraged as wholesome evening leisure activities for young people and the necessary skills were acquired and encouraged. Girls probably received more training in this area than boys, though the boy with a good ear and a natural voice was welcome anywhere. Women were accepted, not only as singers and actresses, but as pianists, violinists, and composers, especially in amateur and church groups.

Music was one of the best-endowed cultural activities. Wealthy citizens funded travelling scholarships for promising young musicians. Chairs of music were established at both the universities of Melbourne and Adelaide in the names of their benefactors—[Sir] Francis Ormond, a keen Presbyterian and notable friend of education in Victoria, and Sir Thomas Elder, whose gift of £20000 in 1897 made possible the Elder Chair in Adelaide. Earlier Elder had endowed a travelling scholarship in South Australia, inspired perhaps by Sir William Clarke's gift in 1882 of such a scholarship to Victoria. The quality of musical activity in Adelaide at the time was remarkable for the size of its population. Elder's patronage was important but he was by no means the only local financial supporter of serious music. In 1884 as the result of some energetic fund-raising by W. R. Cave, a lectureship in music was established at the University of Adelaide. Joshua Ives, B. Mus. Cantab., arrived the next year to fill the position and he was able to gather a strong group of local performers and students, thus laying the foundations for the conservatorium which emerged at the end of the century, also supported by Elder.[171]

A similar willingness on the part of the community to contribute financially to the training of young musicians was evident in Melbourne. When in 1891 Marshall Hall arrived to take up the Ormond Chair, he perceived the need for an associated music school. The council of the University of Melbourne was unwilling to fund such a school, but Marshall Hall was able to finance the necessary building by canvassing the people of Melbourne.[172]

From the creation of teaching institutions, interest in standardized and recognizable examinations and qualifications followed naturally. In Sydney Hector McLean, organist at St James's, was appointed secretary for Trinity College, London, in 1878, to facilitate the participation of Australian music students in the London examinations, though it was not until 1895 that the first practical examinations were held here. The next year Ethel Pedley returned from a trip to England with authority to examine for the Royal Schools of Music in Sydney.[173] Thus piano playing was regulated and made professional. For many young ladies at least, the annual examinations in music theory and practice became their first experience of the system of competitive entry into the professions.

The piano had already achieved the status of a cliché in Australian history as a symbol of bourgeois civilization and its preoccupation with status and a particular kind of culture.[174] As a mechanical means of making music it had no equal in this period; its versatility was incomparable alongside the other mechanical musical sources, the musical boxes and mechanical organs. In many ways it prefigured the gramophone in social life and musical enjoyment.

Some idea of the cultural and economic importance of the piano can be sensed from the way in which piano makers rushed to exhibit and display their instruments at the major international exhibitions in Australia. Rather in the manner of modern car manufacturers hiring champion drivers to show off their products in races and competitions, the British, German, French and other piano and organ manufacturers hired top players to perform on their instruments, both in grand concerts in the specially constructed concert halls, and daily in the courts devoted to national exhibits. The outstanding performer was said to be Madame Summerhayes (no first name given) who gave 120 recitals during the six months of the 1879 Sydney exhibition, mainly on British Brinsmead pianos.[175] Whether this contributed to the reputation or sale of Brinsmeads is not known, but German pianos designed and built for the mass market were beginning to challenge those of British manufacture. According to Oscar Comettant, who was representing France on juries at the International Exhibition in Melbourne in 1888, the 210 German pianos exhibited outnumbered British and American models, while only five French pianos were on display.[176] The German pianos, 'vulgarly over-decorated and for the most part very mediocre', were arranged in a vast square in the concert hall which seated a thousand. Beyond them was a stage with selected pianos and harmoniums on which virtuosos performed. Some other instruments were also exhibited, but the 'Pianopolis' dominated the musical offerings of the exhibition.

The everyday trade in pianos and sheet music was conducted at large and, by the 1880s, impressive music shops had been established in the best streets of the main cities. Tuning, repairs, and all kinds of music lessons could be arranged at these establishments which were also involved in some cases in the manufacture of their own instruments. As early as 1872 there were nine piano warehouses in Victoria, making, mending, and tuning pianos, while George Fincham had a factory in Richmond devoted exclusively to the building and maintenance of pipe organs, mainly in churches, though some were in civic buildings like town halls.[177] In Sydney Octavius Beale was laying the foundations of a fortune with his piano factory and warehouse, while the name of W. H. Paling, a Dutch violinist who mixed professional engagements with teaching, concert management, and every other form of musical entrepreneurial work, became synonymous with sheet music and instrument sales.[178] In Melbourne George Allen, a singing teacher and organist who perceived the possibilities of mass musical education through the schools, went on to become a leading music salesman and entrepreneur.[179]

Music provided employment for a host of teachers, especially piano teachers. 'You only have to tap your foot on the ground and they come out', said Comettant.[180] The music studios of the Strand Arcade in Sydney as described by Louis Stone in *Betty Wayside*,[181] their bohemian characters, the hint of old Vienna or Leipzig, of passion, romance, intrigue, were an alternative (an unhealthy alternative some thought, since it led to idle dreaming) to down-to-earth materialism or the healthy outdoors. For some music was a means of escape. Ethel Richardson left Melbourne in 1888 to study music in Leipzig and found her metaphorical way home writing *Maurice Guest*.[182] Music allowed relaxation from the preference for realism in colonial society. Yet because it was associated eventually with both high culture and religious observance, because even the tunes of the most bawdy or subversive songs were harmless without words, it remained respectable and acceptable. Neither theatre, art, nor literature could so easily occupy this middle ground between the respectable and the stimulating, the traditional and the modern. Thus music was considered the most 'innocent' and suitable of all cultural forms for girls and women.

Dancing also attracted large numbers of female students and performers. Most teachers, arrangers, and choreographers in this period were women though the entrepreneurs were men. Dancing was to many women what sport was to many men, an outlet for energy and physical well-being. Yet it was also a legitimate expression of sensuality, aestheticism, romanticism or athleticism. By the end of the century the difference between dancing and gymnastics for training girls in graceful movement and physical culture was fading. For entertainment and pastime, dancing was as popular as music. 'They all dance very well. It is the passion of Australia', Charles Trevelyan reported to his family apropos a Masonic Ball in Dalby in 1898. A dance was convivial and easy to arrange. A large room and a piano or fiddle would get things going.[183]

Unlike the situation with music, there was no connection between dancing for pleasure and professional dancing, of which there was very little. Theatrical performances often contained dances, or a dancer performed an entr'act. Till about 1870 Australia was included in the tours of leading European and American dancers and dancing troupes who brought classic works like *La Vivandière*, *Giselle*, and *La Sylphide*. Some Australian dancers became famous in their own time, like Julia Matthews, her rival Tilly Earle, and Joseph Chambers, the premier dancer who successfully impersonated the amazing Lola Montez in her scandalous 'Spider Dance'. After 1870 most ballet was engulfed for a decade by the versatility and popularity of comic opera and pantomime song-and-dance routines. The skirts of dancers became shorter and their draperies more diaphanous, introducing a titillating or mildly pornographic element into the performance. Paterfamilias in the family circle however had probably taken his wife and children to see a comic dance pantomime like *Jocko, the Brazilian Ape*. Europe's classical dance revival reached Melbourne's Princess Theatre in 1893 when two of the stars of J. C. Williamson's touring Italian opera, Catherine Bartho and Enrichetta d'Argo, danced a ballet, *Turquoisette*, choreographed by Rosalie Phillipini to a potpourri of music arranged by Melbourne's resident composer, Leon Caron. Classical ballet wore its skirts at a respectable length, though not for long.[184]

Beyond natural anxieties about the quality of art, theatre, music, and dance being reproduced in the colonies and whether Australian styles might emerge to enhance the traditional forms, a unique indigenous culture persisted quite beyond the reach of city and small town audiences and critics. Writers had been using Aboriginal Australia as subject matter continuously (and rather more extensively than the canon of literary history allows).[185] The absence of a written tradition among the Aborigines made it difficult for this process to be anything but one-sided. A musician with the creative imagination of Isaac Nathan perceived something of the interest and quality of Aboriginal music. His attempts to transcribe and render into more popular and contemporary European form some of the Aboriginal songs he heard in the Monaro in the middle of the nineteenth century foreshadow the later uses to which folksong was put by composers like Grainger, Bartok and Kodaly.[186] There were some visiting scholars and journalists, as well as a few local students of Aboriginal culture, who sought and were granted the privileged experience of observing corroborees. Spencer and Gillen on their 1901 expedition not only recorded Aboriginal songs on a 'phonograph', 'the first ever thus recorded in Australia',[187] they also carried 'a cinematograph with which to secure records of native dances and ceremonies'. Despite considerable difficulty with heat, dust, and dryness and the primitiveness of their equipment, the results 'were sufficiently good to justify this, which was the first attempt ever made to obtain moving representations of the ceremonies and dances of the Australian aborigines'.[188] Those flickering images could hardly be seen as a creative encounter between Aboriginal and white culture, though they must be among the earliest attempts at documentary film-making. Percy Grainger later listened to the recordings of the songs and attempted to transcribe two of the 'airs . . . as correctly as it is possible to play them on a piano',[189] but there is no evidence that his own music was influenced by what was after all a remote experience. Later he did use recording equipment to great effect, but this was in his research into English folk music.[190]

Aboriginal art, which made its way as artefacts in huge quantities into museums both here and abroad, was the most accessible aspect of Aboriginal culture, though if it was admired, this was for its decorative possibilities rather than its deeper meaning. Tommy Macrae never went to art school. His handsome illustrations for *Australian Legendary Tales* (1898) were drawn with a pointed stick dipped in ink. His collaboration with Kate Langloh Parker was unusual in crossing the gap between the two cultures, for Tommy Macrae was an Aborigine from Corowa aided and encouraged by the local doctor, W. H. Lang, himself a writer on horsey subjects and a brother of English folklorist Andrew Lang.[191]

Different questions produce different assessments of the nature and extent of cultural development in the Australian colonies in the second half of the nineteenth century. Those who are interested in the growth of 'high culture' frequently find this period disappointing, provincial, imitative. Nevertheless, the institutional foundations to channel culture remorselessly into 'high' forms had been well and truly laid.[192] Nor can there be much doubt that it was a creative and vigorous period for the growth of the masculine egalitarian Australian culture celebrated by Arthur Jose and Vance Palmer.[193] At mid-century the term 'culture' was still associated with privilege, a classical education, a knowledge of the great traditions of British (and to a lesser extent, European) civilization. Leisure and education, prerequisites of culture, were available only to a few though both became generally accessible in Australia before the end of the century. Culture retained its traditional forms and classical connotations.[194] Indeed, both were strengthened by education. But it was coloured by Australian experience, increasingly so, as a majority of its practitioners had neither a classical education nor first-hand acquaintance with the traditional forms. By the end of the century there was a tendency to think that Australian settings of themselves produced originality, that all else must be derivative. The suspicion that colonial culture was still inferior was countered by asserting that at least it was 'manly' in comparison with the effeminate stuff of the old culture. Egalitarianism disguised intellectual thinness, emotional insecurity masqueraded as nationalism, and progressivism concealed a ruthless and impatient approach to social relations.[195] Apart from the environment, sport, mateship, alcohol and unhappy marriage were the chief sources of inspiration, though while unhappy marriages drove men to the bottle or to seek solace with their mates, those same marriages became a major theme in women's novels of this period. Rosa Praed, Jessie Couvreur ('Tasma') and Barbara Baynton wrote from their own experience of unfulfilled romance and sexual longing, of drunken husbands, and loneliness. Not one of the romantic stories Rigby and his mates swap while they wait for the fish to bite has a happy ending, and Rigby, as he listens, forgets his promised meeting with the woman he loved and who has travelled half way round the world to find him again.

5
POWER

CONSTITUTIONS ADOPTED in New South Wales, Victoria, South Australia, and Tasmania in 1856 and in Queensland in 1859 established the institutional framework in which political development occurred during the next forty years. Western Australia was governed by a wholly nominated legislative council until 1870, when it became partially elected. In 1890 when it also became self-governing under its own constitution, Western Australia was still the smallest of the colonies in population—48 502 plus an unknown number of Aborigines. Its political development, therefore, is not automatically included in generalizations about the other colonies.

The one thing Western Australia had in common with the other colonies was its political evolution from the constraints imposed by the Colonial Office and the role of the governor as mediator between the colonists and Westminster. In all the other colonies, by 1860, the framework of government had come to resemble the British model. Governors represented the monarch and played a greater or lesser constitutional and social role depending on their personalities, predelictions, and experience. The earlier governors were, on the whole, more explicitly and intrusively representatives of the Colonial Office. They saw their prime responsibility as the upholding of British values and traditions of government.

Towards the end of the period, and after the main constitutional battles over the authority of the governor and his right to intervene in colonial politics had been won, the governors ceased to be career colonial administrators. They were more likely to be minor aristocrats or semi-retired senior army officers, chosen for their social skills or their ability to afford the style of government house.[1]

At the same time the focus of negotiation between Britain and the colonies shifted from the colonial capitals to London. Colonial agents-general were more used as links in the chain of communication. This change was emphasized by more frequent visits of colonial leaders to Britain, also by a shift in British policy from simply managing the colonies to contemplating complex schemes of imperial government.

The British model of government was continued in two houses of parliament, an upper and a lower house, in each colony. As governors maintained colonial office surveillance over colonial attempts at self-government, so the upper houses were intended as a check on hasty or ill-considered legislation emanating from more democratic lower houses. Just who might sit in a replica of the House of Lords in a country where there was no hereditary aristocracy remained a perplexing problem. In New South Wales, and in Queensland where the constitution was based closely on that of New South Wales, the old nominated advisory Council became the basis of an upper house nominated by the governor in consultation with his ministers in the lower house. Victoria, South Australia, and Tasmania all elected their upper house, though on a much more restricted franchise than the lower house.

Property qualifications for both members of parliament and electors, and the way in which the electorates were drawn up, ensured upper houses which represented property above all. In South Australia, for example, where the least restrictive electoral qualifications were adopted (£50 or £20 leasehold, or £25 per annum rent), the colony voted as a single electorate for the upper house until 1881 when four districts were created. This meant that successful and well-known figures in Adelaide were more easily elected than candidates whose reputations were localized. Names were more important than policies, and although the system un-

THE OPENING OF THE PARLIAMENT—HIS EXCELLENCY THE GOVERNOR
DELIVERING THE OPENING SPEECH

The opening of parliament was an important social occasion, as can be seen by the ladies seated not only in the gallery but in the body of the House

doubtedly encouraged civic spirit and benevolence among aspirants, it also ensured that the Council was made up of the colony's business, professional, and landed élite. In Tasmania, though upper house members were elected from the different regions, a restricted franchise ensured that they came, almost invariably, from well-established and dominant families. A small population and a closely knit rural society gave Tasmanian politics an eighteenth-century feeling.[2]

During the second half of the nineteenth century, the elected upper houses gradually emerged as the most effective forces for conservatism in government. Nominated upper houses were obviously unrepresentative and potentially conservative. The threat, however, of swamping them with new nominees of a different political colour was always available. It was not so easy to show that elected upper houses, however limited their franchise, were entirely unrepresentative. Nor could they be changed, except through the ballot box. Still, sometimes broadly based shifts in public opinion showed here. The first acknowledged members of the Labor Party in an Australian parliament were elected in 1891 to the intensely conservative South Australian Legislative Council. Similarly, in the twentieth century, women have found the upper houses easier of access because of the more extensive electorate and a different approach to choosing candidates.

The constitutions of the 1850s acknowledged the lower houses as the chief source of initiative in government. By 1860 all were undoubtedly more democratic and accessible than the House of Commons on which they were modelled. This was partly due to their methods of election, partly to the kinds of members available for election, partly due to limits on their powers. Colonial parliaments were small in size and as yet without much in the way of tradition to uphold, so they tended to be informal. They had no responsibility for foreign policy which was still administered from London, so much of their earnestness was expended on minor local matters.

By 1860 voting by secret ballot had been adopted in all colonies (except Western Australia where it was adopted in 1878). Manhood suffrage was introduced for lower house elections in South Australia in 1856. Property qualifications were retained elsewhere, although they were so small—a miner's right in Victoria, a £10 p.a. rental or leasehold in Queensland, or status as a permanent resident in New South Wales—that the right to vote was fairly widespread in the male population. Some groups excluded by the property or rental qualifications were gradually incorporated into the system, such as itinerant male workers. Tasmania was the last of the colonies to lower the property barrier to voting qualifications. There, in 1890, property of a rateable value of £40 p.a. or wages of £60 p.a. were still required. Manhood suffrage came to Tasmania with federation in 1901.

Property qualifications came to be considered an evil, not simply because of their symbolism or the groups they excluded from the electoral system, but mainly because of the extra votes they bestowed on the owners of extensive property. S. W. Griffith was on the roll for four different Queensland electorates in 1883, and by 1892 the number had risen to six. It is quite possible that he voted in all six.[3] In New South Wales G. N. Hawker estimated[4] that plural voting as a result of property ownership accounted for 15 per cent of the total vote. In Tasmania in 1897, 500 electors were known to have two votes, 98 to have three, and one person to have had as many as seven.[5] Property qualifications additional to those required for electors for nomination to parliament had been abandoned in most colonies by 1860. Only Tasmania retained them till the end of the century.[6]

Property qualifications were no guarantee, however, of the right to vote. Women property owners were entitled to vote in municipal elections in Victoria from 1854 and did so. However, their votes were not always accepted by electoral officials, and once the precedent of rejecting women's votes was established it was easy to maintain. In other colonies women were also theoretically entitled to vote in municipal elections, but how often they did is unclear. In 1893 an Adelaide headmistress, Eliza Kelsey, described her experience as a municipal voter as a half-way step to full suffrage.[7] If women ratepayers' votes were acceptable at some municipal elections, it seems strange that more was not made of this, both by those in favour of and those against women's suffrage.

Most men learned political behaviour in the voluntary organizations proliferating in this period, the church committee, the lodge or friendly society, the trade union or professional association, the debating society. Such groups were better placed than individuals to assert their views. As early as 1859, trade unions operating in Melbourne had formed themselves into a Trades Committee which then became the Trades and Labour Council. Similar councils soon followed in all colonies. By 1879 it was possible to arrange the first intercolonial trade union congress in Sydney. These organizations created their own power structures and opportunities for advancement both internally and onto the wider stage. (Conversely, the political experience of women was severely limited by the absence of opportunities to participate in them.) For most men newly admitted to the status of voter, however, municipal politics was their introduction to full citizenship. Matters that really affected ownership of real estate, the quality of the immediate environment, or the success of small enterprises were settled at that level.[8] Rates were the most commonly encountered form of taxation and ratepayers' organizations the simplest political form. Municipal politics were a logical starting point for many a political career as well. It cost least, both in time and money, to enter municipal politics. In theory, any man over the age of twenty-one who was entitled to vote in lower house elections (except in Tasmania) was also able to offer himself as a candidate for parliamentary election. In practice, however, the barriers against candidature were its cost—estimated in 1887 at an average of £300 for a successful campaign in New South Wales[9]—and, until payment for members was introduced, the loss of time and income while attending to parliamentary duties.

Payment of members has been seen as crucial to the entry of significant numbers of workingmen into parliament, but pressure for some kind of subsidy came also from rural representatives. Unlike their city colleagues who could attend to their own businesses in the morning and to their parliamentary responsibilities in the afternoons or at night, rural members were forced to spend most of the week, or even the whole session away from their properties, farms, or country businesses. Often the strain, as Friedrich Krichauff discovered, whether physical, or financial, was too much.[10] Travel subsidies and attendance allowances helped to improve the quality and continuity of rural representation. They were so successful when introduced in Queensland in 1886 that by 1892 it was decided they could be cut by half, partly as an economy measure.[11] However, it was Victoria which began the experiment of payment for members in 1870. There the financial sacrifice of attending to one's parliamentary duties when other men were making money at a great rate became marked. It was 1889 before New South Wales made similar provision, and 1890 in South Australia and Tasmania; Western Australia followed in 1900. There were considerable differences in the value of such 'emoluments'. Initially, £300 p.a. was paid in Queensland with a railway fare to and from Brisbane for the session. Even when cut to £150 in 1892, this still compared well with Tasmania's £100.[12] The restoration of the original allowance in Queensland in 1896 helps to explain the extent of the working-class movement in that colony. At a time when memberships of many trade and other voluntary organizations had fallen as a result of depressed conditions, parliamentary salaries were useful as a subsidy to the unions.

As well as church or charitable committees, self-education or mutual improvement associations, trade unions or friendly societies, certain newspaper offices and legal firms provided political apprenticeships, especially for those who were too ambitious to begin in a small way at the municipal level. Politics at mid-century demanded a high level of involvement from those who wished to participate. The number of voters represented by any one member was small while property qualifications, voluntary registration and voting, and the sex barrier, all contributed their own disqualifications and discouragements. In New South Wales in 1860, seventy-two members represented 300 000 men, women and children, that is, one member for each 4167 of the population. The average number of enrolled electors in each New South Wales electorate, however, was about 1500. In Queensland it was even easier for a politician to know his electors. There, twenty-six representatives dealt with the problems of 28 000 people in 1860, an average of 1077 each, though some electorates had fewer than 100 on the roll. This situation changed during the next four decades as the population grew and new centres developed. New electorates and redistributions became necessary from time to time. By 1900, New South Wales had 141 representatives, Victoria 95 (as compared to 78 in 1860), Queensland 72, South Australia 54 (36 in 1860), Tasmania 36 (30 in 1860). Western Australia began in 1890 with 30.[13] Some of the disparities eventually reproduced in the federal constitution were neatly illustrated by Henry Parkes's comment in 1890.

The tally board outside the Age *office for elections in Victoria in 1894*

I live in a Sydney suburb, which is separated from the City by one of the arms of Port Jackson. This suburb [Balmain] contains more than two thirds of the number of people in Western Australia. Its 30 000 people send only four members to our Parliament, whereas 44 000 persons in Western Australia have a Parliament of their own.[14]

The spread of settlement tended to maintain the power of rural voters against the metropolitan centres. Extra electorates were added as they became necessary in areas of recent settlement without compensatory changes in the older rural or urban electorates, though some of these became multi-member electorates, and a new member was added each time the number of enrolled electors passed a certain point. In a multi-member electorate, the voter was entitled to as many votes as there were members to elect. This encouraged a more varied selection of members and fostered minority interests and candidates, just as Senate voting practice does today. Multi-member electorates may have helped the early labour candidates to establish themselves. It was common for voters to 'plump' for only one of the candidates and give him all their votes so that second and subsequent members were elected by a tiny number of votes. This problem was partially overcome by voter education, and by the best-known candidates running 'tickets'. But by the end of the century multi-member electorates were thought to be less than fair, so they were gradually redistributed out of existence. By then, however, the map of settlement was filled in and a rough balance between the size of urban and rural electorates had been established in most colonies.

When it is remembered that in addition to lower house representatives, only slightly smaller numbers sat in the upper houses, and many more were involved in municipal politics, or on committees responsible for schools, hospitals, and other local institutions, it will not seem surprising that there was often difficulty in finding a willing candidate. Some members were returned unopposed at almost every election. At the first election in Western Australia in 1890 only eleven out of thirty seats were contested. John Forrest was returned unopposed by the 187 registered electors of Bunbury to become the first premier.[15] Those men who had taken the trouble to register as electors often took a personal interest in the possibility of nominating a candidate, securing his election, following his speeches, and monitoring his performance in parliament. Because of the numbers involved it was not improbable that the local member would be known personally to most of his electors. For many of them, participation in the political process was a mark of their new-found economic security and respectability. Registration as a voter implied self-respect, a sense of civic responsibility, an active interest in the business of managing the affairs of society. Much political debate was aimed at educating and involving potential electors. So too was the gradual move towards political organization.

This emphasis on the status and the sense of self-worth conferred by participation in politics had a marked effect on those who were excluded by the system. Their inequality was thus sanctioned, institutionalized as political dependence, and easily extended. Women early felt their exclusion, especially when the subjects debated were close to their own concerns, for the debate itself was accessible in every newspaper. Wives and daughters could feel justly aggrieved when the men supposed to be representing them would not listen or act on their views. The status of nomadic Aborigines was worsened by their exclusion. It was not unknown in the 1860s for the representatives of urban electorates to point out that they carried a heavy burden because of the disproportionate number of unenfranchised women and children they had to represent. Likewise, it was sometimes said that sheep must have votes in rural electorates because there were so few people. But no one claimed responsibility for an unknown number of Aborigines. So South Australia, Queensland, and Western Australia entered the twentieth century disproportionately burdened with non-voting, and therefore uncounted, populations whenever anything was assessed on an electorate basis.[16]

An analysis of the membership of the New South Wales Legislative Assembly shows that in the 1860s a majority were immigrants.[17] Men born in England, Scotland and Ireland dominated the Victorian parliament too, until the last years of the century.[18] In Queensland, immigrants outnumbered the native born throughout the period. Indeed, constant immigration produced a parliament in which two-thirds of the members had been born outside Australia, mostly in England, though Scotland produced 88 out of a total of 627 members, Ireland 77, and Wales 13. Of the 36 other members of the Queensland parliament with birthplaces outside Australia, most were from within the Empire—from the West Indies or India—or from Germany. To be native born, however, came to imply moral purpose and youthful vigour which seemed to compensate for lack of numbers.[19]

Anglicans were present in the New South Wales parliament in disproportionately large numbers, but Presbyterians, Methodists, and other Protestants were strongly represented.[20] In Victoria, Presbyterians were conspicuous among MPs.[21] Methodists and other Protestants were similarly in evidence in South Australia.[22] There were some Catholics among the

Irish-born members of the Victorian legislature, notably Charles Gavan Duffy and John O'Shanassy. About 9 per cent of New South Wales representatives were Catholic by 1890. Patrick Jennings became the first Catholic premier of New South Wales in 1886. The proportion of Catholics was probably higher than this in Queensland, though many of those with Irish birthplaces were in fact from northern Ireland, and probably Protestant.

The men who dominated colonial politics seem mostly to have been lawyers and pastoralists, thus giving the impression that there was a battle between these two groups, further identified as representing urban and rural interests. But the occupational experience of members of parliament really encompassed a much larger range, and touched most aspects of colonial life. Pastoralism was but one of the occupations of many a successful immigrant, while the law was often the first or last occupation of a nineteenth-century educated man. Between 1860 and 1900 in New South Wales at least, members drawn from banking, commerce, the professions, journalism and medicine as well as law, always outnumbered the squatters and farmers. Despite the growth of farmers' organizations, there were few farmers in the New South Wales parliament before 1900.[23] Rural electorates centred on towns like Orange or Armidale were more likely to return a local journalist, lawyer, or businessman than a farmer, though the member who did not remain loyal to the local interest, or conversely, bring prestige and power through his ministerial successes, could not look forward to a long career in politics.

Joy Parnaby's analysis of the Victorian parliament shows a shift from urban to rural occupations among members by 1881, with farmers beginning to outnumber pastoralists, though the upper houses continued to represent squatting.[24] In South Australia the interests of Adelaide and the country were closely identified until the late 1880s when the urban challenge of the labour movement consolidated the affection of the traditional power-brokers for their country connections. The labour movement found country support too among the small farmers, the blockies who were as likely as not tradesmen forced on to the land by their inability to find other work.[25]

In contrast, rural interests were strongly represented among Queensland MPs. Squatters, pastoralists, graziers,[26] sugar planters, cane-growers,[27] flour millers, stock and station agents, miners, mine owners and investors, produce merchants and timber millers all made their appearance in Brisbane's George Street and in numbers large enough to hold their own against the lawyers, merchants, storekeepers, newspaper proprietors and assorted small businessmen who represented Brisbane and the other towns. Until the 1890s, even a miners' representative like J. M. Macrossan was more inclined to support the squatters and graziers against the schoolteachers, clerks, butchers, bakers, chemists and drapers of the towns. The growth of rural unionism and the example set, especially by the shearers at Longreach and Barcaldine in 1893, helped to clarify the conflict of interest between rural workers and those who merely invested in, owned, or controlled conditions on stations and remote mining settlements.

The range of occupations represented in the Queensland parliament and their continuing rural bias are reminders of the most significant contrasts between the colonies during this period, their differing dependency on rural exports, industrialization and urban growth. Queensland's diverse representation was partly a function of size, of widespread, though sparse, settlement. But the absence of senior representatives of commerce, banking, and industry was notable. For example, Burns Philp, which dominated the coastal shipping trade and commercial agencies in most Queensland ports, was a Sydney-based company. Robert Philp, who had been the senior partner based in Townsville, lost his place in the firm soon after he entered politics. Though in 1899 he did become premier of Queensland, he had ceased any connection with the firm that bore his name. James Burns's 'strict and unsentimental control' of the firm might as well have been from Edinburgh as Sydney.[28] Battles with Burns Philp, or AML & F, or CSR could not be settled in the Queensland parliament. Legislation passed in New South Wales or Victoria could affect the lives of workers in the smaller colonies as much as any they approved for themselves. The introduction of strict controls over factory conditions in Victoria, for example, produced a sudden crop of badly managed factories in South Australia. Employers simply moved. The events of the 1890s showed clearly that capital was intercolonial in its operation though politics were not. There was never any doubt among the constitution-makers of the 1890s that 'trade and intercourse' between the colonies must be 'absolutely free'.[29] The remoteness of 'capital', however, may have had a liberating effect on the labour movement in the smaller colonies. Labour involvement in politics was less contested in Queensland and South Australia than in New South Wales. Certainly perceptions of 'capital' were less affected by paternalism than in Victoria. The Queensland labour movement in particular brought a touch of idealism to intercolonial labour conferences.

The typical member of the New South Wales parliament entered politics in his prime (in the 1860s and 1870s this meant his late twenties or early thirties) but stayed only about five years. In South Australia, six years was the average length of a parliamentary career.[30] This suggests that the effort and cost of looking after an electorate were not worth the rewards of prestige or power. Few politicians died on the job. It is likely that until members were paid for their time, most regarded politics as a duty or a mark of esteem which once fulfilled might be surrendered to someone else. The disadvantages of this otherwise useful sharing of power and experience were that there was little continuity from parliament to parliament. Only a handful of men built up experience and expertise in the art of government and administration. They became the more influential with each group of neophytes. Often the experts were lawyers whose training and work outside parliament was closely akin to the work in parliament, though Henry Parkes and Alfred Deakin found journalism an equally sympathetic profession. Distance also effectively disadvantaged many members. Unless they could afford to be absent from their businesses for the whole of the parliamentary session, or to maintain a second home in the city, they had to choose between two careers.

There seems to have been a handful of political families whose names were almost a guarantee of a place in parliament—the Dangars, Stephens, Suttors, Windeyers in New South Wales, the Thornes, Bells, Weinholts, Archers in Queensland, and the 'first families' of Western Australia. More commonly, however, ambition, energy, and versatility rather than heredity, culture, or leisure were the qualities necessary for a successful career in colonial politics. S. W. Griffith tried to read and minute every sheet of paper that passed through his department as premier. At the same time he appeared in court as a barrister, both for and against the government.[31] John Hirst has written of businessman John Hart who introduced the title 'premier' into South Australia when he filled that role in 1870–71:

It is difficult to visualize Hart as a fervent land reformer as he moves each day after a morning's work at the Treasury to the bank to discharge his duties as a director, and then to the Adelaide Club for lunch before reappearing in his public role again at two o'clock as leader of the House of Assembly.[32]

For a lawyer, politics could be good publicity; for a businessman, his politics might be a natural extension of his entrepreneurial activities, as in the case of Melbourne land-boomers James Munro or Thomas Bent, or Queensland developer William McIlwraith; for most men who still had a living to earn however, politics was an opportunity or a luxury they could afford only briefly. Payment of members was perceived as a way of improving the quality of representation, of lessening inequalities between the few who became professional politicians and the many amateurs. It also helped to discourage dubious practices by which men who were without means but nonetheless effective politicians kept their seats. Thus, much of the manoeuvering of colonial ministries in the 1860s and 1870s which may be read as a sophisticated tale of conflict and allegiance over policy and position, could have an equally valid but simpler economic explanation. Speakers and cabinet ministers were usually provided with salaries so that they could devote themselves properly to their responsibilities. Yet these were hardly steady sources of income, given the frequency with which cabinets were rearranged. At times, the need to hang on to a paid post outweighed principle and policy. The amazing flexibility of Henry Parkes's political position was frequently related to his closeness to bankruptcy.[33] So probably was the reputation of Arthur Macalister, Queensland's 'Slippery Mac' of the 1860s and 70s. Duncan Gillies managed to survive the best part of twenty years in Victorian politics without an alternative source of income (admittedly without a wife or family to support), though this gained him a reputation for 'pliancy'.[34] Between 1883 and 1886 Edmund Barton calculated that he earned more than half his annual income from the speakership. This was worth £1500 in 1884 compared with £634 8s from professional fees—yet Barton was said to be 'indolent' as a lawyer.[35] (Barristers were said to be able to earn £3000 a year or more.[36]) Though the ideal of the 'independent' member was still held dear, in truth most members were dependent on a group of loyal constituency supporters, on business associates who carried on in their name, or later, a union or labour league.

In the lower houses 'independent' members gathered in what were generally known as 'factions', though there were some more colloquial names for these groups, such as the various 'bunches', the 'Deakinites'. The basis around which a faction formed might be a simple thing like old friendships, loyalties, obligations, shared interests or values. It might be, as some of the 'bunches' were, a regional grouping, related to some shared local need for a railway extension, a bridge, harbour facilities. Before long these fairly spontaneous factions were incorporated into a sophisticated network of debts and alliances. Parliamentary majorities were constructed from carefully managed groups of factions. The man who could command the greater number of votes through his own faction and the support of other faction leaders became the head of government. Most often his cabinet was formed to include the heads of those other factions. Cabinet re-shuffles were as often as not aimed at reorganizing the overall voting strength of the government by realigning factions represented by their leaders in the ministry.

This system enhanced the appearance of independence for the individual member. In theory he could bring down a delicately balanced government by deflecting his vote. In practice, much political energy and ingenuity was devoted to keeping such crises at bay, though they could not be staved off indefinitely. Three (or five) years had been set as the maximum time between elections, but elections occurred at more frequent intervals, and government re-shuffles which themselves required ministerial elections (the practice whereby a potential holder of an office of profit had to submit himself to his electors once more to have his original election confirmed) were necessary sometimes more than once during a parliamentary session.

Such furious activity at least helped to sustain interest and a sense of opposition, though most groups were broadly in agreement on ends. Crises were seized upon by the press. Public meetings were frequent and well attended. There was much 'superficial violence' or name-calling and general rowdyism at political meetings, often as a substitute for real knowledge or careful analysis, but also in recognition that those present had the right to be heard.[37] Michael Davitt was shocked by what he thought to be quite inappropriate and unnecessary vehemence and vituperation in the politics of the otherwise admirably self-governing labour colonies on the Murray,[38] but George Black could write nostalgically of the organized interruptions at political meetings in New South Wales in 'the roaring days' of the labour movement. Edmund Barton kept an old coat specially to wear for political meetings; he called it his 'ova coat' because it was so stained with egg yolk.[39] Sectarian questions, class relations, motives, personal and political, all were subject to creative insult, leading logically to the high alliterative inuendoes of John Norton's *Truth*.[40] Much of the racism of this period seems more offensive than it may really have been simply because we are no longer accustomed to the verbal violence which was ordinary political discourse then, though we expect it on questions which nineteenth-century politicians dared not mention. Thus it was said that politics were too rowdy for women, or conversely, that politics needed the civilizing influence of women. In fact, the familiar rough style of Australian politics with its mixture of self-confidence, arrogance, aggression, insensitivity, and ignorance was already established. Politeness and sensitivity were unnecessary when electors and their representatives regarded each other as equals, spoke the same language, shared values and aspirations.

From the late 1850s to the early 1890s, ministries were mainly short-lived. Twenty-five months in Queensland, eleven in South Australia, and seventeen and eighteen months in New South Wales and Victoria was the average duration. In all colonies from 1893–94 however, the ministry survived two full terms, enduring till the end of the century.[41] The search for economic security had something to do with this, the need for co-operation rather than competition, for consensus and greater unity. With the growing size and scope of governments, something like a party system emerged to regulate the complexity and fluidity of the factions. Members decided to act in concert to achieve short-term aims in alliances between factions. Then they pledged themselves to a permanent faction for the sake of long-term goals, as in the young labour parties of New South Wales, Queensland, and South Australia. But, as many a reluctant labour representative discovered, the power his pledge gave the caucus did not necessarily compensate him for the power he lost as an 'independent' member. There were many defections before party solidarity could be shown to be the powerful weapon it later became. Meanwhile the new labour parties contributed to the stability and credibility of colonial governments.[42]

Many observers distrusted the growth of party politics. David Syme thought the political party an abrogation of democracy. Ideally, representatives really represented the views of their electors, consulted them frequently, and were answerable entirely. Party politics reduced representation to a class or sectional interest. The extent to which a member represented his electors was accidental or a technicality. Underlying the federal Constitution was an expectation that regional alliances would continue to be more important than party politics, even though 'state rights' were intrinsically less democratic than nation-wide parties. Only much later did it become apparent that there were also groups whom not even the parties actively or effectively represented—the 'minorities' of the mid-twentieth century, women, for example, and various ethnic communities.

Andrew Dawson was born in Rockhampton in 1863. When both his parents died he was placed in a Brisbane orphanage. He lived for a while with an uncle in Gympie, then, like

MR. DAWSON'S MINISTRY LEAVING GOVERNMENT HOUSE GROUNDS
AFTER THE SWEARING-IN

Sarah Campion's fictional hero 'Mo Burdekin', who may have been based on Dawson, he went to Charters Towers, working at anything thrown up by the goldfields town.[43] In 1886 he was in the Kimberleys, but the next year he returned to Charters Towers to marriage, journalism, and to serve as president of the local Miners' Union. He changed his name to Anderson (his father's name) and was elected in 1893, along with John Dunsford, his co-proprietor of the Charters Towers *Eagle*, as a labour member for the town. In late November 1899, the Governor of Queensland, Lord Lamington, asked him, as the apparent leading opposition member, to form a government to replace that of (Sir) James Dickson who had resigned. Dawson's short-lived ministry—it lasted only a week as he really did not have enough support to command a majority of votes in the house—is interesting as the first labour government to take office anywhere. Of the seven, two were Australian-born, two were English, two Scots (Andrew Fisher and William Kidston) and one from Dayton, Ohio (H. F. Hardacre). Four were union officials, and like Dawson had worked at a variety of occupations including mining and butchering. Joseph Turley had been a wharf labourer, Kidston a bookseller. C. B. Fitzgerald, the attorney-general, was necessarily a lawyer. He was also the only Catholic among three Anglicans, two Presbyterians, and one Congregationalist. Catholics and men of Irish backgrounds were slightly better represented in the parliamentary party from which this cabinet was drawn. So too were immigrants from elsewhere in Australia, mainly Victoria.[44] Just as a little later, labour organization in Western Australia was to draw considerably on the talents and experiences of men who moved with the mining frontier, so Queensland in the 1890s was the gathering place for restless, ambitious, idealistic types. The chance to move, to another place, to another job, to some new ideas about the meaning of life—these freedoms were essential to the labour movement's view of the good society, though it was not obvious how they related to steadiness, security or the more mundane home and family preoccupations of the majority of the population.

The labour members in New South Wales after the first election in which they were successful (1891), were a 'coalition of disparate elements, often linked more by hostility to the forces of Australian conservatism than by clear bonds of common policy'.[45] Fifteen came from Sydney electorates, twenty from outside Sydney. They certainly had more entrenched conservative forces to contend with than their fellows in Queensland. The New South Wales group was also different in that it contained a higher proportion of immigrants within its 'band of unhappy amateurs . . . made up somewhat as follows: several miners, three or four printers, a boilermaker, three sailors, a carrier, a few shearers, a tailor, and—with bated breath—a mine-owner, a squatter, and an MD.'[46] In South Australia, the only other substantial group of labour members in an Australian parliament before 1900 contained no shearers and few miners. Englishmen and tradesmen were well represented, and the railways contributed further to the sense of 'industry'.[47]

La Nauze calculated that on the basis of their representation in the colonial assemblies by the end of the century, there should have been at least six labour members of the 1897 Federal Convention in addition to W. A. Trenwith from the Victorian Trades and Labour Council. He speculated that a labour bloc would have encouraged a more democratic constitution and been less preoccupied with states' rights than the other delegates.[48] But as L. F. Crisp wrote, 'the Federation issue was being rushed to decision just ten or fifteen years too soon for Labour'.[49] Though power in the lower houses had become a possibility, the labour movement was only beginning to understand the many ways in which an upper house could frustrate that power.

Australian democracy was admired for its achievements—adult male suffrage, the secret ballot, payment for members of parliament, the election of workingmen, the appearance of labour parties, even the granting of votes to some women. Although the colonies surpassed Westminster in the development of democracy, Westminster remained the model. The quality of oratory in Australian parliaments was unfavourably compared with Westminster, for example. Some of the strengths of colonial democracy—intercolonial rivalry, experimentalism, provincialism, fluidity—were seen as weaknesses, or became weaknesses in the larger entity created by federation. Political experiments, feasible with six small and comparable systems, became unwieldy with federation. In order to emulate Westminster it was necessary for Australian politics to become more like British politics, more cautious, more conservative, more class conscious. To add a new level of government was to create a new hierarchy of parliament, courts, and bureaucracy. Labour parties, essential in class-bound British society, grew easily in the freer atmosphere of the colonies. They also fostered a heightened sense of class consciousness and a weakness for paternalism in colonial society.

There has been a tendency to see New South Wales as the model of colonial political development, with the other colonies merely as variants.[50] In fact, a different political style emerged in each colony according to the reforming ideas most popular during its period of most intensive development. Thus in New South Wales, the values of the would-be old landed aristocracy were pitted against the idealism of Edmund Burke and the radicalism of Cobden and Bright. Victoria's post-gold-rush expansion was already improving the liberalism of J. S. Mill in the 1860s. The expansion of the Queensland economy in the 1880s and 1890s took place in a climate of ideas influenced by Bellamy, Hyndman, and Marx. At the same time, the sequence of political fashions suited some colonies more than others. Free selection as devised in New South Wales was inappropriate for Queensland, much as they tried to have it, and irrelevant in South Australia. The trade union movement as it arose in New South Wales, the colony which was most like Britain in entrenched ownership of land and capital, was timely and necessary, but in Victoria, avant-garde liberalism (that is, colonial liberalism) had already taken the logical step of concerning itself with the liberty of the workman as well as of his employer to make a decent income. Unionism in Queensland may have 'lagged behind that of the other Australian colonies', but this probably freed the labour movement to apply the latest ideas from England and America more aggressively in politics.[51] Intercolonial differences were played down for the sake of federation, though the structure of the Senate remains a powerful reminder of former rivalries. Queensland was unwelcome while the use of Pacific Island labourers continued. For New Zealand, despite economic, social and political development in parallel with the six mainland colonies, the accommodation required for federation was simply too great.

Colonial politics could be uninhibited. The responsibility and expense of real independence need not be taken seriously. In Britain, the fortunes of the political parties in the later stages of the nineteenth century were shaped by problems of imperial policy completely outside the scope of colonial constitutions. The Empire, Disraeli decided, was the perfect vehicle with which to renew conservatism, threatened as it was by the rising demand for democracy. Gladstone found in the Irish quest for home rule a perfect and conveniently distanced image of liberalism. In the Australian colonies, however, the Empire was a fact of life, sometimes useful, sometimes not. Colonial conservatives were reduced to self-serving assertions of the privileges of property, while colonial liberals without a high imperial ideal as their duty frequently appeared dangerously self-interested.

'We gave to our Australian colonies a noble dowry,' wrote C. A. Dilke, 'in handing over to them all their lands.'[52] The land was not only a 'noble dowry' on which, as we have seen, the colonists borrowed extravagantly. Since its original owners had been dismissed as irrelevant and without compensation, the land provided an unprecedented opportunity for collective decision-making about the distribution of power and wealth within that society. As part of the dowry, Britain had bequeathed a working set of administrative arrangements for colonial lands. Three categories of land could be distinguished in 1860: land which had already been granted, sold, or otherwise alienated from the Crown; land which had been distributed as leasehold under various conditions; and the balance still under the Crown which was unexplored or unsettled. As settlement expanded the demand for land rose. So did the value of land already granted, leased, or sold. There was a corresponding shift in government policies. Land was no longer treated as an incentive for settlement, though for most of the nineteenth century the Queensland government continued to offer land as a subsidy for the fares of intending settlers. Land became a major source of revenue, either through sales, through rent for leasehold land, or as security for overseas borrowing. Once the land was no longer to be given away, its value became important, whether that value was to be realized immediately by sale, or continuously through leasing. Land values were established slowly through the operation of the market, but so much land had yet to reach the market or was coming on the market for the first time in the 1860s that a 'fair' or 'just' price was largely a matter of theory.

The relative merits of leasehold and freehold to governments—the people—and to those who would use it—also the people—provoked much discussion. From 1860 to at least the mid-1880s, those who supported the widest possible distribution of land consistent with adequate economic returns to governments argued against those who sought selective distribution to the graziers and farmers who had already demonstrated their skills and their commitment to the land by becoming successful lessees. The main argument was that effective use was more important than mere distribution. Debate was emotionally influenced by the experience of closer settlement in the United States and tenant rights movements in Ireland and Scotland. The yeoman farmer with his individualism and independence by 1860 was already established as an antithesis to wage slavery.[53] The unsettled lands of Australia seemed ideally suited to continue the American experiment with Jeffersonian democracy. The expansion of small farms would provide useful and productive employment to retain gold-rush immigrants and ensure self-sufficiency in basic foodstuffs. Land would repay dispossessed farmers from Ireland and Scotland for loss of their rightful inheritance and foster their sturdy morality, if not their peasant ways.

Land laws devised in New South Wales and Victoria and carried into operation in the early 1860s seemed novel and were highly contentious, yet the most fiercely debated aspect of the land reform programme of the 1860s—free selection before survey it was called in New South Wales—was little different in principle from the regulated squatting practised by an earlier generation. It seems likely that New South Wales farmer and grazier John Robertson, who remembered the economic and social excitement of the 1840s when the squatters went forth (himself among them) to seize the ancient Aboriginal hunting grounds for their sheep, hoped that in the 1860s a similar movement of farmers into areas underutilized by grazing would have a comparably stimulating effect. It certainly stimulated a sense of injustice among lessees who saw the best parts of their runs ruined by roads, clearing, cultivation and the erection of unsightly huts. In Victoria, perhaps because of Charles Gavan Duffy's links with the Irish Land League, the selection movement seemed a deliberate assault on the squatters and their privileges, though in fact Victoria's safeguarded selection in specially designated agricultural reserves was more orderly and controlled than free selection as implemented in New South Wales. It was thus correspondingly less effective as economic and social policy. Much energy was wasted in seizing and defending small parcels of land for political rather than productive ends.

In both Victoria and New South Wales the strongest opposition to the land reforms of the 1860s came from the legislative councils. In the upper houses the argument that this legislation was no more than a means of economic development or job creation did not disguise its equalizing potential, its undermining of privilege in the distribution of property, and the radical, if not revolutionary meaning of such slogans as 'unlock the lands'. Conflict emerged early in New South Wales where the forces of old landed conservatism had been entrenched in the nominated Legislative Council and where they had been unashamedly outspoken in their hostility to the coming of democracy. During the session of 1860–61, the Legislative Assembly went so far as to pass a Bill to make the council elective. This was, needless to say, rejected by the council itself. The council's further rejection of free selection, even though John Robertson's Bill had received specific electoral support when he had taken both the Bill and his government to the people in a dissolution in 1860, provided grounds for drastic action. Robertson sought and received support from the governor (Sir William Denison) for 'swamping' the council with twenty-one new nominees sympathetic to the principles of free selection. Still the old councillors resisted. Because they absented themselves from the chamber there was no quorum available for swearing in the new members. However, when, a few days later, the five-year appointments of existing members of the council came to an end, they were simply replaced.[54]

Thereafter the New South Wales Legislative Council remained sensitive to the power of the assembly and governor to alter its composition. In Victoria, however, an elected council was able to assert its own version of the wishes of the people long after its successful interference in the land reforms of the 1860s. It has been estimated that between 9 per cent and 11 per cent of all bills sent from the Victorian assembly to the council in the second half of the nineteenth century were rejected or allowed to lapse.[55] George Higinbotham complained in 1873 of the obstructive behaviour of the council:

> We suffer shame and humiliation in the feeling that we are called night after night to sit and discuss public measures when we know that the whole of our discussion is fruitless, and that our talk is idle and purposeless.[56]

Land reform was not the only legislation rejected by the Victorian upper house. Most bills relating to tariff protection or aimed at reforming conditions in factories, especially regulating the use of out workers (sweated labour which was largely female) were lengthily contested by the council before they reached the statute book. Anti-sweating legislation was seen by members of the council as 'interference with the right of the poor to earn their bread in their own way'.[57] The self-confidence of the self-made men of the Victorian Legislative Council succeeded also in blocking electoral reforms in that colony. Payment for members of parliament was rejected four times before a three-year trial was permitted in 1870. Abolition of plural voting was rejected six times, and the female franchise, eleven times.

Despite hostile upper houses, additional farming land was thrown open in New South Wales and Victoria by 1862. This produced a new crisis of equality in the neighbouring colonies, Queensland and South Australia. Already suffering population losses to the goldfields of the older colonies, they began to lose farmers as well to the apparently cheap and available land promoted through the selection debates. There was a great irony in this, especially in South Australia where farm land had always been available to those who were willing to pay the upset price. This price, fixed too high to attract speculators who had no intention of using the land, generally attracted earnest, even knowledgeable farmers seeking land which they judged to be good value for their money. From time to time there was agitation that land should be made available more cheaply and easily under some kind of selection arrangement as in Victoria, yet the accidental combination of South Australia's geography and Wakefieldian theory had effected closer settlement. There was little of the corruption in land administration which made closer settlement so costly elsewhere. Relations between farmers and pastoralists were fairly amicable. They were not in direct conflict for land. The pastoralists were big capitalists, often diversely engaged in business, farming and politics as well. Their status was more secure than that of squatters, graziers or pastoralists in the other colonies, and their power therefore greater.[58] By the 1870s, only the old landowners of New South Wales who had acquired early freehold grants could compare with them. Unlike other pastoral tenants and lessees who could be subjected to selection or resumption, South Australian pastoralists had paid the full price for their land. This made it difficult to accuse them of privileged access. Such security and wealth were of course cumulative and eventually produced 'the Adelaide Establishment' and an upper house and political system more conservative than any in Australia. The wisdom of maintaining Wakefieldian pricing principles in South Australian land policies was debated often. At £1 per acre, land was always available, but the prizes went to those with money. A fair market and an equitable system of distribution are quite different things. Strenuous campaigns were mounted by politicians like Ebenezer Ward and Charles Mann to reform land sale procedures, but there could be no quarrel with the legitimacy of outcomes produced by the market itself.

In Queensland it was believed that only the magic of free selection would entice and hold settlers in competition with the southern colonies. In fact, with its agricultural reserves and land order system, Queensland was, from the early 1860s, virtually paying immigrants to take up farms. As it became clear that free selection in New South Wales was not producing miracles of settlement, and when the Victorian legislation was revised almost annually to try to make it work, then the clamour for free selection died down in Queensland. In any case, gold discovered at Gympie in 1867 began working its own magic.

Despite the many experiments of the 1860s, by 1880 only about a third of the land in Australia had been alienated as freehold. It had also become clear that regardless of the rhetoric of land reformers, ownership of a small farm did not lead automatically to contentment, prosperity, and equal citizenship. On securing his freehold title after fulfilling the conditions of residence, cultivation, and instalment payments, the selector was just as likely to sell to a neighbouring farmer or grazier as he was to plough for the next season. Then he would use the money to pay off his debts before becoming an itinerant labourer again while searching for another selection 'further out'. John Shaw Neilson's father did this several times in his life, always hoping to improve his financial position. The profits went to others while the back-breaking work of clearing and settling was done by men like Neilson senior and his family who were desperately poor. Their health showed it.[59] This was their chance for independence—or so they thought. The fallacy in the theory of free selection lay in the assumption that all acres were equal and every man was a born farmer, just as every woman was a natural mother.

By the end of the 1870s it could be seen that although land was only released to the market for the benefit of small farmers, most of it passed eventually into the hands of large ones. Most criticism was directed at such practices as 'dummying' in which squatters selected land of strategic value on their existing runs in the names of family or friends, or 'peacocking' in which selectors 'picked the eyes' out of a run, for example, the waterholes and river flats, making the rest of it practically useless, and thus holding the squatter to ransom. Some consolidation of small selections or their gradual incorporation into larger mixed farms or grazing properties was inevitable if holdings were to be economically viable.

As mere lessees of Crown land, the squatters had proved a powerful interest group. The move towards more intensive land use and closer settlement during the 1860s and 1870s had the unwanted side-effect of creating large freeholdings of the best land. Would not those who became outright owners of large tracts of the best land be more powerful than the leaseholding squatters of old? A partial answer may be seen in the structure and function of the upper houses outlined above. Land conferred status, and ownership was not essential. Many squatters bankrupted themselves or became impossibly indebted to financial institutions in their determination to outwit selectors. The fact that they were retained only as managers does not seem to have affected their standing, though in the long term, consolidation of many runs into small empires managed for banks, pastoral companies, and

The Government of Western Australia

DESIRES THAT THE

FOOD SUPPLIES

SHALL BE

GROWN IN THE COLONY,

AND THEREFORE, AS AN INDUCEMENT TO THIS END, OFFERS NOT ONLY

A FREE GRANT OF 160 ACRES

TO EVERY PERSON WHO CHOOSES TO AVAIL HIMSELF OF IT,

BUT ALSO WILL ADVANCE, SUBJECT TO CERTAIN EASY CONDITIONS, AT £5 PER CENT. INTEREST PER ANNUM,

£800 for IMPROVEMENTS

ON THE LAND.

If the area which is given free is not sufficient, a FURTHER ACREAGE MAY BE HAD on payment of the nominal sum of Sixpence per Acre per Annum for 20 years, at the expiration of which term it becomes the Freehold of the Settler.

EVERYONE IS GIVEN AT LEAST ONE OPPORTUNITY IN LIFE, AND WHETHER IT IS TAKEN ADVANTAGE OF OR NOT LARGELY DEPENDS MATERIAL SUCCESS.

IS THIS MY OPPORTUNITY ?

SHALL I PASS IT BY, OR SHALL I CONSIDER THE MATTER MORE CLOSELY?

IF THE LATTER.

Further particulars, with Handbooks and Maps relating to the Colony of Western Australia and its capabilities, may be obtained at the Office of this Paper, at the Office of the Commissioner of Crown Lands, Perth, or at the Office of the Agent-General for Western Australia, 15 Victoria-street, London, S.W.

READ, MARK, LEARN, AND INWARDLY DIGEST

THE ABOVE FACTS.

individual operators like James Tyson or Sidney Kidman, reduced the number of identifiable landholders as well as their collective authority. The idea of a big landholder was more potent than the reality. Images and fears based on memories of relations between landlords and tenants in Ireland and Scotland played their part. So did ingrained deference. Though he was no longer the local MP and his property and his next clip were already the property of the bank or the pastoral company, the squatter/manager still acted as magistrate or justice of the peace. He was more likely to be the patron than the president of the local race or show committee. His home was expected to be a superior centre of civilization and culture. In much of Queensland, Western Australia, western New South Wales and South Australia, station homesteads were synonymous with settlement and they were marked thus on maps. Travellers and itinerant workers moved from one to the next. They were the only places where accommodation, stores, even fresh horses might be found. All this added greatly to the squatter's authority. Those of his workers who lived permanently on the station were beholden for board and lodging. Those who were itinerant had limited choice in the vast distances of the places they might try and those they wished to avoid.

In his study of the squatters of northern New South Wales in their capacity as justices of the peace, David Denholm found that they had subtle and far-reaching authority, especially in relation to the lowliest of the persons who came before them, the Aborigines.[60] G. C. Bolton has described the relations between Aborigines and pastoralists in north-western Western Australia late in the nineteenth century as 'quasi-feudal'.[61] Terms like 'gentry' or 'squirarchy' have also been used to describe the position of the squatters. In part the range of terms reflects regional difference. In the Western District of Victoria, South Australia, the south of Western Australia, and in southern New South Wales, the squatter was treated as a squire. Despite disgust with the extreme conservatism of the Victorian Legislative Council, there was often no opposition to the nomination, and therefore automatic election, of the local landowner at council elections. The less cultivated or cruder version of the squatter to be found in the harsher parts of New South Wales and southern and central Queensland might be considered 'gentry'. Conflict between these squatters and the AWU in the 1890s was touched with resentment of a kind predating industrialization. Living conditions for the workmen, though considered temporary, were both rough and lacking basic dignity. Sidney Webb, an expert on working conditions in Britain, was surprised by the primitive conditions he observed in the pastoral industry.

The accommodation leaves much to be desired. Even the shearers sleep, eat and cook in the same room—a long shed, with the fireplace at one end, a long table down the middle, long forms on each side of the table, and two tiers of bunks along the walls. Washing and sanitary accommodation was conspicuous by its absence. The hut used by the rouseabouts was a degree rougher in its arrangements.[62]

In the remote north of Australia, relations between the squatter and his employees who were mostly Aborigines, might be described as pre-feudal. Not even labour relations in the deep south of America were comparable, since they involved much more responsibility. Only in central Africa was it possible to observe anything like the lack of responsibility for and degree of exploitation of a completely powerless workforce. Perhaps fortunately, both remoteness and the relatively small scale of operations prevented this abuse of power from becoming a more important part of the national system, but its existence should not be overlooked.

Inquiries were launched in the 1880s in New South Wales, Victoria, Queensland, and South Australia into the effectiveness of the previous decades' closer settlement policies. These inquiries showed that the market was failing to distribute land equitably, but they also hinted at government dependence on land as a source of finance, as well as a means of diffusing tension or dissatisfaction with other policies. Land taxation, introduced in Victoria in 1877 and South Australia in 1884, won the approval of the many enthusiasts for Henry George's theories. The otherwise puzzling spectacle of colonial governments repossessing large areas of land at considerable expense confirmed their dependence on land as an instrument of social and economic policy. There was no serious shortage of land for small settlers. The land available in every district was not of the best quality or the most convenient for transport and water supplies, or always cheap. Idealists and land reformers still envisaged popular control or ownership of the bulk of the land, and at worst, widespread small proprietorship. This the market had not achieved. Repossessions, though expensive, seemed suitable gestures in the direction of popular ownership.

In Victoria and South Australia a series of irrigation settlements on the Goulburn and Murray rivers were planned to strengthen the position of small independent farmers working in co-operation with others to establish and maintain watering systems and pumping plant. Such plans for independent yet co-operative forms of land use became more important with deteriorating economic conditions in the 1890s. Legislation was introduced in most colonies to permit village, group, or co-operative settlement where before only individuals or families had been considered suitable. Partly a reaction to intense individualism, this was also a response to urban unemployment. For those without the initiative to join a group settlement, Colonel Goldstein's somewhat controversial labour colony at Leongatha promised farming experience and rehabilitation in the work ethic.[63]

In another gesture against land monopoly, the government of S. W. Griffith in Queensland brought in legislation which encouraged the establishment of central co-operative sugar crushing mills, thus also encouraging small sugar-cane farms as an alternative to the large plantations. The co-operative mills were a triumph for small proprietorship. No longer would small farmers be dependent on the plantation mills or be treated as mere tenants by plantation managers.[64]

At the other extreme, William Lane's New Australia experiment was the ultimate co-operative, wishing to remove itself entirely from the contamination of existing societies to a remote part of Paraguay.[65] As in the group irrigation settlements on the Murray, the settlers at Cosme found human nature less easily perfected than political theory, but unlike the Murray colonists who were beneficiaries of a sophisticated economy, the New Australia settlers found subsistence farming a distraction from the dynamics of co-operation.

The land reformers of the 1860s had been influenced in a

general way by the ideas of Edward Gibbon Wakefield, by the need to maintain a balance between the price of land and the kind of settlement that price produced. By the 1880s, priorities were changing. Karl Marx's prediction of the 1850s was proving to be uncannily accurate.

By its very nature, small holding property forms a suitable basis for an all-powerful and innumerable bureaucracy. It creates a uniform level of relationships and persons over the whole surface of the land. Hence it also permits of uniform action from a supreme centre on all points of this uniform mass. It annihilates the aristocratic intermediate grades between the mass of the people and the State power. On all sides, therefore, it calls forth the direct interference of this State power and the interposition of its immediate organs. Finally, it produces an unemployed surplus population for which there is no place either on the land or in the towns, and which accordingly reaches out for state offices as a sort of respectable alms, and provokes the creation of state posts.[66]

The bureaucracy had certainly expanded, so that in most colonies an inquiry into its size and efficiency seemed necessary before the century was out.[67] Nor was land an infinite resource. Land revenues stabilized or began to diminish. However the idea of land as a source of power and the key to independence and self-sufficiency was as strong as ever. It was given new authority by the writings of the American political economist, Henry George, especially through his *Progress and Poverty* (1879).[68]

George's exposition of land as the fundamental source of wealth in any society seemed obvious in its application to Australian circumstances. Further, his idea that a 'single tax' should be levied on the unimproved value of land suggested a way of maintaining land revenue which was both equitable and a continuing assault on inequality and the growth of monopolies. Not surprisingly, the 'single tax' on land became an important component of working-class movements in the late 1880s and 1890s in all colonies. Property was condemned alongside capital.

This was easily done in the abstractions of newspaper columns and debating societies. Most of the discussion about access to and ownership of property in land, and the role of land as a source of wealth and power to individuals or to the people as a whole, was based on land used for grazing or farming. Yet in the towns and cities, the need for housing was gradually and subtly transforming the terms of discussion. From the 1860s the political franchise was broadly based, initially because it was difficult for a man to avoid being a householder paying the requisite £10 p.a. in rent. During succeeding decades the trend was for increasing numbers of these householders to build or buy their own homes. In practice the aspirations of the land reformers of the 1860s for free selection were really worked out in the suburban subdivisions of the 1870s and 1880s. The Australian dream of the 1860s, a small farm in a fertile valley, by the 1880s was a small cottage on its own block of land not too far from a suburban railway line. And here, the market seemed to work. The original land grants surrounding the cities were steadily subdivided. Houses were built and, most importantly, sold. Torrens titles made such transactions little more complicated than buying or selling a horse and buggy. The cash flow generated by such sales allowed development and building with only modest outlays of capital. Even cooperation in the form of building societies seemed viable. Widespread home ownership was seen as wholesome by those who denounced monopoly and abhorred landlordism. The owner-occupier joined the small farmer as the ideal citizen and democrat. Land taxes were deemed undesirable on such modest holdings. Where land taxes were introduced on rural land, they frequently did not apply to small holdings, and it was with great reluctance that rates were imposed on urban housing. Business premises should be taxed, but homes were another matter.[69]

It was occasionally suggested that there need be no individual ownership of land, that the land should be held collectively and made available on life tenancies to those who would use it. These ideas were pursued most consistently in Queensland where in 1900 only about one-fortieth of the colony's land had been alienated, but here too the desire for a secure inheritance, something to pass on to the children, overcame all theory. Otherwise, it was said, people had no incentive to work, to care for property which was not really theirs. Land for housing was never a part of these discussions. It was beyond politics except at local government level where politics were about little else. Nor were differences of opinion over land policy and administration ever strong enough to give rise to separate parties advocating distinctly opposed views. Rather, the 'dowry' was energetically and enthusiastically exploited. All government intervention and management of economic policy flowed from that fact. State railways, irrigation, water supplies followed easily on the experience of managing the Crown lands. But as Henry George had pointed out, the supply of land was finite. That realization was rapidly approaching by 1900. Opening up new farms (as for returned servicemen after the First World War), or building new houses could continue as government's response to social or economic problems, but never on the same scale or with the ease experienced between 1860 and 1890.

The aim of land administration was to moderate collective and individual access to the basic source of wealth, but education promised more effective equality. In this period, education is usually seen as a symbolic battleground of sectarianism, but all attempts to make it a party matter failed.[70] There was basic agreement that it was necessary. The problem was to foster equality while maintaining choice. Education and opportunity went hand in hand. Those who came of age in the 1860s and 1870s placed immense value on education and they sought something better than they had received for their children. As this had probably been minimal or haphazard, they were not in a position to question what was provided by the authorities, be they church or state. Here was a duty most parents were pleased to relinquish, an expense they could do without. State schools had the advantage of being free (or almost). They seemed to offer progress, and maintained quality control through the inspection of teachers and examination of children. Whatever the churches might threaten for the soul and the hereafter, this was the key to survival and advancement in the immediate future.

The reformers of the 1860s were building on a tradition which already accepted the need for systematic education to remedy possibly inherent colonial moral and social weaknesses. For men like Richard Heales (1821–64), the architect of the Victorian Common Schools Bill of 1862, education was the basis of a liberal democratic society. Heales was a typical successful immigrant. He had arrived in Melbourne as a young man and established himself as a coach-builder in a small way. In addition to a general system of education, he advocated total abstinence, reform of the Legislative Council as a prelude to land reform, payment for members of parliament, and protection for industry.

Heales' Common Schools Act placed all schools receiving financial assistance from the state under the control of a single board of management. The practice of tying financial assistance to some form of supervision or quality control was adopted subsequently in all other colonies. Nineteenth-century governments were not anxious to vote large sums for activities that were not obviously 'reproductive'—that they could not envisage eventually paying for themselves. Though it was expensive and could show no direct profits, the generalized benefits of education were understood. But this made public financial accountability the more necessary, and recommended the inspection of teachers and the public examination of children. In return for sums expended, each year there was a new crop of young workers with certified levels of attainment.

As the state system became more comprehensive and more widely accepted by parents, so the need to subsidize church schools seemed less imperative, and this expense was more resented. Indeed, the need for non-specific religious education was questioned. By the 1870s 'free, secular, and compulsory' education was accepted in New South Wales, Victoria, Queensland, and South Australia in principle, though it took a little longer to implement.[71] This was a severe blow to the power and authority of the churches. Most church leaders were bewildered at the apparent ease with which they were superseded by anonymous boards of education. But they had never been well placed to resist the change. Since the churches had no private system of patronage or official representation in the structure of government in any of the Australian colonies, they had no influence other than through lay spokesmen. As institutions they were being pushed to the periphery of power. An Anglican bishop was appointed by the Queen, but he had no place in a colonial upper house or in any official capacity outside the church. His position gave him precedence in official processions (though not when the Catholic archbishop was present), but only sheer ability or personality could command respect such as Charles Perry received in Melbourne. The other churches, both Roman Catholic and Nonconformist, found it refreshing to be on equal terms with their Anglican confrères, but the harsh fact was that all churches were diminished by the absence of an established church and by the persistence of sectarianism, becoming pawns in a suburban game of status.[72]

When Cardinal Moran offered himself as a candidate for election to the Federal Convention of 1897 he came fourteenth in a list of fifty candidates for ten positions in New South Wales, with something over 40 000 votes.[73] (Catherine Helen Spence stood for election to the South Australian delegation at the same time, the first woman to stand for election to public office in Australia. She came twenty-second in a list of thirty-three, with 7383 votes).[74] Moran hoped by his candidature to break down the 'Orange bigotry' which was very intense, but the response was an even more intense outburst of anti-Irish or anti-Catholic feeling.[75] The story of Moran's candidature illustrates most succinctly the kind of powerlessness with which church leaders had to contend. Only mobilization among the laity as eventually occurred with the federal constitutional conventions, would really influence the political process.

The daily preoccupation of most people was with wages or profits. Working conditions and the money they had in their pockets or bank accounts from week to week instantly defined power and powerlessness. The organization of labour and industry affected too many people too closely to be treated as theoretically as the land question or as optimistically as education. So economic management proved more divisive than land policy or education. The aristocratic and sometimes naïve Harold Finch-Hatton was scathing of the motives of colonial politicians whom he saw as 'a succession of selfish, sordid adventurers'. 'It is absurd to distinguish the members of either party as Conservatives or Radicals, as it is to call them politicians, since the transparent motive of all of them is to plunder the colony.'[76] 'Rival syndicates' was Charles Dilke's description of the politics of New South Wales.[77] Each colony seemed to behave like a syndicate too, especially in the competition for loan money, immigrants, and rail freight.

In the 1860s the notion that any local industry should be protected by tariffs in order to encourage its growth as a reliable source of both goods and employment seemed interfering with the laws of nature. Competition, as Adam Smith taught, produced the survival of the fittest in any economic structure. Who wanted a weak economy? What was a strong economy? Was it self-sufficient, or did it rely on a strong export sector balanced by massive imports? Was it one in which full employment produced vigorous consumption and increasing opportunities for expansion and diversification? Or did it rely on the wealth of a minority and the exploitation of the rest?

While exports of wool, gold, and wheat continued to be highly profitable, and labour in short supply, the Australian colonies could afford a little of both worlds, subsidizing their rising standard of living from their most profitable exports. This was, as David Syme never tired of pointing out to the readers of the *Age*, short-sighted.

The object of industry, or that of labour by which men live, is not the greatest development of foreign trade, it is the comfort, well-being, and moral progress of the masses of each separate nationality. Under no circumstances therefore can it be the duty of any Government to give up the care of the labour, that is of the labourers, of the country.[78]

Competition favoured the strong. The essence of free trade, of the doctrine of *laissez-faire*, was not competition, but eventual dominance. In relation to the British economy and British manufactures and markets, most Australian enterprises were weak and poorly equipped to compete, except where they were protected by distance or special local requirements. Small manufacturers came quickly to favour protection. So did most of their employees. The labour movement officially adopted protection as one of its policies.

Free-trade advocates were in industries which could compete in export markets, such as wool, or were already in a dominant position in intercolonial trade, like New South Wales coal and Queensland sugar. Shippers, importers and retailers, the 'calico jimmies' whose profits lay in buying cheap in Britain and selling dear in the colonies, were naturally free traders.[79] New South Wales with its strong export base was for free trade. The argument for protection was greatly assisted in Victoria (and the other smaller colonies) by the fear of losing population. While Victoria appeared to be flourishing during the 1870s and 1880s, her tariff policies were advanced as explanation, but Sydney became increasingly a beneficiary of protective measures taken in other colonies. Actress Nellie Stewart commented in her memoirs, 'Sydney then [in the 1880s] had most of the initial productions, for, by importing the wardrobe there, a lot of Customs charges were evaded and when the worn costumes went into Victoria they were no longer dutiable.'[80] In the 1890s the need for protection became desperate as Victorian manufacturers, especially, sought to expand their markets within Australia. Under George Turner, premier 1894–99, Victoria led the push for federation as a sound business arrangement, though not without the assistance of the ambitious young natives-in-a-hurry.[81]

Once accepted in Victoria, the doctrine of protection expanded rapidly.[82] Much of the labour and industry legislation of the later decades of the nineteenth century was devised to ensure more equal access to the economic competition in which all were engaged. Regulation of shop trading hours may serve as an example. Voluntary moves towards early closing were never more than partially successful because there were always a few who could profit best by standing out against the many. This invariably drove the weakest, poorest, or least efficient shopkeepers to stay open late at night and all weekend in hope of catching the shreds of trade at those hours. This imposed a burden on their families or their employees. Their costs for lighting and labour were rarely justified by their takings. Order and reasonable conditions could not be introduced into the competitive jungle of retailing without some losses. Undercapitalized businesses, often a desperate bid for economic independence by a widow or the wife of an invalid husband, were the first to fail. The hardest task, according to the shop inspectors who policed the early closing regulations, was reforming the habits of housewives. In poorer neighbourhoods

> years of slipshod household management have built up a system of casual night trade that is difficult to reform. Here the shopkeeper and his assistants are the white slaves of housewives, whose petty purchases extend over the whole day, late into the night, and cease only with sleep.[83]

Modern wisdom admits functions other than simple economic ones fulfilled by the 'corner store'. Yet the work done to systematize trading hours was a step in the direction of a 'fair go' for all concerned. Elaborate consideration and weighting of all manner of special cases did not produce equality or even fairness, but some advantages and disadvantages were evened out. This was already a long step from a belief in the survival of the economically fit.

Regulation of wages, arbitration of industrial disputes, acceptance of special conditions for those workers who were slower, weaker, or less productive than the ideal young man in the prime of life can all be traced to a dawning perception during this period that unrestrained economic competition was not necessarily in the national interest. The immigrants of the 1850s ceased to be young men in the prime of life by the 1870s. A society which loves winners must have losers. Even the trade unions, strong though they were in comparison with unorganized workers, and vocal about their own requirements, found themselves at the mercy of employers who would not negotiate. The most powerful employers' organizations, the Steamship Owners' Association, the Newcastle Coal Owners' Mutual Protective Association, and the Pastoralists' Federal Council could afford to wait till the men were starved into submission or faded away and others were brought in.[84] Strikes only worked where there was some rough mutual dependency between employers and employees. When the employers were graziers, coal owners and shipping companies, as they were in the decisive strikes of 1890–93, there was no need for them to talk or to make any concessions. The work would keep. C. C. Kingston, the South Australian attorney-general, felt that the community should interfere to see that strikes were settled. Striking and resort to force were barbarous weapons and victory in the struggle was decided by strength and not necessarily by justice.

Kingston further argued that if the contestants would not agree to have their dispute judged by a third party they should be compelled to do so. Awards should be made in settlement of disputes and obedience to the terms of such awards should be legally binding.[85]

Now seen as a system for controlling the union movement, the industrial arbitration system devised during the economic depression of the 1890s was aimed at forcing employers, whatever their strength, to listen to the grievances of their employees. (It was not intended that it should be as legalistic as it has become. The Constitution adopted in 1901 left no choice but to develop nation-wide arbitration though the courts.[86]) Arbitration tribunals appointed by governments included worker representatives and expressly excluded lawyers. Here again the contrast between New South Wales and Victoria was marked. The worst disputes originated in New South Wales (and Queensland, by extension as the branch office). Protective legislation and a system of wages boards in Victoria already had surveillance of working conditions and wages for a large proportion of workers. As well, few Victorian employers were strong enough to survive indefinite strikes. So it was in New South Wales that the need to force employers to listen was greatest, and it was there that unionism became most necessary as a way of proving to employers that they were dealing with officially recognized groups of workers. New South Wales therefore instituted compulsory arbitration with unions an essential component of the machinery.

Our arbitration laws applied to labour are company law. When you allow capital to organise, it is subject to certain State requirements, and you submit the incorporating individuals to special legal liabilities and restrictions in return for the rights you give them; so should you do with labour if you allow it to organise. You require capital to incorporate in order to exercise certain capitalistic powers; you should require labour to incorporate in order to exercise certain collective labour powers. Every argument based on social grounds that you can advance for the one, is equally applicable to the other.[87]

At the end of the century there were approximately 80 000 unionists in New South Wales. Victoria managed much longer with various forms of mutual or supervised agreements, and as a result the union movement remained small. There were about 8000 unionists in Victoria in 1900. (Compare this also with 8000 in Queensland, 5000 in South Australia, and 12 000 in Western Australia where mining had brought strong unions and industrial legislation).[88]

Visitors to the Australian colonies in the second half of the nineteenth century almost invariably made some comment on the democratic tendency or the egalitarian tone of society. In his saga of colonial life, *The Recollections of Geoffrey Hamlyn*, first published in 1859, Henry Kingsley wrote bitterly of 'Australia, that working man's paradise'. His phrase, like its modern equivalent, 'the lucky country', became so misquoted that by 1890, William Lane could use it with heavy irony.[89] 'An Australian National Song' sung to music composed by G. W. L. Marshall Hall at federation rallies more simply defined the working *man's* paradise.

Australia, Australia, thou land of liberty
Where each man is himself a king, each home a monarchy.[90]

Writing to his sister Elizabeth on 2 January 1881, Patrick McMahon Glynn, then twenty-five, described the structure of colonial society with an incisiveness which explains his successful career in politics.

The working classes are well off, their houses cozey, with verandahs interlaced with vines, flowery gardens, their daughters pianists, their children disobedient, their food cheap (mutton 1½ and 2d. per lb.) their working hours eight, their holidays many . . . I can't see how the middle classes are better off than at home—*now*—they *were*. Too many shops are being opened, too many towns built. The Upper classes—but who are they, perhaps bankers, squatters, book makers and some doctors,—are some very rich, some not, but they are rich because they came here early. Lawyers. The old solicitors are well off, because they encroach on barristers' business. Barristers are a great many briefless, a few monopolise the business. A man of middle class ability, without a fluent tongue and having no connections here, had better stay at home. If I get on—the devil thank the colony. I can't see how I am to do so without becoming a politician; and as such I ought to succeed if I get a chance. I spoke recently at a Reform League... If I see my way clear to getting soon into Parliament through them, I will, as this league belongs to the conservative, or anti-democratic side, and consequently wealthy side of Politics... As I *can* speak... I feel confident of emerging from the rank and file.[91]

Glynn identified not only the improved condition of the working classes but also the still fluid relationship between status and opportunity in the colonies. Comparison with the more deferential and hierarchical society of England or Europe was implicit. Egalitarianism was Australia's distinguishing characteristic, a sign of social and economic progress, an element of the growing national consciousness. In political and social rhetoric, egalitarianism became the basis of democracy and the classless society.

In fact the egalitarianism of the late nineeenth century was merely a framework of economic opportunities within which individuals and groups manoeuvred to assert traditional forms of authority or to invent new ones. Those who fell outside the economic structure or were excluded by sex or colour were irrelevant to the rhetoric of egalitarianism. Even so, colonial society was more open than the British society implicit in comparisons. Contemporaries frequently noted the absence of an upper class when trying to explain this openness. Yet it may not have been the absence of that class so much as the weakness of traditional institutions, or even the absence of tradition itself, which encouraged a sense of egalitarianism.

It was also easy for contemporaries like Glynn to overlook the power of the upper class, or ruling élite, for it was largely absentee, composed of investors, land and company owners, Colonial Office bureaucrats, those members of the British Establishment who made decisions or approved decisions made in Australia. To a point these groups could be ignored and seemed of little immediate consequence to those so powerless as to be ignorant of their existence. But they had their representatives in the colonies—the governors, the managers, those employees whose loyalties and values were firmly located in the hierarchies of 'home'. Equality-loving colonials might mock at vice-regal ceremonies and expectations of deference. Others were only too willing to accept invitations to balls and receptions, drop their curtsies and carry off their KCMGs. Though George Higinbotham refused a knighthood because he wanted nothing to do with imperial honours, other politicians were proud to be represented in *Burke's Colonial Gentry* complete with slightly improbable coats-of-arms.[92] Of the fifteen delegates representing the Australian colonies at Queen Victoria's Golden Jubilee ceremonies in London in 1887, nine were already 'Sir'. Mortimer Franklyn (an American) thought the fondness he perceived for formal observation of status and procedure, especially in Masonic lodges, friendly societies and trades unions simply typical of English-speaking people.[93] Michael Davitt could not understand why 'this pitiable hunger for handles to names has afflicted the Sir Lancelot squatters, the Sir Godefroy fatmen, and the others who chance to become the premiers of colonies or mayors of Melbourne or Sydney'. His explanation was too easy—'Cherchez la femme', although he may not have been entirely wrong.[94]

Rosa Praed in various novels described the status-conscious débutantes she observed as the daughter of a widowed Queensland cabinet minister in the 1860s and 1870s.[95] In the segregated world of colonial politics, wives, as J. A. Froude remarked of Lady Loftus, worked as hard if not harder than their husbands to secure social advantages.[96] Audrey Tennyson's condescending descriptions for her mother's amusement at the presumptions of successful colonials, especially their wives, illustrates and explains something of the female preoccupation with status.[97] Men earned their rewards in society through their money or their politics. For them, status was synonymous with wealth or power. Defending their rough ways as democratic, the author of an 1888 article on 'Nouveaux Riches' remarked, 'Australia is a country that owes so much to her self-made men, that the British idea of despising them, ought not to prevail here. . . .'[98] Women, on the other hand, earned their rewards through their husbands or the marriages of their children. 'The unpleasant *nouveaux riches* are oftenest women, who are apt to take credit to themselves of the faculty of their husbands or fathers for money-making.'[99] The most 'ladylike', that is acceptable, colonial wives were those who stayed quietly at home, having nothing to do with either society or politics. Women's traditional exercise of power had been in 'hatching, matching, and dispatching'. In the Australian colonies in this period, their importance as producers of the next generation was acknowledged, though as befitted developmental attitudes, appreciated more for its quantity than its quality. The modern, egalitarian society had largely democratized courtship or consigned it to the market, so that women's traditional role in this area became redundant and appeared frivolous. Only death remained as a reluctant and residual area of responsibility in a society which could boast of its longevity and afford the services of professionals such as doctors, nurses, and hospitals. Women's work in the new society was to be 'civilized', to instil standards of behaviour in the young, to maintain social distinctions on a broader front, to exercise the compassionate and human touch. The egalitarian society had no place for them. They were patently not equal to men. And whereas men could pretend equality in that they were all dignified as citizens or voters, the common denominator for women was service (not reproduction—reproduction was limited to married women, but housework was universal for women from age five to age ninety-five).

Men did not touch their caps or pull their forelocks in Australia. There were, to Anthony Trollope's regret, no harvest homes over which squires presided.[100] Clergymen were as likely to be the objects of their parishioners' charity in Australia, Ada Cambridge discovered, as the reverse.[101] Yet this absence of deferential behaviour, which seemed to characterize Australian attitudes, was by no means a conscious development. There was plenty of deference in nineteenth-century Australia. It was the deference of wives to husbands, daughters to their fathers, maidservants to their masters. The serving class was almost entirely female. Deference therefore was subsumed into the traditional duties and responsibilities of women towards men, seen not for what it was, but as a natural attribute of femininity.

In a society without tradition, tradition could have a beguiling face. The traditional power attributed to broad acres has already been noticed, but among urban professionals, the natural way to measure achievement was against traditional standards. The desire to preserve tradition for its stabilizing effects was certainly present in daily matters such as style of dress, eating habits, interior decoration. Government House was not only the symbolic apex of government. It was the source of authority in matters of style and behaviour. Governors and their wives were a real link with 'home', representing the 'best circles', or reflecting the fashionable world of London. The Government House people usually accepted their role as leaders of society, whether it was being firm with colonial politicians, chaperoning their motherless daughters, bestowing honours, setting a good example with charities for unmarried mothers and badly fed babies, or providing jolly encouragement for troops setting off to serve the Empire. Governors were required to observe the forms and maintain a proper distance in a friendly fashion. Then, as Lady Stephen, wife of Sir Alfred who was several times acting-governor of New South Wales, observed, life at Government House could be dreadfully boring.[102]

Immigration blurred the significance of their origins for many settlers, and sharpened their interest in material success. Some hoped to crown their colonial achievement by re-entry into British society at a higher level than they had left. Others became passionately opposed to the British class system. The few who sought to enter British politics found themselves perpetually identified as 'colonial'. More effective were the retired bankers, journalists, and wealthy businessmen who constituted a semi-professional colonial lobby and information network in London.

Far from creating a new society in a new world, successive generations of immigrants found themselves required to choose between two kinds of authority legitimated by two powerful traditions. The older (and stronger) was that derived from Britain—the rule of law reinforced by landowning interests and moderated by freedom of the press and the right to public assembly. People who identified with this tradition increasingly gravitated to the Government House view of Australian society. The newer (and increasingly vigorous) Australian national tradition emphasized equality, opportunity and independence. It was easily conflated with the transplanted anti-establishment attitudes of the English and Irish working class, but it lacked the memory of a long struggle for survival. Nor did it need to assert its right to political expression after the 1850s. So it lacked also the cohesion and the more thorough-going class consciousness of the Old World and the discipline of material and moral adversity.

Native-born Australians had three choices. They could be captivated by a rosily perceived idea of 'home', reject its blackened version, or assert their own special egalitarian identity. By the 1890s, the native born had become the group most scrutinized for their social origins, identified with them, and trapped by them. They had risen above their lowly 'currency' status by their enthusiastic espousal of nationalism and by their pursuit of Australian 'traditions' of the kind, for example, that Russel Ward has traced back to convict times.[103] Still, they were invariably stamped as colonial. The more assertive their Australianness, the more colonial they seemed. Since they could not aspire to status in the imperial context, status and patronage in the colonies became increasingly important. As the native-born generations assumed prominence in the professions and in government, loose arrangements based on an assumption that the colonies were merely an extension of home ceased to be acceptable.

The chief difficulty in describing class or status in the late nineteenth century is that meanings have altered so much. Our understanding of class or status now inevitably includes theories which were in the course of development during this period. Nor did the theories necessarily have Australian conditions in mind. Neither press nor politicians in late nineteenth-century Australia had any qualms about using the word 'class' frequently in pejorative references to the policies of their opponents.[104] They were as yet uninhibited by the writings of Karl Marx and unaware that they should be more carefully defining their terms. So what they said might mean anything. It was often meant to mean as much as possible. The fact that it was possible to condemn legislation as emanating from one class or designed to benefit only one class suggests that listeners were also accustomed to thinking of themselves as belonging to one class or not belonging to others.

Class was also a product of the nineteenth century's preoccupation with scientific ordering, with 'classification'. The division of populations and occupations into classes ordered according to a notion of income or contribution to the national wealth placed some people at the top of the scale and others at the bottom just as effectively as the most blatant 'class legislation'. Out of such classification came the justification for real discrimination between social groups. Furthermore, bureaucratic classifications made it easier to treat the members of each designated group uniformly. This may have produced an appearance of order and equality within categories, but it also produced unfairness between categories, for those marginally excluded were relegated to the common denominator of a lower category. In a sense, the real architects of class consciousness were the early census takers and statisticians, the bureaucrats whose definitions provided both a scale for measuring success and a series of hurdles which became barriers as well. The nature of opportunity changed as well during the second half of the nineteenth century. The economy became less flexible, and the institutional framework of society was fixed.

Writing to Friedrich Engels on 8 October 1858, Karl Marx predicted a powerful role for an Australian bourgeoisie:

The specific task of bourgeois society is the establishment of the world market, at least in its outlines, and of production resting on its basis. As the world is round, this seems to have been completed by the colonization of California and Australia and the opening up of China and Japan. The difficult question for us is this: on the Continent the revolution is imminent and will also immediately assume a socialist character. Is it not bound to be crushed in this little corner, considering in a far greater territory the movement of bourgeois society is still in the ascendant?[105]

Though he may have been quite right about California, it seems unlikely that the Australian bourgeoisie has contributed as Marx predicted to the demise of the socialist revolution in Europe.

In New South Wales, the oldest, largest and most stable of the colonies, the census in 1861 categorized the occupations of the people as shown in table 5.1. More than half the population (58.51 per cent) consisted of women and children at school or under school age (i.e. occupations not stated). About 30 per cent can be seen to be employees or small business people. The 'middle classes' hardly make an impressive showing, even if their wives and children are included.

Table 5.1: Census listing of occupations, 1861

Class	Males	Females	Total	Per cent of total population
1 Government Service—				
Naval and Military	532	—	532	0.15
Civil Officers and subordinates	764	—	1 752	0.50
Police, including Police Magistrates	988	—		
2 Learned Professions—				
Legal	432	—		
Medical	534	—	1 293	0.37
Clerical	327	—		
3 Educated Professions—				
Professors and Teachers	942	1 128	2 584	0.73
Other Educated Professions	484	30		
4 Scholars under tuition—				
At Home	3 748	4 277	45 953	13.10
At School	19 477	18 451		
5 Trade and Commerce	7 325	1 135	8 460	2.41
6 Providers of Food, Drinks, and Accommodation—				
Producers	1 987	183		
Distributors	2 377	378	7 325	2.10
Providers with accommodation	1 697	730		
7 Skilled Workmen and Artificers—				
In the superior arts	1 129	—		
In the metals	3 097	—		
In wood	5 550	—	18 454	5.26
In stone and earth	2 871	—		
In leather and skins	3 502	—		
In other materials	2 305	—		
8 Mining—				
In the precious metals	20 365	—		
In the inferior metals	38	—	21 382	6.10
In coal	979	—		

Table 5.1 (*cont.*)

9	Agriculture— Proprietors employing men Tenant farmers ditto Hired farm servants	4 503 8 862 18 337	1 167 2 157 2 342	37 368	10.65
10	Pastoral— Proprietors of sheep and cattle farms Lessees and licensees Shepherds, stockmen, and other servants	456 2 282 10 838	16 157 1 058	14 507	4.13
11	Horticulture and Vinedressing— Proprietors and masters Hired servants	399 1 539	51 38	2 027	0.57
12	Unskilled workmen	13 047	—	13 047	3.71
13	Domestic Duties— Persons not hired Hired servants	2 150 4 481	63 638 13 189	83 458	23.78
14	Seafaring Persons	3 141	—	3 141	0.90
15	Paupers and persons receiving public support	1 134	879	2 013	0.57
16	Miscellaneous Occupations	6 598	5 047	11 645	3.32
17	Occupations not stated	39 571	36 321	75 892	21.63

Source: *NSWVP*, 1862, vol. III, Census Report 1861, p. 11.

By the end of the century the census-takers were even more anxious to classify occupations according to whether they were producing wealth or consuming it, and it is not evident who was earning disproportionately in each category, or what proportions might have been self-employed.

Both these classifications were based on a hierarchy of occupations arranged according to exercise of power rather than income, and therefore undoubtedly applicable to a society derived from the British class system, but not so obviously suited to a society devoted to the pursuit of material

Table 5.2: Census listing of occupations, 1901

Class	Occupation	Persons	Males	Females
I	Professional—Embracing all persons, not otherwise classed, mainly engaged in the Government and defence of the country, and in satisfying the intellectual, moral, and social wants of the inhabitants.	38 065	25 035	13 030
II	Domestic—Embracing all persons engaged in the supply of board and lodging, and in rendering personal services for which remuneration is usually paid.	68 825	20 965	47 860
III	Commercial—Embracing all persons directly connected with the hire, sale, transfer, distribution, storage, and security of property and materials, and with the transport of persons or goods, or engaged in effecting communication.	103 185	96 500	6 685
IV	Industrial—Embracing all persons not otherwise classed, who are principally engaged in various works of utility, or in specialities connected with the manufacture, construction, modification, or alteration of materials so as to render them more available for the various uses of man, but excluding, as far as possible, all who are mainly or solely in the service of commercial interchange.	167 890	145 600	22 290
V	Agricultural, pastoral, mineral, and other primary producers—Embracing all persons mainly engaged in the cultivation or acquisition of food products, and in obtaining other raw materials from natural sources.	175 330	160 150	15 180
VI	Indefinite—Embracing all persons who derive incomes from services rendered, but the direction of which services cannot be exactly determined.	12 745	5 685	7 060
VII	Dependents—Embracing all persons dependent upon relatives or natural guardians, including wives, children, and relatives not otherwise engaged in pursuits for which remuneration is paid, and all persons depending upon private charity, or whose support is a burthen on the public revenue.	798 550	265 060	533 490

Source: *W and P*, 1900–01, p. 704.

wealth. According to R. A. Gollan, in 1891 approximately 76 per cent of the people were employees, 14 per cent engaged on their own account—that is, self-employed—and the remaining 10 per cent employers (it is not clear where women and children fitted into this classification).[106] No classification, however, can measure the extent of opportunity, real or perceived. Yet nothing was more important in late nineteenth-century Australia. Most fields were still open to talent, or energy, or education. Money and/or education could overcome traditional barriers of class or status. His genteel English relatives might feel embarrassed by their Uncle Piper, but they needed his money even though he had earned it as a butcher to buy their way into respectable society in the colonies.[107] The old class-based standards of behaviour were well supplied by the flow of immigrants already well-educated in the proprieties, especially into professional positions. As Sir G. F. Bowen wrote to the English solicitor-general in 1862 seeking a suitable chief justice for Queensland,

What we want is a *gentleman* rather than a mere lawyer. The principal officials in a Colony form a sort of social aristocracy. The Chief Justice ranks next to the Governor, and his influence extends far beyond the sphere of his judicial duties—indeed, whether for good or for evil, it is immense.[108]

One of the aims of a new civil service act in South Australia in 1874 was 'to treat the Civil Servants as gentlemen'.[109] Behaviour mattered in the best circles. Henry Parkes's second wife was not invited to functions at Government House because she was known to have been his mistress before his first wife's death.[110] John Douglas, an otherwise capable politician and administrator could not be appointed president of the Queensland Legislative Council because in that capacity he might be called upon to deputize for the governor. The fact that he had married his servant after his wife died was enough to rule him out of polite society, and banish him to a thoroughly safe distance administering Thursday Island.[111] Had Parkes or Douglas also been wealthy, their behaviour might have mattered less. Even in the professions, wealth became an indicator of success. H. S. Chapman, author of the Victorian secret ballot in 1856 and correspondent of J. S. Mill, had worked in many capacities as a clerk and spare-time writer before qualifying as a lawyer and marrying at the age of thirty-nine. He accepted legal appointments in New Zealand, Tasmania, and Victoria in an attempt to earn enough to support his family. Though his political principles sometimes got in the way of his legal advancement, by 1860 he was established in a large house in St Kilda and earning £5000 a year, a judge, member of the Legislative Assembly, and lecturer in law at the University of Melbourne.[112] James Balfour began his career in the colonies on £200 a year as a clerk. During the land boom his firm was converted into a limited liability company with an authorized capital of £250000, nor did his subsequent failure tarnish his reputation as a man of financial ability.[113] H. B. Higgins was only a lad when he arrived in Victoria from Ireland with his family. He managed to support himself by teaching while he qualified as a lawyer at the University of Melbourne. By the 1890s he was earning in the vicinity of £5000 a year as a leading equity barrister.[114] Graeme Davison has shown how in part the growth of 'Marvellous Melbourne' was due to the expansion of professional and managerial activity from the 1860s. Not only lawyers and accountants, doctors and dentists, but importers and exporters, bankers, agents, and bureaucrats provided services as intermediaries between rural producers and their markets on the other side of the world.[115] At the same time an expanding and financially comfortable society required and could afford more of such professional and managerial goods and services. A similar expansion of these activities took place, though on a smaller scale, in each of the colonial cities which served as port, market, or seat of government and administration. Here, among the managers, the professionals, the public servants, were the people most likely to hand on advantage to their children.

There was as yet no need to limit the numbers entering the skilled professions, though the power their exclusivity bestowed was already understood, if only in terms of the fees they could command. In 1863 there were 44 barristers in New South Wales. There were 65 a decade later, and 140 in 1900.[116] When S. W. Griffith was admitted to the Queensland Bar in 1867 he became the twenty-sixth practising barrister, five of whom had already been members of parliament. As a whole the Queensland legal profession numbered about a hundred.[117] Skilled tradesmen similarly capitalized on their scarcity and experience to create and sustain a place in the structure of power. Master and servant relations underwent a dramatic change.[118] General prosperity before the 1890s was partly responsible, but so too was the shortage of skilled or qualified men. The capable engineer or carpenter soon ceased to be a servant and became his own master. As the report of a Royal Commission into Accounts and Departments—Southern Side in Tasmania in 1863 wisely observed,

It is indispensible that some inducements should be offered sufficient to weight against the greater and more rapid gains of the Professions or Commercial and other speculative pursuits. These may be looked for in the importance of the Civil Service, its fixity of tenure, and the provision which it has always been the policy of wise Governments to make for its old and superannuated servants.[119]

Not all talent prospered straightforwardly. Constance Jane Ellis grew up in a gentle, literary environment in London where her grandfather was editor of the *Kentish Independent*. She came to the colonies, ostensibly to join her brother, but in fact to seek her own adventure, and she found it as a 'lady companion' on Kyabra station near Quilpie in south-west Queensland. There in 1889 she married the storekeeper, another escape from her ladylike upbringing. For the next ten years the Ellises lived in tents between one casual bush job and the next, settling at last in Mt Morgan, then in Brisane so that their six children could be educated.[120] The considerable contribution of the Ellis children to Australian intellectual and political life in the first half of the twentieth century suggests a complicated interaction between education, opportunity and initiative in late nineteenth-century Australia, one which is only occasionally glimpsed in references to a mother's influences on her children. It also shows some of the ways in which the idea of the open society was kept alive. Social mobility was a real possibility, at least within those groups who were not excluded from the workforce or education system. Yet opportunity was created and maintained by various forms of exclusion. Furthermore, exclusion gave a deceptive impression of solidaridity and equality to the membership of most professions and organizations. Both democracy and equality were illusions resting on narrow definitions. They would require more strict interpretation or greater self-deception as the population grew larger and more complex.

When women were first admitted to the University of Adelaide in 1874, special imperial sanction was required for such a dramatic departure from British practice. Though women qualified as doctors in South Australia, they were not accepted by their professional association, the local branch of the British Medical Association.[121] In such small ways was the power of Britain felt in the colonies. And the Empire grew in significance in the second half of the nineteenth century, easily matching the rise of colonial aspirations for independence. The old queen aged and grew more marvellous, revered for sheer survival. India was secured. The conquest of Africa proceeded with violence. The Pacific remained a glittering possibility.

Some of the Australian colonies' oldest links with Britain were maintained through India, especially in trade, communications, and through the transfer of imperial administrative personnel. The strong black tea in the bushman's billy was of Indian or Singalese origin. The breeding of 'Walers' for use in India, often superintended by old India hands, was an important industry in the Hunter Valley, parts of the Western District of Victoria and in Western Australia. The squatter lounging on his verandah, the workman in his dungarees, the housewife stitching at her chintz, cambric, or muslin were indebted to India, while retired Indian public servants, merchants, and planters in Australia demanded chutney to enliven their mutton and believed that curry powder in soups and stews both disguised a multitude of culinary sins and was eminently suited to the climate.

The opening of the Suez Canal in 1867 and the symbolic elevation of Queen Victoria as Empress of India in 1887 brought India even more sharply into focus in the south.

India was the real edge of Europe, the last relay station on the overland telegraph from London, the last port of call before landfall in Australia. Between Galle in Ceylon and Fremantle, Albany or Adelaide was the longest leg of the voyage out of touch with the telegraph system. After 1867, ships using the Suez Canal called at Galle, not only for coal, but to pick up the latest news from Europe for transmission to Australia. Even after completion of the East India undersea cable and the Australian overland telegraph from Adelaide to Darwin in 1872, India was a crucial link in the communications system, especially for maintenance of the undersea cable.

The voyage via India produced some interesting acquaintances for the travelling classes. When Rudyard Kipling visited Hobart, Melbourne, Sydney (briefly by train from Melbourne), and Adelaide in 1890 on a cruise for the sake of his health, he was joined on the return voyage to Ceylon by General Booth and his party of Salvationists who, having evangelized New Zealand and Australia, were continuing on to work through India.[122] The same year Alfred Deakin visited India with David Syme's son, Herbert, to report on irrigation, religion and poverty there for the *Age* in a series of articles published in 1893 in book form as *Irrigated India* and *Temple and Tomb in India*, the one keenly read by believers in irrigation, the other by the growing number whose interest in India was missionary or mystical.[123] Religious business brought Annie Besant from Madras to Australia in 1894,[124] but for the majority of Australian-born travellers, a visit to a temple in Ceylon, or a stop-over in Bombay harbour was their first taste of foreign sights, sounds, and customs. At the highest levels of society—in Government House circles—not to have seen some service in India was unusual. Thus Audrey Tennyson felt obliged to warn her successor, Lady Northcote, that Australians would not show the same deference towards the governor-general and his wife as they had become used to in India.[125]

India continued to hold a significant place in the strategy of Empire. 'The defence of Australia', it was said, 'begins on the hills outside Herat.'[126] The notion that Australian-based troops or British ships in Australian waters could be the first on the scene in the event of another crisis in India was never far distant. Charles Dilke discussed the idea in his *Problems of*

Greater Britain, and dismissed it as too expensive to be practical. He was thinking no doubt of some of the difficulties raised by the Sudan expedition,[127] though a similar idea was put into practice during the Boer War.

Between the Indian Mutiny at mid-century and the South African campaigns at the end, the Australian colonies began to discover that there was no simple complementarity between British or imperial strategic concerns and their own growing external preoccupations. Though they flung themselves enthusiastically into the service of Empire (1475 volunteers to the Maori Wars in 1863, 750 infantry plus artillery to the Sudan in 1885, 16 175 men to South Africa between 1899 and 1902, a South Australian gunboat and naval contingents from Victoria and New South Wales to China in 1900 during the Boxer Rebellion), their own security remained uncertain.[128]

In 1860 the colonies depended for their defence on the British garrisons that remained and the Royal Navy. There were also small detachments of Native Police guarding the frontiers, mainly in Queensland, but these were somehow never considered part of the security forces. By 1870 the last of the British garrisons had been withdrawn, leaving the colonies to build their defence forces from volunteers. Most were spare-time soldiers who drilled at weekends, though a nucleus of officers and other professionals was maintained. From 1881, in order to improve both numbers and quality, a 'militia' system which paid a small annual allowance (£10 or £12) plus arms, uniforms, and equipment was instituted. At the end of 1900, the defence forces of all the colonies totalled just under 26 000.[129]

Few of the colonists were unhappy to see the British legions depart. Though required to contribute to their upkeep, they had no say in the kind of troops sent or the manner of their deployment. In Victoria especially, knowledge that the governor might have sole command of these British troops exacerbated memories of Eureka and did nothing to soothe tensions between the Legislative Assembly and the Colonial Office. Victorians believed they could and should defend themselves, and organized their police force to supplement the volunteers.[130] There were naturally thoughts of a navy as well. Here relative British and colonial authority became acute. Put most simply, Britain was reluctant to permit the colonists to play with such dangerous toys as gunboats. Though they raised substantial armies, the colonial governments had no legal authority over them once they left Australian shores. The volunteers in New Zealand in 1863 all went as individuals and fought in British regiments. A special act of the New South Wales parliament was required in 1885 to place the Sudan contingent under the authority of the British Army Act and the whole episode was so legally confused that the British government was reluctant to repeat it.[131] Authority or control over necessarily very mobile naval units based in colonial ports was even more troublesome. Further, as became clear during the 1870s and beyond, the interests of the colonists in the Pacific region had been developing at a different pace and with objectives sometimes at odds with those of the British War and Colonial offices.

In the second half of the nineteenth century Britain experienced not only severe financial strain on her capacity to defend her expanding Empire (hence her withdrawal from relatively secure places like Australia) but also the expensive transformation of her navy, the main agent of British strategy, from sail to steam. The new steam-powered navy was no longer vulnerable to the vagaries of wind and waves. Instead it was closely tethered to its bases and coaling stations. The Pacific with its watery distances and infrequent coaling stations was a severe trial to the new technology. It was also remote from London and unimportant in trading and strategic priorities. With the annexation of Fiji in 1874, a token presence was established. Any other strategic needs, it was thought, could best be served by a British naval squadron operating out of a series of well-fortified Australian bases where coal was plentiful.

The main question in the Pacific of concern to Britain at this time was not a strategic but a moral one, the policing of the labour trade. For this, a base in Fiji, the creation of the Western Pacific High Commission, and regular patrols by the deputy commissioners, Peter Scratchley, and his successor, H. H. Romilly, were the best that could be done within reasonable limits of expenditure. Responsibility for control of the labour trade was a matter shared with the government of Queensland, and for Britain to interfere more than this was undiplomatic and probably ineffectual.

In Australia, however, Britain's annexation of Fiji and the creation of the Western Pacific High Commission were seen as the first steps toward a stronger British presence and through it, greater Australian influence in the south Pacific. Writing to her father Henry Parkes, mainly about family and money matters on 17 October 1874, Menie Thom remarked, 'I note too, the cession of Figi [sic] with satisfaction. Will any special advantage accrue to N. S. Wales from its share in the matter?'[132] Like many another fairly ordinary citizen at the time, Menie Thom had slightly more than a passing interest in the Pacific. Married as she was to a Presbyterian minister, there was always a prospect of missionary work on one of the islands which might be more enticing and worthy than the tedium of the manse, first at Pambula on the south coast of 'N. S. Wales', then at Ballan near Melbourne. For members of the Presbyterian church, especially in Victoria, concern about the fate of their missions in the New Hebrides was such that they kept considerable pressure on the government through the late 1870s and into the 1880s.[133] There was less overt interest in 'N. S. Wales' where attitudes on economic, defence, and strategic questions were more in sympathy with British policy, though the employees of such firms as Burns Philp, Robert Towns and Co., and W. R. Carpenter could hardly have been disinterested. In Queensland where the treatment of labourers from the Pacific islands had become a divisive issue, however, church people could rally support by reference to the work of their fellows in Victoria. Queensland like Victoria had had its differences with the Colonial Office and the government of Queensland was very sensitive on any issue which came under the scrutiny of the Colonial Office.

In 1876 a Bill to prevent Chinese working on any goldfield in Queensland during the first three years of its operation (the Goldfields Amendment Bill) was vetoed by the Crown on the ground that such restrictive legislation was repugnant to Britain's treaties with China. The constitutions of all the Australian colonies contained a provision for the governor to reserve his assent for certain classes of legislation, especially anything incompatible with British laws and treaties. These powers were used sparingly—only five Bills were ever disallowed.[134]—though it should be remembered that the colonial constitutions were themselves Acts of the British parliament, and that even in the case of the Bill to create a federated Australia, skilful negotiation was required so that the wishes of the Australian people could be made acceptable to the British parliament.

Thanks to the remarkable eccentricities of Mr Justice Boothby who as acting-governor of South Australia in 1865 rejected all legislation passed that session as repugnant to English law, largely on the grounds that he found the South Australian constitution and parliament repugnant to English legal precedent, the British government brought in the Colonial Laws Validity Act. This strengthened the position of colonial legislatures against extremism of the Boothby variety. The tendency also of advice given by subsequent British secretaries for the colonies to colonial governors was that the governors should treat the wishes of duly constituted local governments with all possible respect. Thus Governor Bowen's cautious, even deferential approach to Victoria's constitutional crisis of 1878 when the upper and lower houses could not agree on supply, earned him a rebuke from the Colonial Office for his extreme leniency regarding the colonists' wishes, as well as the opprobrium of the Victorian government itself, because he had presumed to explain to them what he thought the Colonial Office reaction might be. Colonial governments were not as powerless as the Victorian attorney-general of the day, George Higinbotham indignantly suggested, namely, that 'the million and a half Englishmen who inhabit these colonies, and who during the last fifteen years have believed they possessed self-government, have really been governed during the whole of that time by a person named Rogers'.[135] (Frederic Rogers was the colonial law officer in the 1850s who alerted the British government to the real potential for independence contained in early drafts of the New South Wales and Victorian constitutions. They were subsequently modified in favour of the Colonial Office.)

In the end, Britain called the tune. Victoria, for example, was not permitted to vary her tariff so that she charged her neighbour colonies less than she charged Britain on imported

TORRES STRAITS

goods. This had a silly and sometimes detrimental effect on certain industries in the other colonies. Laws on shipping, immigration, the issue of coinage or any other form of legal tender, defence, and military discipline were all reserved for royal assent. Britain retained control of the prerogative of mercy which could only be exercised by the governor. Resentment of the Colonial Office's power to supervise and interfere is illustrated not only by Higinbotham's outburst, McIlwraith's introduction of another version of the vetoed Goldfields Amendment Bill, and the battle in 1888 over another round of Chinese exclusion Bills, but also by the colonies' refusal to co-operate in the introduction, for example, of uniform divorce legislation which Britain requested in 1858.

Britain's reluctance to act when and where the colonies thought it imperative was also a source of frustration. In 1879, at the urging of Premier Thomas McIlwraith, Queensland annexed those islands lying between the continent of Australia and the mainland of New Guinea (the Coast Islands Act) without repercussion. McIlwraith argued that the security of the Torres Strait was essential to maintain traffic and communication along a route which had become as busy as the Bass Strait route, especially since the introduction of steam. Queensland had always suffered from the fact that Sydney was the major port of entry. Queensland, unlike Victoria, did not impose heavy tariffs, yet Queenslanders invariably paid additional charges for being bypassed by the major shipping lines. Sydney was the main location for many firms with extensive businesses in Queensland, and there was much trans-shipping and re-routing of goods through Sydney. Then the P. & O. Company, which carried mail as well as passengers and cargo from Britain, relocated its terminus from Sydney to Melbourne. The incentive to secure an alternate service using Torres Strait and calling first at Queensland ports before terminating in Sydney was doubled. The complicated story of McIlwraith's battle to secure the contracts and push the necessary legislation through an upper house which was as opposed to entrepreneurs as it was to workingmen, amidst the shifting rivalries of intercolonial businesses and power-brokers, is only one of many such stories from these years. Intercolonial differences created advantages and openings for some as they caused damage and irritation to others. In attempting to secure the Torres Strait for Queensland's advantage, McIlwraith strengthened the Sydney lobby trying to regain their lost dominance over Melbourne and the mails. Simultaneously he gave unwitting encouragement to another group in Melbourne seeking greater Australian influence in the far north with a view to extending missionary activity there.

Colonial ministers met from time to time to sort out problems caused by conflicting policies and attitudes. There were some twelve such meetings between 1863 and 1891.[136] Politically it was often expedient to sustain the appearance of difference though compromises were necessary. The long railway platforms at Albury and Wallangarra with a set of different gauge tracks on either side were symbolic of the kinds of solutions found. *Ad hoc* adjustments of the effects of intercolonial tariffs which seemed tidy in the *Votes and Proceedings* took no account of the inconvenience and irritation to ordinary people going about their business from one colony into the next. And while co-operation could be secured in the case of a major crime, tracing a minor thief or deserting husband in another colony was practically impossible. Regulation of navigation on the Murray River could be arranged by compromise, though as irrigation came to be seen as the key to further settlement in the interior (and as the drought of the 1890s began to affect the volume of water available, especially in the lower reaches of the Murray–Darling system), the co-operation of four colonies was required.

Intercolonial negotiations were the nearest most colonial politicians came to international diplomacy, at least until journeys to London to discuss defence, tariffs, and trade, or imperial government and federation became more common. The few politicians who stayed in office long enough to become known through intercolonial diplomacy acquired additional standing in this way. Yet when the federal conventions began, few of the members knew each other, even by repute. Sizing up the other delegations and adjusting to the composition of conventions with changing electoral fortunes certainly added to the length and complexity of their deliberations.

On Monday, 16 April 1883, *The Times* of London as well as several other British morning papers carried the following announcement: 'The Queensland Government has taken formal possession of New Guinea.'[137] The Queensland government resident at Thursday Island, Henry M. Chester, had in fact raised the flag on land near modern Port Moresby on 4 April, nearly two weeks before the news arrived in Britain, but it had taken him several days to reach a telegraph station from whence to transmit his news to Brisbane. From there it travelled to London with reasonable contemporary speed. Victoria, New South Wales, and South Australia all responded within the week by sending telegrams of support to Queensland. It was early July before the British government decided that the annexation must be disavowed.[138] Queensland was reassured that should there be signs of any moves by foreign powers, Britain should be informed by cable 'and if the circumstances had justified immediate action it could have been taken without delay of more than a very few hours'.[139] Given the time it clearly took to get news in and out of New Guinea, and since McIlwraith's action was the climax of several years of rising anxiety about German designs on New Guinea, evidence of which was plainly visible in the comings and goings of German ships and personnel in Australian ports, such assurances from Britain could not be taken seriously. McIlwraith had no choice but to accept Britain's alternate suggestion that the Australian colonies should raise the £15 000 needed to pay for the gradual establishment of a British protectorate in the area.

A conference of representatives of all interested Australasian governments including New Zealand and Fiji was called in November 1883. Unusually for any gathering of politicians at that stage, seven of the fifteen delegates had been born in the antipodes.[140] Even at the first Federal Convention in 1891 only sixteen, perhaps seventeen, of forty-six delegates were colonial by birth.[141] At the 1883 conference it was resolved that a Federal Council should be set up, formally to manage such questions as defence co-operation. Thus it became the precursor to federation. Though handicapped by its lack of legal authority and without a source of regular funding, its very problems exposed the difficulties in the way of a more substantial form of intercolonial co-operation. Not least of these were differences especially between New South Wales and Victoria arising from their different trading and economic policies. Because of Sydney's position as the major south Pacific trading centre at this time, New South Wales felt more confident of her influence in the Pacific than Victoria. Furthermore her free-trade policy caused her to welcome the presence and strength of the British navy without which free trade was a dubious concept anyway. Victoria's protective policies naturally distanced her more from Britain and inclined her to favour a self-reliant attitude towards defence with greater intercolonial co-operation. An *Age* editorial of 29 May 1883, probably written by C. H. Pearson, went so far as to suggest that the Australian colonies, regardless of British policies, should adopt their own version of the Munro Doctrine for the South Pacific.[142] Queensland was drawn towards the Victorian position through her dependence on Pacific island labour, though her concern about German activity gave high priority to continuing British involvement, especially the British navy.

With the encouragement of the Federal Council, uniform Bills to enable greater federal activity were introduced into each colonial parliament. Without popular enthusiasm, there was no political advantage in pushing them, certainly not while there were more obviously vote-winning measures on the agenda, so they took their chance in the administrative housekeeping. Meanwhile, German plans to secure the islands of New Guinea as reservoirs of labour for Samoan plantations were well advanced. From September 1883 a German warship was stationed in New Guinea waters to protect German labour traders from Queensland ships. Within the year, Germany had declared possession of both New Britain and New Ireland as well as the northern coast of New Guinea itself, thus forcing Britain to a comparable gesture on the southern coast or forgo the whole of New Guinea.[143]

Such were the origins of Australia's lingering resentment and distrust of German imperialism which re-surfaced so easily and with such violence in 1914. These events were important also for at least two more immediate reasons. To some observers this was a turning point in the history of British imperialism. The failure of Britain to absorb this latest conquest made on her behalf indicated a fatal infirmity of purpose. Though it was said that Queensland wished only to secure the New Guinea labour trade for herself and that her record in black labour administration left much to be desired, the strategic motives put forward at the time continued to preoccupy Australian politicians while the islands to the north could be seen as stepping stones to invasion, namely until after long-range aircraft did away with the need for refuelling and supply bases. The historian of the Queensland parliament went so far as to describe McIlwraith's annexation of New Guinea as 'his coup d'état . . . a statesmanlike and far-sighted act'.[144] In Sydney, the Catholic *Freeman's Journal* (23 June 1883) declared, 'It is as the first independent act of a son announcing . . . that he has come of age.'[145] As far as Britain was concerned, the age of Empire was rapidly degenerating into the white man's burden. Nor were New Guinea and the Torres Strait among the 'strong places of the world' to which the British navy required access to sustain its mobility. For the Australian colonies, however, these events proved their powerlessness in the councils of the mother country and raised proper doubts about the automatic availability of British protection. Henceforth there was a growing willingness for the colonies to co-operate with each other on defence matters, as in 1889 they participated in a general survey of existing defence facilities and requirements by Major-General J. Bevan Edwards.

The abiding uncertainties of defence policy were purpose and adequacy. It might be thought that the oft-demonstrated reluctance to spend money on defence seemed natural to a remotely situated society which had no reason to expect or provoke hostilities. That there were many of this disposition can be inferred from the mixed response to W. B. Dalley's precipitate action in dispatching a New South Wales contingent to the Sudan in 1885.

I do not meet many people [Menie Thom wrote to her father on 9 March 1885] but every man or woman I have spoken to has expressed unqualified reprobation of Mr. Dalley's action, and decisive fear as to its result. All *I* say is—take the verdict of six months hence. I have seen a little boy blow up a paper balloon until it was very big, and fled high, and I have seen that balloon go crash against the rock, and I have seen the result. I shall see some things the same in this case.[146]

When the contingent returned without having seen any action other than labouring on the Suakim to Berber railway, Menie asked acidly, 'What are they going to do with the Contingent now? Embalm them?'[147]

That there were many more who endorsed Dalley's impulsive loyalty can be seen in the New South Wales parliament's acceptance of his action and the thwarted efforts of the Victorian, Queensland, and South Australian governments to emulate it. But were the General Gordon statues and memorials and the crop of baby Gordons which sprouted notably in Victoria symbols only of loyalty to the cause of Empire?[148] They may also have been manifestations of repressed militarism, or simply in admiration of futile heroism.

What kind of defence did the colonies need? For what purpose—to fend off the enemies of Britain in Europe, Africa, or Afghanistan, to protect British and Australian interests in the Pacific, to defend Australia's far-flung resources? The advice of 'experts' ranged from suggesting improbable medieval coastal fortifications to acquiring the very latest in expensive naval architecture and weaponry. Britain's own

FORTIFICATIONS, SOUTH HEAD

perception of imperial defence based on the navy coloured most colonial thinking. Certainly the model of Britain's army was too difficult to adapt to Australian conditions. Britain had traditionally drawn her defence forces from the disadvantaged classes officered by a small élite of upper-class professionals, younger sons of the establishment. In Australia the Native Police with their white (often British) officers and their poorly paid but armed and mounted Aboriginal troopers, who were thus infinitely better-off than their victimized fellows on reserves and stations, approximated the British army 'at home' or on the frontiers of India. High employment levels, good wages, and adult male suffrage, as well as legislative reluctance to vote too much money on salaries for soldiers whose duties were usually ornamental, all combined to defeat the British army tradition in Australia.

American democracy had elevated the concept of bearing arms in self-defence to a constitutional right and duty. There too, a successfully weathered civil war showed what could be done with modern weapons, modern medicine and modern attitudes to force unity out of division. A developing tendency among Australian governments to use the armed forces or specially sworn constables as a supplement to the police in maintaining order during civil disturbances, especially those connected with strikes, strengthed Australian distaste for the concept of a standing army.[149] Yet the volunteer movement failed to create anything like the nucleus of a citizen army. By the end of the century, Australia was turning in its anxiety about security towards universal military training for boys before they were old enough to resist, vote, or sink into sloth and hedonism.[150]

Were such steps necessary? The kind of national spirit, love of country, which would ensure an automatic response to a real attack was beginning to appear among the younger generation. Levels of paranoia were fuelled by the regular publication of fantasy invasion stories in popular magazines, sturdy local variants of the genre becoming fashionable in England and Europe, though appropriately in their Australian setting, often involving race war or war to defend the Empire. In George Darrell's immensely popular play, *The Forlorn Hope*, first performed on 26 December 1879 in Melbourne, an Australian contingent defended the Empire at great cost. There were embarkation scenes, battle scenes and at the end a public welcome for the returned survivors outside Parliament House.[151] When Dr William Fitchett, headmaster of Melbourne Methodist Ladies' College, wrote his *Deeds that Won the Empire* (1897) and *Fights for the Flag* (1898), or when Henry Lawson prophesied in 'The Star of Australasia',

There are boys out there by the western creeks, who hurry away
  from school
To climb the sides of the breezy peaks or dive in the shaded pool
Who'll stick to their guns when the mountains quake to the tread of
  a mighty war,
And fight for Right or a Grand Mistake as men never fought
  before,[152]

they struck a chord already familiar to many of their readers. The colonial habit of ignoring or discounting violence, either latent or realized in ways other than donning uniforms and marching to military bands, led to a serious misreading of the aggressive potential of the population as a whole.

As with so much else, two separate standards were applied to the defence of Australia during the second half of the nineteenth century. In late 1859, journalist Frederick Sinnett, describing the latest phase of white settlement during the gold-rush to Canoona (near modern Rockhampton), had no doubt that he was witnessing a war.

I am unfortunately safe in saying that the ordinary relation between the black and white races is that of war to the knife. The atrocities on both sides are perfectly horrible, and I do not believe the Government makes any effort to stop the slaughter of the Aborigines. A native police force is indeed actively engaged, but exclusively against the blacks who are shot down by their bloodthirsty brethren at every opportunity. I believe the blacks retaliate whenever they can, and never lose a chance of murdering white man, woman or child... The number of blacks killed it is impossible to estimate. They are being killed officially by police and unofficially by settlers and diggers every day, nor are women and children spared when murders are being revenged by the whites... I believe the border warfare about the FitzRoy, the Dawson, and the adjacent districts to be as savage to this day as any war with the Aborigines that in any part of Australia ever darkened with disgraceful incidents the history of our progress.[153]

Pioneer anthropologist A. W. Howitt wrote in 1880 that a line of blood marked the advance of European settlement in Australia.[154] Noel Loos's cautious estimate is that 4000 Aborigines were killed on the north Queensland frontier between 1860 and the end of the century.[155] Richard Broome, drawing on the work of Loos and Henry Reynolds, suggests that 'somewhere between 1000 and 1500' Europeans died fighting Aborigines on Australian frontiers. The number of Aboriginal deaths may have been 20 000 or more.[156] These figures compare strikingly with those for the official military engagements of the period, the Sudan campaign, the Boer War, the Boxer Rebellion. Yet the myth persisted that we were a peaceful people who had never been guilty of offensive action, who needed the baptism of blood. It would perhaps be more honest to see ourselves as a people accustomed to wage war by stealth, in a spirit of meanness and self-deceit about our activities and our motives.

The hunting, shooting, and other violent tendencies of certain sections of the Australian population during this period cannot be ignored here. Nor can the weakness for heroics, and the bushranging cult which reached its apotheosis in Edward Kelly. In 1892 it was estimated that upwards of 70 000 men were involved in the strikes and lock-outs of 1890.[157] The employers were provocative. The union leaders were pugnacious. Governments and citizens who were not involved or were pacifist by nature were fearful of a widespread breakdown of law and order. Many of the protagonists believed the rhetoric about war between labour and capital, and not only because they wished to see labour win. Ernie Lane did think there was a chance for revolution.[158] His elder brother, to judge from his later career, had a natural propensity for militaristic or totalitarian solutions. With William Lane's encouragement, Henry Lawson and E. J. Brady wrote battle hymns for the 1890s. Australia seemed edgy. For some it was on the brink of greatness. The course of industrial confrontation in the 1890s was a welcome change both for capital and labour. Henry Lawson summed up the feelings of many when he wrote that they were rotting in 'a peace that lasts too long'. The future he hoped for would bring violent release:

if the cavalry charge again as they did when the world was wide,
'Twill be grand in the ranks of a thousand men in that glorious race to ride
And strike for all that is true and strong, for all that is grand and brave,
And all that ever shall be, so long as man has a soul to save.[159]

Lawson's romantic feeling for war was almost medieval in its belief in chivalry, fealty, honour, and its assumptions about the trappings and rituals. A contemporary critic discerned something more primitive, however, in the yearning for blood. Annie (Mrs Charles) Bright thought that Lawson was 'apparently anticipating the redemption of his country, through the development of the savage war spirit which grew in times of tribal conflict when human brotherhood was undreamed of'.[160] Brotherhood, mateship, were they spiritual, uplifting, intellectual ideals, or were they dark, primitive bonds sealed by violent encounters and blood?

Even the imagery of federation and nationhood was often militaristic as in 'Australia's Cherished Dream' written by Marion Miller to be sung at federal meetings:

They are gathering for the Roll-call from the East unto the West!
Now, rise up, every slumbering one and follow with the rest!
I hear the bugles sounding, and the trumpet's steady blare,
And a Nation's mighty shouting with its thunder rends the air!

306 THE OXFORD HISTORY OF AUSTRALIA

AN INCIDENT OF THE STRIKE IN THE SHIPPING TRADE—UNLOADING THE COAL UNDER POLICE PROTECTION

The strike of 1890

They are mustering fast and faster! From the hilltops to the sea
They chant the glorious tidings of a well-won victory!
They gather not for bloodshed, though their weapons brightly gleam,
But with joyful pride, to realise Australia's cherished dream.

Refrain
They come, they come, their banners stream on every wind that blows,
And Federation's holy name in every fold disclose.
Now halt! O standard bearer! and face the risen sun—
The Nation's first great festival, its joyance hath begun.[161]

When the opportunity for a 'real stoush' came in South Africa in 1899, there was none of the hesitation of 1885 in the colonies or from Britain. In quick succession colonial contingents were offered and accepted, costs of equipment and transport to South Africa were found by their originating governments, with wages and expenses while under British command to be paid from imperial funds. The New South Wales government supplemented the pay of its men as imperial rates were thought too low for Australians.[162] The few who dared to question involvement became victims of patriotic scorn. H. B. Higgins, for example, lost his seat in parliament for opposing the despatch of the Victorian contingent. 'It seems to me to shock all conscience for us to venture to go into war with a light heart, and without inquiring closely into the justice of it.' He had seen plenty of soldiering as a boy in Ireland and thought Australia would have problems 'dealing with the merits of a war in which we have no voice and no control. We are dealing with information which comes only from one source.'[163] Pacificism, conscientious objection, or pro-Boer sympathies quickly became synonymous with anti-British sentiments.[164] Information 'only from one source' produced a simple enthusiasm to stamp out the Boer, and then the shocking court-martial and execution of Henry 'Breaker' Morant for the murder of Boer prisoners, women and children. Morant's callousness was a typical product of the Australian frontier experience and fostered by the wider struggle for economic security in Australian society. It thrived on ignorance and was protected by its self-imposed sense of inferiority. Morant was not unusual among his contemporaries, a little more dashing and confident perhaps. His case exemplified the complexities of nationalism and independence. Morant became a hero because he was an Australian martyr to imperial law. Soon after, the Commonwealth parliament passed legislation entrusting the discipline of Australian troops to their own command, and forbidding the death penalty except for traitorous acts. Yet the Australians were in South Africa because they supported the Empire.[165] Or was it because they could not resist a fight?

EPILOGUE

CENTENNIAL PARK, SYDNEY, was Henry Parkes's less than fully realized 'monument' celebrating one hundred years of white settlement in Australia. With its token Victorian gardens, formal drives leading to still wild patches of bush and swamp, it was the ideal outdoor setting for the large crowd expected to hear the proclamation of the Commonwealth on 1 January 1901. The preparations went on for weeks. Elaborate arches were built along the route of the ceremonial procession from the Domain to Centennial Park. Stands were erected at vantage points. The Centennial Park Grand Stand adjoining the Queen Street entrance to the park was advertised as being erected 'under Government Supervision', and designed to carry ten times more than the weight required. It had a special overhead canopy for protection against rain or heat, also

attendance and ample accommodation for Horses and Vehicles. Refreshments of every description, also Musical Selections within the Stand. The position commands a Panoramic View of the whole of Centennial Park, overlooking the *Swearing In Ground*. Comfortable Seats with Reclining Backs 18 in × 2 ft 8 in. All seats Numbered and will be Reserved throughout the day. Ample sanitary accommodation. Entrances to the Stand from three sides; no necessity to wait for hours.

Seats cost 6s, 7s 6d and 10s. In comparison 1s secured entry to a massed military band concert with Christmas Carols at the Sydney Cricket Ground, while in the same columns of the *Sydney Morning Herald*, gramophones were advertised at £7 with discs at 2s 6d, the going rate for a set of artificial teeth was £1, and the Bayswater Hotel was offering 12s 6d a week for a waitress.[1]

The ceremonial procession on 1 January was a perfect mirror of colonial status. Its order was only fully agreed on 30 December. First (that is, last in the order of precedence) came representatives of trades and labour unions. The shearers, for example, had a decorated car with 'a sort of goddess' and a 'representation of the process of sheep shearing'. They were followed by thirty members of the Australian Workers' Union 'all mounted on good bush horses... fine-looking backblocks men with a certain freedom of bearing and suggestion of capability that was effective'. The trade unions were followed by representatives of friendly societies in regalia, representatives of the press, district court judges, heads of various churches, ex-members of executive councils, consuls, the mayor of Sydney and other visiting mayors, members of the Legislative Assembly of New South Wales and other states 'according to population', then members of legislative councils. The formula, 'according to population' determined the precedence of all interstate representatives, including visiting governors. After the legislative councillors came 'gentlemen holding the various orders of knighthood, according to the precedence of those orders', ex-executive councillors allowed to retain the title of 'honourable', chancellors and senators of universities, judges of the supreme courts, speakers, presidents of legislative councils, ministers of the new Commonwealth and of the various states and colonies, the prime minister and the premiers, the cardinal, the primate of Australia, the chief justice, visiting governors, His Excellency Admiral Pearson, the naval commander-in-chief, and the lieutenant governor, with the governor-general, his suite and escort in pride of place. Most of the procession rode or drove in carriages, wives and daughters accompanying some of the visiting premiers. Apart from 'goddesses' and other symbolic female figures on the 'decorated cars' they were the only women included. There may have been some symbolic Aboriginal or other non-white characters on the 'cars' but they were not mentioned in the report. Military and civilian brass bands, many of the latter brought in from the country for the occasion, accompanied the procession. So did a large contingent of troops. Detachments from some of the more colourful British and Indian regiments had come specially for the occasion. Most colonial volunteer forces were also represented.[2] 'On a typical sunny Australian day, this country was proclaimed a Commonwealth, by the Earl of Hopetoun, our first Governor-General', composer Alfred Hill recalled.

At that precise moment a choir of 10 000 children, 1000 adults and 10 brass bands thundered forth the National Anthem, and I had the honour of wielding the baton, or rather, a broomstick, which had to be large to be seen by that vast assembly of singers and musicians.[3]

Melbourne by comparison was very quiet. A salute of 101 guns was fired and the city was decorated with flags, but there were no public functions. It was just an ordinary holiday. The weather was 'exceedingly pleasant, but rain set in shortly before 8 o'clock and fell heavily'.[4] Melbourne's turn came in May when the first federal parliament was opened in the Exhibition Building in a manner reminiscent of earlier grand openings staged there. Nellie Stewart, the darling of the musical theatre, sang an ode to Australia specially written for her by Charles Kenningham, and George Musgrove earned a silver inkstand to commemorate his splendid management of the occasion.[5]

A more spontaneous celebration of what federation was about occurred in May 1900, when three middle-aged and solidly built statesmen 'seized each other's hands and danced hand in hand in a ring' round a room in London when they heard that the British Secretary for the Colonies, Joseph Chamberlain, had finally agreed to a version of clause 74 of the Australian Constitution Bill which 'gave them back all they wanted—all they could define as distinctly Australian'.[6] Alfred Deakin, Charles Cameron Kingston, and Edmund Barton, were all three Australian born. Australia was for them, as for many others who opted for federation, the only country they had ever known. Like Kingston till he made this trip to London, most Australians in 1900 had never been outside Australia. Unlike their immigrant parents or grandparents, they did not have the choice of thinking of somewhere else as home, or evading responsibility for its future. Unlike those who could recall another life elsewhere, and knew they had done well in Australia, the native born had only their ideals to pose in comparison with what they saw about them. Like the generations of Aboriginal people with whom they had in common much more than they knew, they had begun 'imaginatively to possess the land'.[7] Writing to Tom Roberts in 1891, the young Arthur Streeton caught the essence of the country near the railway construction works he immortalized in *Fire's On*.

To the right sou'east the plain rises a little, and thousands of strong gum trees stand in a line, up to which comes the civilised side of things—crops of maize, lucerne and prairie grass, all so soft and peaceful and yet gradually edging their way into the stately dominion of eucalyptus. Looking from here the gums look like thousands of bronzy grey ants coming over the gentle rises—just like that. Then I rise and go back a mile to my lodging at Glenbrook. So interesting and Australian altogether, men and all.[8]

Edmund Morris Miller, the future bibliographer of Australian literature, recalled how as a nineteen-year-old at a New Year's Eve party in Melbourne in 1900,

At the zero hour I withdrew from the company to a corner of the verandah where the Southern Cross constellation was visible. I gazed at it in silent rapture, almost worshipping it as a priest-like guardian of our New Commonwealth and wondering what the future portended for Australia.[9]

Despite the optimism of federation and the mood of satisfaction evoked at official ceremonies, the 1890s marked the end of a phase in Australian history and the beginning of a more sombre period. The glittering morning light had begun to harden. Never would there be such natural resources for the taking, the endless supplies of land for pasture, for crops, the goldfields, the opportunities for investment with certain profit. The optimism which had fuelled the wilder speculations in Australian development had come to an end. It was a

PRIVATE STUDY OF NELLIE STEWART IN 1901

different kind of sober capital which sought out federated Australia. Its expectations were cautious, its requirements very conservative. Growth itself created tougher circumstances. As the scale grew larger, so the competition became more intense. A boy born in the Australian colonies in 1875 was unlikely to know many of the opportunities on which Australia's generous reputation and expectations were based. He would be lucky to gain entry to a professional career with the same ease or educational qualifications as his father. Most likely he would spend his adolescence struggling to preserve the family farm through drought and depression. Or he would see his hopes of steady work, a modest home, and a family of his own at risk in a capricious economy. The girl born in 1875 who was still not married in 1895 was unlikely to find a husband. In Australia that made her a failure. If she had married, the chances were that by 1918 she would have lost a son on Gallipolli or in France. For this generation, the first generation in which native-born Australians dominated the adult population, the world was changing beyond their imagination. In his history of England 1870–1914, R. C. K. Ensor quoted a line of Robert Browning's to describe this period, so often seen as the height of Empire, but in reality the beginning of decline—'there was never glad confident morning again'.[10] In 1946 Brian Fitzpatrick used the phrase to describe the close of the nineteenth century in Australia: 'What took place was like the ending of a childhood: the curtain's fall on wide-eyed expectation, the entrance instead of uncertainty, doubt and mistrust: "never glad confident morning again".'[11] R. A. Gollan also quoted Ensor in his chapter in *Australia: A Social and Political History* (1955).[12] And Geoffrey Serle used the Fitzpatrick version to introduce his chapter on the 1890s in *From Deserts Prophets Come*.[13] The years from 1860 had been 'glad, confident morning', the best of times for those who prospered, and there were many. For the rest, the unemployed, the aged, the infirm, the unmarried mothers and deserted wives, those who did not conform to the requirements for prosperity or success, they were the worst of times. So it has always been. We have yet to produce a social system which does not inflict its own suffering, unhappiness, and unfairness.

Despite depression and drought in the last decade the dominant image of this period has ever been of progress, energetic development, social reform, cultural achievement, and growing national pride. In comparison with the period that went before, this was a time of social stability, of steady growth, of hope, and until near the end, prosperity. In comparison with the period that came after it seemed a time of peace, of security, of building, with bricks and iron, with beliefs and ideas. It has become a kind of golden age, noted for its great beginnings, for the foundations which were so truly laid. Yet these images overshadow costs which are still being paid for the achievement of those years. Our social, political and cultural institutions are still rooted in the values we espoused then.

British investors had dreamt of endless profits in the unexplored regions of the furthest colonies. Now they knew there was nothing magical about the ordinary pursuits of railway building and sheep-farming, that even Australian gold-mines were limited and hungry for capital investment. Australian governments had believed in spending money on development or its image as the soundest way to raise more money. Now it was plain that there was a limit to the colour of money they could borrow. Like the governments they supported, the modest Australian capitalists had believed that money boldly spent attracts more money to buy. They found there was an end to buyers with money for the number of houses available in Melbourne by 1893, regardless of ingenious schemes for lending and borrowing. Distance enforced a physical limit to the export markets available to the boldest or most imaginative producer of local goods. Furthermore, the Australian worker had become convinced of his right to a fair wage and a reasonable standard of living, which after all added to demand. One of the lessons of the last two decades of the nineteenth century in Australia was how delicately balanced, how interdependent were the interests of workers and capitalists, producers and consumers. Those margins which elsewhere could be found in exploiting the unemployed or the hapless immigrant, or switching to another market, were denied in Australia by a small and isolated population, and a fairly uniform one. Neither class nor geographical markets were available to the Australian producer, so in failing to meet local demand, he failed altogether. This is not to say that there were no inequalities. Nothing was clearer by the end of the 1890s than who did not count, who was not seen.

The Commonwealth of Australia emerged, so it seemed, without violence, without bloodshed, without social upheaval. The real upheavals had occurred long before—when the original inhabitants were banished, and in the ruthless destruction and exploitation of natural resources. The new nation's greatest strength was its ability to provide for the material well-being of its people. To do this it had accepted regulation of the population and the labour market, and imposed a code of self-discipline and conformity. Its dangerous weakness was an attitude of mind which sought praise for achievements but avoided the truth of their significance.

Often we have turned to this period to trace the origins

of national traditions we would uphold and cherish. We are proud of our spirit of democracy, sense of fair play, our modest but distinctive contribution to culture, to social organization, to the enjoyment of leisure. Perhaps the time has come to recognize the less attractive aspects of our heritage emerging at this time—our easy assumption that the real responsibility for our economic welfare lies elsewhere; our tendency to ignore what is inconvenient; to exploit whatever is weak or easily bought; to distrust hard-won ideas and high principles; to use equality as an excuse for ugliness, or expediency; to admire and reward self-proclaimed success while suspecting modest excellence; to believe that we have a right to maximum gain for minimum effort.

History is not a balance sheet nor a company report. It is not possible to show just how the management of one year has succeeded or failed in the next. Nor can a society be wound up or sold off if it fails to perform according to expectations. Its people may be unemployed but they do not go away. Neither the problems of the past nor the solutions found there may seem relevant or illuminating to the present. Our institutions change as our circumstances change, how quickly we are sometimes too well aware. The slowest things to change are our social habits, our mores, the ways in which we think of ourselves as human beings. That may seem hard to believe. We all think things have changed so much. We are all so different, even from our own parents. Have we changed that much? We are bound to the past by some very simple ties. We are born. We die. We experience hunger, cold, greed, desire. The emotions which moved people then, move us still today. Henry Kendall's bell-birds can still be heard (sometimes) in those same coastal gullies. The light that shines in Arthur Streeton's *Still Glides the Stream* (1889) still shines on the right kind of day in the right place.

We are inclined to think that we may be both better and worse than they were then. We are more progressive, more enlightened, but they were more hopeful, more idealistic. History filters out the petty, the ugly, the gross, the mere irritants. In our own memories we recall the good times and the very worst. Yet we know ourselves to be the products of much more than that, a mass of habits, beliefs, assumptions and reactions. So it is that the people of the late nineteenth century have their relevance. Their times were a little like our times. They were not stirred by great national emergencies. They seemed to be times of peace, of prosperity, ordinary times, normal times. Such times are the best times to see into the soul.

NOTES

ABBREVIATIONS

ADB	*Australian Dictionary of Biography*
AEHR	*Australian Economic History Review*
ANU	Australian National University
ANZ	Australian and New Zealand Book Company
APCOL	Alternative Publishing Cooperative Limited
CUP	Cambridge University Press
HS	*Historical Studies*
JIR	*Journal of Industrial Relations*
JRAHS	*Journal of the Royal Australian Historical Society*
JRH	*Journal of Religious History*
LH	*Labour History*
ML	Mitchell Library
MUP	Melbourne University Press
NLA	National Library of Australia
NSWVP	*New South Wales Votes and Proceedings*
OUP	Oxford University Press
RHSVJ	*Royal Historical Society of Victoria Journal*
SUP	Sydney University Press
UNSWP	University of New South Wales Press
UQP	University of Queensland Press
UWA	University of Western Australia
W and P	*Wealth and Progress of New South Wales*

PROLOGUE

1 Stanley Lane-Poole, *Thirty Years of Colonial Government*, Longmans, Green and Co., London, 1889, vol. 1, pp. 87–8.
2 ibid., p. 90.
3 ibid., p. 101.
4 ibid., p. 85.

CHAPTER 1: MATERIALISM

1 William Westgarth, *Half a Century of Australasian Progress: A Personal Retrospect*, Sampson Low, London, 1889, p. 137.
2 G. C. Bolton, *A Thousand Miles Away*, Jacaranda Press, Brisbane, 1963, p. 20; see also Jean Farnfield, *Frontiersman: A Biography of George Elphinstone Dalrymple*, OUP, Melbourne, 1968.
3 Stanley Lane-Poole, *Thirty Years of Colonial Government*, Longmans, Green and Co., London, 1889, vol. 1, p. 193.
4 On Leichhardt, see Alec H. Chisholm, *Strange Journey: The Adventures of Ludwig Leichhardt and John Gilbert*, Rigby, Adelaide, 1973; Edgar Beale, *Kennedy of Cape York*, Rigby, Adelaide, 1970; also Ernest Favenc, *The History of Australian Exploration from 1788 to 1888*, Turner and Henderson, Sydney, 1888, esp. chap. XI on the Jardines.
5 Mary Durack, *Kings in Grass Castles*, Corgi Books, London, 1967; Gordon Buchanan, *Packhorse and Waterhole. With the first overlanders to the Kimberleys*. Angus and Robertson, Sydney, 1933.
6 Favenc, *History of Australian Exploration*, chap. X.
7 Wendy Birman, *Gregory of Rainworth: A Man in his time*. UWA Press, Nedlands, 1979; F. K. Crowley, *Forrest*, vol. 1, 1847–91: *Apprenticeship to Premiership*, UQP, St Lucia, 1971.
8 Mary Howitt Walker, *Come Wind. Come Weather: A Biography of Alfred Howitt*, MUP, Melbourne, 1971.
9 P. F. Donovan, *A Land Full of Possibilities: A History of South Australia's Northern Territory*, UQP, St Lucia, 1981; Alan Powell, *Far Country: A Short History of the Northern Territory*, MUP, Melbourne, 1982; Ross Duncan, *The Northern Territory Pastoral Industry 1863–1910*, MUP, Melbourne, 1967.
10 Ann Moyal, *Clear Across Australia: A History of Telecommunications*, Nelson, Melbourne, 1984.
11 K. T. H. Farrer, *A Settlement Amply Supplied: Food Technology in Nineteenth Century Australia*, MUP, Melbourne, 1980, esp. chap. 7.
12 Noel Loos, *Invasion and Resistance: Aboriginal-European Relations on the North Queensland Frontier 1861–1897*, ANU Press, Canberra, 1982, chap. 4.
13 Farnfield, *Frontiersman*, p. 15.
14 Roy Connolly, *John Drysdale and the Burdekin*, Ure Smith, Sydney, 1964.
15 See Glen Lewis, *A History of the ports of Queensland: A Study in Economic Nationalism*, UQP, St Lucia, 1973, pp. 57–62.
16 H. S. Russell, *The Genesis of Queensland*, Turner and Henderson, London, 1888.
17 G. C. Bolton, 'The Valley of Lagoons: A Study in Exile', *Business Archives* 4 (1964), pp. 99–116.
18 David S. Macmillan, *Bowen Downs 1863–1963*, A Centenary Publication issued by The Scottish Australian Company Limited, Sydney, 1963.
19 J. D. Bailey, *A Hundred Years of Pastoral Banking: A History of the Australian Mercantile Land and Finance Company 1863–1963*, OUP, Clarendon, 1966.
20 Donovan, *Land Full of Possibilities*, pp. 121–2.
21 Bolton, *A Thousand Miles Away*, p. 30; Dorothy Jones, *Cardwell Shire Story*, Jacaranda Press, Brisbane, 1961; Margaret G. Kerr, *The Surveyors: The Story of the Founding of Darwin*, Rigby, Adelaide, 1971.

22 C. T. Stannage (ed.), *A New History of Western Australia*, UWA Press, Nedlands, 1981, pp. 105–9.
23 Mrs Aneas Gunn, *We of the Never Never*, Hutchinson, Melbourne, 1982 (1908).
24 Alfred Searcy, *In Australian Tropics*, Kegan Paul Trench Trubner, London, 1907, p. 173.
25 See Alan Barnard, *The Australian Wool Market 1840–1900*, MUP, Melbourne, 1958, for a detailed examination of these movements.
26 J. D. Lang, *Queensland, Australia: a highly eligible field for emigration, and the future cotton-field of Great Britain: with a disquisition on the Origins, Manners, and Customs of the Aborigines*, Edward Stanford, London, 1861.
27 H. Mortimer Franklyn's article on R. Goldsborough and Co. in his *A Glance at Australia in 1880*, The Victorian Review Pub. Co., Melbourne, 1881, p. 201.
28 S. H. Roberts, *History of Australian Land Settlement 1788–1920*, MUP, Melbourne, 1924 is still the best general account; see also Alan Barnard (ed.), *The Simple Fleece: Studies in the Australian Wool Industry*, MUP, Melbourne, 1962; Margaret Kiddle, *Men of Yesterday: A Social History of the Western District of Victoria 1834–1890*, MUP, Melbourne, 1961; R. L. Heathcote, *Back of Bourke: A Study of Land Appraisal and Settlement in Semi-arid Australia*, MUP, Melbourne, 1965; Eric C. Rolls, *They All Ran Wild: The Story of Pests on the Land in Australia*, Angus and Robertson, Sydney, 1969; Geoffrey Bolton, *Spoils and Spoilers: Australians Make Their Environment 1788–1890*, Allen and Unwin, Sydney, 1981, chap. 8.
29 Barnard, *The Australian Wool Market*, pp. 229–30.
30 See W. H. Hall, *Statistics. Six States of Australia and New Zealand 1861–1904*, Government Printer, Sydney, 1905, pp. 20–3.
31 Anthony Trollope, *Australia*, edited by P. D. Edwards and R. B. Joyce, UQP, St Lucia, 1967, p. 662.
32 R. B. Walker, *Old New England: A History of the Northern Tablelands of New South Wales 1818–1900*, SUP, Sydney, 1966.
33 Commonwealth of Australia, *Official Year Book*, 1901–7, p. 413; Geoffrey Blainey, *The Rush that Never Ended*, MUP, Melbourne, 1978, p. 131.
34 Randolph Bedford, *Naught to Thirty-three*, MUP, Melbourne, 1976, p. 22.
35 Elena Grainger, *Hargrave and Son: A Biography of John Fletcher Hargrave and his son Lawrence Hargrave*, UQP, St Lucia, 1978, chap. 5; Dorothy Jones, *Cardwell Shire Story*, chap. XVIII.
36 Loos, *Invasion and Resistance*, pp. 16–19, chap. 4.
37 Mary Durack, *The Rock and the Sand*, Constable, London, 1969.
38 Monica E. Turvey, 'Missionaries and Imperialism: Opponents or progenitors of empire? The New Guinea case', *JRAHS*, 65 (1979), pp. 89–108; also N. Miklouho-Maclay, *Travels to New Guinea: Diaries, Letters, Documents*. Progress Publishers, Moscow, 1982; E. M. Webster, *The Moon Man: A Biography of Nikolai Miklouho-Maclay*, MUP, Melbourne, 1984.
39 R. I. Crivelli, 'History of the Coal Industry in Australia', *JRAHS*, XVI (1930), pp. 151–60.
40 Robin Gollan, *The Coalminers of New South Wales: A History of the Union 1860–1960*, MUP, Melbourne, 1963, p. 31.
41 ibid., p. 10.
42 Crivelli, 'History of the Coal Industry', p. 160.

43 Lionel Wigmore, *Struggle for the Snowy: The Background of the Snowy Mountains Scheme*, OUP, Melbourne, 1968, p. 66.
44 George S. Baden-Powell, *New Homes for the Old Country*, Richard Bentley and Son, London, 1872, p. 420.
45 Jurgen Tampke (ed.), *Wunderbar Country: Germans look at Australia 1850–1914*, Hale and Iremonger, Sydney, 1982, p. 152.
46 Tony Dingle, *The Victorians: Settling*, Fairfax, Syme and Weldon Associates, Sydney, 1984, chap. 7.
47 Westgarth, *Half a Century of Australasian Progress*, pp. 88–91.
48 *W and P*, 1900–1, pp. 342–3.
49 Farrer, *A Settlement Amply Supplied*, p. 97.
50 Mortimer Franklyn, *A Glance at Australia*, p. 64.
51 I. McLean, S. Molloy and P. Lockett, 'The Rural Workforce in Australia 1871–1911', *AEHR* 22 (1982), pp. 172–81; also Katrina Alford, 'Colonial Women's Employment as Seen by Nineteenth Century Statisticians and Twentieth Century Economic Historians', *Working Papers in Economic History* 65 (1986), ANU, Canberra.
52 A. B. Facey, *A Fortunate Life*, Fremantle Arts Centre Press, 1981.
53 Caroline Chisholm, *Comfort for the Poor! Meat Three Times a Day! Voluntary Information from the People of New South Wales collected in that colony*, John Ollivier, London, 1847; Mortimer Franklyn, *A Glance at Australia in 1880, or Food from the South*.
54 T. A. Coghlan, *Labour and Industry in Australia*, OUP, 1918, p. 1190.
55 F. A. Larcombe, *A History of Local Government in New South Wales*, vol, 2 1858–1906, SUP, Sydney, 1976, pp. 214–15.
56 Alan Barnard, *Visions and Profits: Studies in the Business Career of T. S. Mort*, MUP, Melbourne, 1961, pp. 102–4.
57 Andrew Garran (ed.), *Picturesque Atlas of Australasia*, Picturesque Atlas Publishing Co., Sydney, 1886–88, reissued as *Australia: The First Hundred Years*, Ure Smith, Sydney, 1974, p. 409.
58 Wigmore, *Struggle for the Snowy*, chap. 4; also Dingle, *Settling*, pp. 119–24.
59 Edgars Dunsdorfs, *The Australian Wheat-growing Industry 1788–1948*, MUP, Melbourne, 1956, pp. 147–9.
60 M. E. Robinson, *The New South Wales Wheat Frontier 1851–1911*, ANU, Canberra, 1976, p. 12; also Dingle, *Settling*, p. 113.
61 A. A. Giraud, *The Sugar Cane; its Culture and Manufacture, specially adapted to Queensland*, Watson Ferguson, Brisbane, 1883, p. 8 recommended Meera, Rose Bamboo or Rapoe, Black Java, Striped Singapore, Ribbon, Creole, Guingant and Red cane as the varieties most suited to Queensland.
62 J. M. Powell (ed.), *Yeomen and Bureaucrats: The Victorian Crown Land Commission 1878–79*, OUP, Melbourne, 1973, pp. iii–xxxvi.
63 Frances Wheelhouse, *Digging Stick to Rotary Hoe: Men and Machines in Rural Australia*, Cassell, Melbourne, 1966.
64 See K. R. Bowes, *Land Settlement in South Australia 1857–1890*, Libraries Board of South Australia, Adelaide, 1968.
65 D. W. Meinig, *On the Margins of the Good Earth: The South Australian Wheat Frontier 1869–1884*, John Murray, London, 1963, p. 70.
66 J. M. Powell, *Environmental Management in Australia 1788–1914. Guardians, Improvers and Profit: An Introductory Survey*, OUP, Melbourne, 1976, pp. 87–9.
67 Nancy Robinson (ed.), *Stagg of Tarcowie: The Diaries of a Colonial Teenager (1885–1887)*, Lynton, Adelaide, 1973, p. 51.

68 ibid., pp. 58, 67.
69 ibid., p. 83.
70 Dunsdorfs, *Wheat Industry*, p. 186.
71 See Roland Skemp, *Memories of Myrtlebank: The Bush-farming Experiences of Rowland and Samuel Skemp in North-Eastern Tasmania 1883–1948*, MUP, Melbourne, 1952; also Dingle, *Settling*, pp. 114–19.
72 Westgarth, *Half a Century of Australasian Progress*, pp. 232–3.
73 Steele Rudd, *On Our Selection*, Bulletin Newspaper Co., Sydney, 1899; Michael Cannon, *Life in the Country*, Nelson, Melbourne, 1973; John Ritchie, *Australia As We Were*, Heinemann, Melbourne, 1975, chap. 6.
74 B. D. Graham, *The Formation of the Australian Country Parties*, ANU Press, Canberra, 1966, chaps 1–2; Clarence Karr, 'Origins of the New South Wales Farmers' Union Movement', *JRAHS*, 61 (1975), pp. 199–206.
75 J. Henniker Heaton, *Australian Dictionary of Dates and Men of the Time*, George Robertson, Sydney, 1879, p. 276.
76 John D. Keating, *Bells in Australia*, MUP, Melbourne, 1979, chaps 6–7.
77 N. G. Butlin, *Investment in Australian Economic Development 1861–1900*, CUP, Cambridge, 1964, pp. 215–17.
78 R. V. Jackson, *Australian Economic Development in the Nineteenth Century*, ANU Press, Canberra, 1977, p. 121.
79 ibid., p. 128.
80 Butlin, *Investment in Australian Economic Development*, pp. 204–5; Jackson, *Australian Economic Development*, p. 117.
81 J. M. Freeland, *Architecture in Australia: A History*, F. W. Cheshire, Melbourne, 1968, pp. 84–8, 168.
82 Michael Cannon, *Life in the Cities*, Nelson, Melbourne, 1975, p. 160.
83 Bernard Barrett, *The Inner Suburbs: The Evolution of an Industrial Area*, MUP, Melbourne, 1971; see also W. A. Sinclair, 'Economic Growth and Well-being: Melbourne 1870–1914', *Economic Record*, 51 (1975), pp. 153–73; A. J. C. Mayne, *Fever, Squalor and Vice: Sanitation and Social Policy in Victorian Sydney*, UQP, St Lucia, 1982; Shirley Fitzgerald, *Rising Damp. Sydney 1870–90*, OUP, Melbourne, 1987.
84 Hutchinson's *Australian Encyclopaedia*, London, 1892, *passim*.
85 Butlin, *Investment in Australian Economic Development*, p. 324.
86 Brian Kiernan (ed.), *Henry Lawson*, UQP, St Lucia, 1976, p. 78 (from 'The Roaring Days').
87 Butlin, *Investment in Australian Economic Development*, pp. 305–14; also Samuel Clyde McCulloch, *River King: The McCulloch Carrying Company and Echuca (1865–1898)*, University of Melbourne Archives Board of Management, Melbourne, 1986.
88 Butlin, *Investment in Australian Economic Investment*, pp. 314–20; see also *The Roadmakers: A History of Main Roads in New South Wales*, NSW Department of Main Roads, Sydney, 1976, pp. 42–59.
89 G. J. R. Linge, *Industrial Awakening: A Geography of Australian Manufacturing 1788–1890*, ANU Press, Canberra, 1979, p. 716.
90 Commonwealth of Australia, *Official Year Book*, 1901–1907, p. 434.
91 As discussed by Jackson, *Australian Economic Development*, p. 115; also Jenny Lee and Charles Fahey, 'A Boom for Whom? Some developments in the Australian labour market 1870–91', *LH*, 50 (1986), pp. 1–27.

92 e.g. Mortimer Franklyn, *A Glance at Australia*.
93 J. W. Turner, *Manufacturing in Newcastle 1801–1900*, Newcastle Public Library, 1980; also J. C. Docherty, *Newcastle. The Making of an Australian City*, Hale and Iremonger, Sydney, 1983, chap. 1.
94 Docherty, *Newcastle*, p. 4.
95 Linge, *Industrial Awakening*, pp. 250–61.
96 P. L. Cottrell, *British Overseas Investment in the Nineteenth Century*, Macmillan, London, 1975, p. 43.
97 A. R. Hall, *The London Capital Market and Australia 1870–1914*, ANU Press, Canberra, 1963, pp. 189–90.
98 Hall, *London Capital Market*, p. 188.
99 Stephen Mills, *Taxation in Australia*, Macmillan, London, 1925.
100 Susan Magarey, *Unbridling the Tongues of Women: A Biography of Catherine Helen Spence*, Hale and Iremonger, Sydney, 1985, p. 175.
101 Queensland *Official Year Book*, 1901, p. 145.
102 *W and P*, 1886–87, p. 555
103 *W and P*, 1900–1, p. 860.
104 Victorian *Year Book*, 1905, p. 280.
105 Craufurd D. Goodwin, *The Image of Australia: British Perception of the Australian Economy from the Eighteenth to the Twentieth Century*, Duke University Press, Durham, 1974, p. 182.
106 *W and P*, 1900–1, p. 858.
107 W. D. Rubinstein, 'The Distribution of Personal Wealth in Victoria 1860–1974', *AEHR*, XIX (1979), p. 36.
108 ibid., p. 38.
109 *W and P*, 1900–1, p. 856.
110 See p. 30.
111 C. W. Dilke, *The Problems of Greater Britain*, Macmillan, London, 1890, p. 112.
112 N. G. Butlin and H. de Meel, *Public Capital Formation in Australia*, ANU Press, Canberra, 1954, p. 8.
113 Butlin, *Investment in Australian Economic Development*, pp. 414, 415.
114 See T. A. Coghlan, *Labour and Industry*, pp. 1633 ff.; E. A. Boehm, *Prosperity and Depression in Australia 1887–1897*, OUP, Clarendon, 1971; W. A. Sinclair, *The Process of Economic Development in Australia*, Longman Cheshire, Melbourne, 1976, chap. 6; Geoffrey Blainey, *Gold and Paper: A History of The National Bank of Australasia Limited*, Georgian House, Melbourne, 1958, chap. 10.
115 Scottish Australian Investment Company Papers, Sydney University Archives.
116 e.g. Bailey, *A Hundred Years of Pastoral Banking*; Butlin, *Investment in Australian Economic Development*, pp. 85–124.
117 Boehm, *Prosperity and Depression*, pp. 346–7.
118 G. H. Knibbs, *The Private Wealth of Australia and Its Growth*, Government Printer, Melbourne, 1918, pp. 23–4.
119 Commonwealth of Australia, *Official Year Book*, 1901–7, pp. 707–8.
120 K. S. Prichard, *Child of the Hurricane*, Angus and Robertson, Sydney, 1963, p. 40.
121 G. H. Knibbs, *Trade Unionism, Unemployment, Wages, Prices and Cost of Living in Australia 1891–1912*, Government Printer, Melbourne, 1913, p. 18.
122 *W and P*, 1900–1, p. 858.
123 Shurlee Swain, 'The poor people of Melbourne', in Graeme Davison, David Dunstan and Chris McConville (eds), *The Outcasts of*

Melbourne: Essays in Social History, Allen and Unwin, Sydney, 1985, pp. 91–12; also Ronald Mendelsohn, *The Condition of the People: Social Welfare in Australia 1900–1975*, Allen and Unwin, Sydney, 1979, chap. 4.
124 *W and P*, 1892, p. 841.
125 ibid., p. 843.
126 P. and D. O'Farrell, *Documents in Australian Catholic History*, vol. 1 1788–1884, Geoffrey Chapman, London, 1969, pp. 119–22.
127 R. W. Dale, *Impressions of Australia*, Hodder and Stoughton, London, 1889, pp. 224–5.
128 See J. A. La Nauze, *Political Economy in Australia*, MUP, Melbourne, 1949, pp. 45–97.
129 Mortimer Franklyn, *A Glance at Australia*, p. 302.
130 *W and P*, 1900–1, p. 769.
131 A. G. Austin (ed.), *The Webbs' Australian Diaries 1898*, Pitman, Melbourne, 1965, p. 108.
132 Walter Murdoch, *Alfred Deakin: A Sketch*, Constable and Co., London, 1923, p. 175.
133 *Bulletin*, 14 Feb. 1981, p. 4.

CHAPTER 2: BELIEF

1 William Westgarth, *The Colony of Victoria*, Sampson Low and Son, London, 1864, p. 346.
2 R. Reid Badger, *The Great American Fair*, Nelson Hall, Chicago, 1979, Appendix A, 'The World's Fairs', p. 131.
3 Geoffrey Serle, *The Rush to be Rich: A History of the Colony of Victoria, 1883–1889*, MUP, Melbourne, 1971, p. 285.
4 'J. D. A.', 'Historical Society of Australasia', *RHSV History News*, no. 58, June 1985, p. 2.
5 H. G. Turner and Alexander Sutherland, *The Development of Australian Literature*, George Robertson, Melbourne, 1898, p. 119.
6 P. D. Edwards and R. B. Joyce in the introduction to Anthony Trollope, *Australia*, UQP, St Lucia, 1967, pp. 34–40.
7 Brunton Stephens, *The Poetical Works*, Angus and Robertson, Sydney, 1912, pp. 1, 3, 6.
8 Craufurd D. W. Goodwin, *The Image of Australia. British Perception of the Australian Economy from the Eighteenth to the Twentieth Century*, Duke University Press, Durham, 1974, pp. 110–22.
9 Eric Partridge, *A Dictionary of Slang and Unconventional English*, Routledge and Kegan Paul, London, 1961, p. 222.
10 Marcie Muir (ed.), *My Bush Book. K. Langloh Parker's 1890s story of outback station life*, Rigby, Adelaide, 1982, pp. 151, 171–2.
11 Havelock Ellis, *My Life*, Neville Spearman, London, 1967, chap. 4.
12 Mary Howitt Walker, *Come Wind. Come Weather: A Biography of Alfred Howitt*, MUP, Melbourne, 1971.
13 Elizabeth Salter, *Daisy Bates: The Great White Queen of the Never-never*, Angus and Robertson, Sydney, 1971.
14 B. R. Wise, *The Commonwealth of Australia*, Pitman, London, 1909, p. 231.
15 See Beatrice Webb's comment e.g. A. G. Austin (ed.), *The Webbs' Australian Diary 1898*, Pitman, Melbourne, 1965, p. 76.
16 D. J. Mulvaney, 'The Australian Aborigines 1606–1929. Opinion and Fieldwork', in J. J. Eastwood and F. B. Smith, *Historical Studies*

Selected Articles, MUP, Melbourne, 1964, pp. 1–56; Max Charlesworth, Howard Morphy, Diane Bell and Kenneth Maddock (eds), *Religion in Aboriginal Australia: An Anthology*, UQP, St Lucia, 1986.
17 Baldwin Spencer and F. J. Gillen, *Across Australia*, Macmillan, London, 1912, p. 290.
18 ibid., pp. 6–7.
19 ibid.
20 P. and D. O'Farrell, *Documents in Australian Catholic History*, vol. 1 1788–1884, Geoffrey Chapman, London, 1969, pp. 407–13.
21 Ross Border, *Church and State in Australia 1788–1872: A Constitutional Study of The Church of England in Australia*, SPCK, London, 1972.
22 Jill Roe, 'Challenge and Response: Religious Life in Melbourne, 1876–86', *JRH*, 5 (1968–69), pp. 149–66.
23 See C. R. Badger, *The Reverend Charles Strong and the Australian Church*, Abacada Press, Melbourne, 1971, chaps 4 and 5.
24 Dorothy Scott, *The Halfway House to Infidelity. A History of the Melbourne Unitarian Church 1853–1973*, The Unitarian Fellowship of Australia, Melbourne, 1980.
25 *ADB*, 3, p. 330.
26 *ADB*, 7, p. 85.
27 Renate Howe, 'Social Composition of the Wesleyan Church in Victoria During the Nineteenth Century', *JRH*, 4 (1966–67), pp. 206–17.
28 Arnold D. Hunt, 'The Bible Christians in South Australia', *Journal of the Historical Society of South Australia*, 10 (1982), p. 30; also his *This Side of Heaven: A History of Methodism in South Australia*, Lutheran Publishing House, Adelaide, 1985.
29 Walter Phillips, *Defending 'A Christian Country': Churchmen and Society in New South Wales in the 1880s and after*, UQP, St Lucia, 1981, p. 138.
30 Robert Potter, *A Voice from the Church in Australia: Eight Sermons preached in New South Wales and Victoria with notes on the scientific aspect of Christian doctrines*, Macmillan, London, 1864, p. 23.
31 Robert Easterley and John Wilbraham, *The Germ Growers: An Australian Story of Adventure and Mystery*, Melville, Mullen, Melbourne, 1892.
32 Hans Mol, *Religion in Australia: A sociological investigation*, Nelson, Melbourne, 1971, p. 12, quoting statistics compiled by Walter Phillips.
33 Phillips, *Defending 'A Christian Country'*, p. 136.
34 See e.g, John Stanley James, *The Vagabond Papers*, edited with an introduction by Michael Cannon, MUP, Melbourne, 1969, pp. 107–18; also the series 'Round the Churches' published in the Adelaide *Quiz and the Lantern*, 1894–95, quoted by David Hilliard in *Godliness and Good Order: A History of the Anglican Church in South Australia*, Wakefield Press, Adelaide, 1986, chaps 2 and 3.
35 *W and P*, 1892, p. 756.
36 Mrs Campbell Praed, *Mrs Tregaskiss: A Novel of Anglo-Australian Life*, Appleton and Co., New York, 1895, p. 170.
37 Mary Mcleod Banks, *Memories of Pioneer Days in Queensland*, Heath Cranton Ltd, London, 1931, p. 40.
38 Serle, *Rush to be Rich*, p. 160; also H. R. Jackson, *Churches and People in Australia and New Zealand 1860–1930*, Allen and Unwin/Port Nicholson Press, Wellington, 1987, chap. 5.

39 Phillips, *Defending 'A Christian Country'*, pp. 175-93.
40 pp. 275-6 1st edn, quoted by John Tregenza, *Professor of Democracy: The Life of Charles Henry Pearson, 1830-1894, Oxford Don and Australian Radical*, MUP, Melbourne, 1968, p. 8.
41 Serle, *Rush to be Rich*, p. 174.
42 Walter Murdoch, *Alfred Deakin: A Sketch*, Constable, London, 1923, pp. 167-8.
43 Frank M. C. Forster, 'Birth Control in Australia: Henry Keylock Rusden and Knowlton's "Fruits of Philosophy"', *Victorian Historical Journal*, 50 (1979), pp. 237-44.
44 Frank M. C. Forster, 'The Collins Prosecution, the Windeyer Judgement and Publications on Birth Control', *Australia 1888*, 10 (September 1982), pp. 76-83.
45 Farley Kelly, 'Mrs Smyth and the body politic: Health reform and birth control in Melbourne', in Margaret Bevege, Margaret Jams and Carmel Shute (eds), *Worth Her Salt: Women at Work in Australia*, Hale and Iremonger, Sydney, 1982, pp. 213-29.
46 Sigmund Freud, '"Civilized" Sexual Morality and Modern Nervous Illness' (1908), *The Standard Edition of the Complete Psychological Works*, vol. IX, The Hogarth Press, London, 1959, p. 199.
47 S. McInerney (ed.), *The Confessions of William James Childley*, UQP, St Lucia, 1977.
48 David Walker, 'Continence for a Nation. Seminal Loss and National Vigour', *LH*, 48 (1985), pp. 1-14.
49 Neville Hicks, *'This Sin and Scandal': Australia's Population Debate 1891-1911*, ANU Press, Canberra, 1978.
50 Beverley Kingston, *My Wife, My Daughter, and Poor Mary Ann*, Nelson, Melbourne, 1975; Kerreen Reiger, *The Disenchantment of the Home: Modernizing the Australian Family*, OUP, Melbourne, 1985.
51 Phillip E. Muskett, *The Art of Living in Australia*, Eyre and Spottiswoode, London, 1893, p. 111.
52 Richard Ely, 'Secularisation and the Sacred in Australian History', *HS*, 19 (1981), pp. 553-66.
53 See Peter Coleman, *Obscenity, Blasphemy, Sedition: 100 Years of Censorship in Australia*, Angus and Robertson, Sydney, 1974, pp. 11-12.
54 K. T. Livingston, *The Emergence of an Australian Catholic Priesthood 1835-1915*, Catholic Theological Faculty, Sydney, 1977, p. 70.
55 John N. Molony, *An Architect of Freedom: John Hubert Plunkett in New South Wales 1832-1869*, ANU Press, Canberra, 1973, pp. 191-2; also T. L. Suttor, *Hierarchy and Democracy in Australia 1788-1870*, MUP, Melbourne, 1965.
56 John N. Molony, *The Roman Mould of the Australian Catholic Church*, MUP, Melbourne, 1969.
57 Patrick O'Farrell, *The Catholic Church and Community in Australia: A History*, Nelson, Melbourne, 1977, chap. 4.
58 ibid.
59 O'Farrell, *Documents*, vol. 1, p. 333.
60 Joseph Furphy in Tom Collins, *Rigby's Romance*, Angus and Robertson, Sydney, 1946, p. 164.
61 Marcus Clarke, 'The Future Australian Race' (1877), quoted Ian Turner (ed.), *The Australian Dream*, Sun Books, Melbourne, 1968, pp. 132-3.
62 Commonwealth of Australia, *Official Year Book*, 1901-7, pp. 172-4.
63 C. M. H. Clark, *A History of Australia, IV: The Earth Abideth For-*

ever, 1851–1888, MUP, Melbourne, 1978, p. 142; P. O'Farrell, 'Writing the General History of Australian Religion', *JRH*, 9 (1976), p. 70.
64 C. W. Dilke, *Problems of Greater Britain*, Macmillan, London, 1890, p. 586.
65 Richard Ely, *Unto God and Caesar: Religious Issues in the Emerging Commonwealth 1891–1906*, MUP, Melbourne, 1976.
66 Blair Ussher, 'The Salvation War', Graeme Davison, David Dunstan, and Chris McConville (eds), *The Outcasts of Melbourne: Essays in Social History*, Allen and Unwin, Sydney, 1985, pp. 124–39.
67 Miles Franklin, *Joseph Furphy: The Legend of a Man and his Book*, Angus and Robertson, Sydney, 1944, p. 110.
68 Dilke, *Problems of Greater Britain*, p. 586.
69 Cole Turnley, *Cole of the Book Arcade: A Pictorial Biography of E. W. Cole*, Cole Publications, Melbourne, 1974, chap. 7.
70 Marian Zaunbrecher, 'Henry Lawson's Religion', *JRH*, 11 (1980), pp. 308–19.
71 J. D. Bollen, *Protestantism and Social Reform in New South Wales 1890–1910*, MUP, Melbourne, 1972.
72 Hutchinson's *Australasian Encyclopaedia*, London, 1892, p. 122.
73 Roger B. Joyce, *Samuel Walker Griffith*, UQP, St Lucia, 1984.
74 David G. Green and Lawrence G. Cromwell, *Mutual Aid or Welfare State: Australia's Friendly Societies*, Allen and Unwin, Sydney, 1984, p. 217 suggests an Australia-wide friendly society membership of about 242 000 in 1892.
75 See Ann Mari Jordens, *The Stenhouse Circle: Literary life in mid-nineteenth century Sydney*, MUP, Melbourne, 1979; Brian Elliot, *Marcus Clarke*, OUP, London, 1958.
76 See Joyce, *Griffith*; Sylvia Lawson, *The Archibald Paradox: A Strange Case of Authorship*, Allen Lane, Ringwood, 1983; Axel Clark, *Christopher Brennan: A Critical Biography*, MUP, Melbourne, 1980; Nancy Keesing (ed.), *The Autobiography of John Shaw Neilson*, NLA, Canberra, 1978; Denton Prout, *Henry Lawson: The Grey Dreamer*, Rigby, Adelaide, 1963.
77 N. G. Butlin, *The Life and Times of 'The Australian Economist' 1888–1898*, Working Papers in Economic History 69, ANU, 1986.
78 Joan Radford, *The Chemistry Department of the University of Melbourne. Its Contribution to Australian Science 1854–1959*, The Hawthorn Press, Melbourne, 1978, chaps 1 and 3.
79 J. A. La Nauze, *Political Economy in Australia*, MUP, Melbourne, 1949, pp. 49–97; Tregenza, *Professor of Democracy*.
80 D. J. Mulvaney and J. H. Calaby, *'So Much That is New'. Baldwin Spencer 1860–1929: A Biography*, MUP, Melbourne, 1985.
81 See Margaret Hazzard's introduction to *Flower Paintings of Ellis Rowan*, National Library of Australia, Canberra, 1982; also D. J. and S. G. M. Carr, *People and Plants in Australia*, Academic Press Australia, Sydney, 1981, pp. 325–98.
82 Mulvaney and Calaby, *Baldwin Spencer*, pp. 193, 217–18.
83 Ann Mozely Moyal (ed.), *Scientists in Nineteenth Century Australia: A Documentary History*, Cassell Aust., Stanmore, 1976, pp. 126–7.
84 ibid., pp. 120–3.
85 Craufurd D. W. Goodwin, *Economic Enquiry in Australia*, Duke University Press, Durham, 1966, p. 358.
86 ibid., p. 68.
87 ibid., pp. 330–9.

88 Warren G. Osmond, *Frederic Eggleston: An Intellectual in Australian Politics*, Allen and Unwin, Sydney, 1985, chaps 1 and 2.
89 Zelman Cowen, *Isaac Isaacs*, OUP, Melbourne, 1967; Geoffrey Serle, *Monash: A Biography*, MUP, Melbourne, 1982; Len Fox, *E. Phillips Fox and his family*, Len Fox, Sydney, 1985; Leslie M. Henderson, *The Goldstein Story*, Stockland Press, Melbourne, 1973.
90 F. W. Eggleston, *Reflections of an Australian Liberal*, F. W. Cheshire, Melbourne, 1953, p. 2.
91 J. M. Ward, *James Macarthur Colonial Conservative, 1798–1867*, SUP, Sydney, 1981.
92 H. J. Wrixson, *Democracy in Australia*, Heath and Cordell, Melbourne, 1868, pp. 10–11.
93 W. K. Hancock, *Discovering Monaro: A Study of Man's Impact on his Environment*, CUP, Cambridge, 1972.
94 Archibald Forsyth, *Rapara or The Rights of the Individual in the State*, William Dymock, Sydney, 1897; also David S. Macmillan, *One Hundred Years of Ropemaking 1865–1965: Archibald Forsyth and Company*, A. Forsyth & Co. Pty Ltd, Sydney, 1965, pp. 3–28.
95 Bruce Smith, *Liberty and Liberalism: A protest against the growing tendency toward undue interference by the state, with individual liberty, private enterprise, and the rights of property*, George Robertson, Melbourne, 1887, pp. 241, 245.
96 ibid., p. 327.
97 Bruce Smith, 'Strikes and their Cure', *The Centennial Magazine*, 1 (1888–9), pp. 165–73.
98 Albert Métin, *Socialism without Doctrine*, APCOL, Sydney, 1977.
99 In *The Australian Legend*, OUP, London, 1958.
100 W. K. Hancock quoted in La Nauze, *Political Economy in Australia*, p. 134.
101 R. B. Joyce, 'Samuel Walker Griffith: A Liberal Lawyer' in D. J. Murphy, and R. B. Joyce (eds), *Queensland Political Portraits 1859–1952*, UQP, St Lucia, 1978, pp. 171–2.
102 L. F. Crisp, *Charles Cameron Kingston: Radical Federationist*, ANU, Canberra, 1984, p. 10.
103 P. O'Farrell, 'The Australian Socialist League and the Labour Movement 1887–1891', *HS*, 8 (1958), p. 157; R. N. Ebbels (ed.), *The Australian Labor Movement 1850–1907*, Cheshire-Lansdowne, Melbourne, 1965, pp. 6–7. 'The Report of the Royal Commission on Strikes', *NSWPP*, Second Session 1891, vol. II, pp. 179–209, contains a remarkable bibliography of literature available in Australia at the time on strikes, industrial economics, socialism, etc.
104 Lloyd Ross, *William Lane and the Australian Labor Movement*, Lloyd Ross, Sydney, 1935.
105 Brian Fitzpatrick, *A Short History of the Australian Labor Movement*, Rawson's Bookshop, Melbourne, 1944, chaps 6–8; D. J. Murphy, R. B. Joyce, and Colin A. Hughes (eds), *Prelude to Power: The Rise of the Labour Party in Queensland 1885–1915*, Jacaranda Press, Brisbane, 1970, chap. 7; D. H. Johnson, *Volunteers at Heart: The Queensland Defence Forces 1860–1901*, UQP, St Lucia, 1975, chap. 14.
106 Gavin Souter, *A Peculiar People: The Australians in Paraguay*, Angus and Robertson, Sydney, 1968.
107 Henry Mayer, *Marx, Engels and Australia*, F. W. Cheshire, Melbourne, 1964; Verity Burgmann, *In Our Time: Socialism and the Rise of Labor, 1885–1905*, Allen and Unwin, Sydney, 1985.
108 Quoted O'Farrell, 'The Australian Socialist League...', p. 157.

Higgs's reference to a labour army is reminiscent of Bellamy, whose future society in *Looking Backward* was organized along military lines. Bellamy himself was an admirer of German military efficiency. See Sylvia E. Bowman, *Edward Bellamy Abroad: An American Prophet's Influence*, Twayne, New York, 1962.
109 p. iv.
110 E. H. Lane, *Dawn to Dusk: Reminiscences of a Rebel*, William Brooks and Co., Brisbane, 1939, p. 48.
111 ibid., p. 37.
112 Ebbels, *The Australian Labor Movement*, p. 206.
113 Burgmann, *In Our Time*, p. 6.
114 Bede Nairn, *Civilizing Capitalism: The Labor Movement in New South Wales 1870–1900*, ANU Press, Canberra, 1973.
115 William Pember Reeves, *State Experiments in Australia and New Zealand*, Macmillan, Melbourne, 1969 (1902), vol. 1, p. 68.
116 Christopher Crisp Papers, NLA.
117 Michael Davitt, *Life and Progress in Australasia*, Methuen and Co., London, 1898.
118 First published by Penguin Books, Ringwood, Vic., 1984.
119 *Centennial Magazine*, vol. 1, 1888–89, p. 908.
120 John Miller (William Lane), *The Workingman's Paradise: An Australian Labour Novel*, Edwards, Dunlop and Co., Brisbane, 1892, p. 123.
121 ibid., p. 75.
122 ibid., p. 133.
123 Ian Tyrrell, 'International Aspects of the Women's Temperance Movement in Australia: The Influence of the American WCTU, 1882–1914', *JRH*, 12 (1983), p. 285; also Anthea Hyslop, 'Temperance, Christianity, and Feminism. The Women's Christian Temperance Union of Victoria, 1887–97', *HS*, 17 (1976), pp. 27–49. Compare the figures for trade union membership on p. 277.
124 C. E. Sayers, *David Syme. A Life*, F. W. Cheshire, Melbourne, 1965, p. 91.
125 D. W. A. Baker, *Days of Wrath: A Life of John Dunmore Lang*, MUP, Melbourne, 1985, pp. 489–91.
126 J. A. Froude, *Oceana or England and Her Colonies*, Longmans Green, London, 1886, chaps X–XIII.
127 Murdoch, *Deakin*, p. 167; also J. A. La Nauze, 'The Name of the Commonwealth of Australia', *HS*, 15 (1971), pp. 59–71.
128 Denton Prout, *Henry Lawson: The Grey Dreamer*, Rigby, Adelaide, 1963, p. 62.
129 K. S. Inglis, 'Young Australia 1870–1900: the idea and the reality' in Guy Featherstone (ed.), *The Colonial Child*, RHSV, Melbourne, 1981, pp. 1–23.
130 Ebbels, *The Australian Labor Movement*, pp. 212, 216.
131 John Foster Fraser, *Australia: The Making of a Nation*, Cassell, London, 1910, chap. 2.

CHAPTER 3: SOCIETY

1 Commonwealth of Australia, *Official Year Book*, 1901–7, pp. 162–5; L. R. Smith, *The Aboriginal Population of Australia*, ANU Press, Canberra, 1980, p. 210.
2 Craufurd D. W. Goodwin, *Economic Enquiry in Australia*, Duke University Press, Durham, 1966, chap. 12.

3 Smith, *The Aboriginal Population*, p. 210.
4 ibid., p. 7; also Richard Broome, *Aboriginal Australians: Black Response to White Dominance 1788–1980*, Allen and Unwin, Sydney, 1982, p. 82.
5 Commonwealth of Australia, *Official Year Book*, 1901–7, pp. 144–5; Smith, *The Aboriginal Population*, p. 20; G. Sawer, 'The Australian Constitution and the Australian Aborigine', *Federal Law Review*, 2 (1966), pp. 17–36.
6 This and subsequent discussion based on Commonwealth of Australia, *Official Year Book*, 1901–7, pp. 142–64.
7 ibid., p. 145.
8 All figures from the 1901 UK Census quoted Commonwealth of Australia, *Official Year Book*, 1901–8, p. 155.
9 W. H. Hall, 'The Concentration of Population in the Metropolis and other Urban Districts, the Causes that have led up to it, and a Possible Practical Solution of the Difficulty', *Australasian Catholic Record*, 1908, p. 60.
10 Shirley Fisher, 'Sydney Women and the Workforce 1870–90', in Max Kelly (ed.), *Nineteenth-Century Sydney: Essays in Urban History*, SUP, Sydney, 1978, pp. 102–5.
11 For a sample see Beverley Kingston, 'The Lady and the Australian Girl: Some thoughts on nationalism and class; in Alisa Burns and Norma Grieve (eds), *Australian Women: New Feminist Perspectives*, OUP, Melbourne, 1986, pp. 27–41.
12 See Peter Macdonald and Patricia Quiggin, 'Lifecourse Transitions in Victoria in the 1880s', in Patricia Grimshaw, Chris McConville and Ellen McEwen (eds), *Families in Colonial Australia*, Allen and Unwin, Sydney, 1985, pp. 64–82.
13 Jill Roe, 'Old Age, Young Country: The first old-age pensions and pensioners in New South Wales', *Teaching History*, 15 (July 1981), pp. 23–42.
14 P. F. Macdonald, *Marriage in Australia: Age at First Marriage and Proportions Marrying, 1860–1971*, ANU, Canberra, pp. 105, 107.
15 ibid., p. 96.
16 *W and P*, 1900–1, p. 218.
17 Macdonald, *Marriage in Australia*, pp. 148–9.
18 *W and P*, 1900–1, p. 1000.
19 Ronald Mendelsohn, *The Condition of the People: Social Welfare in Australia 1900–1975*, Allen and Unwin, Sydney, 1979, p. 232.
20 Bryan Gandevia, *Tears Often Shed: Child Health and Welfare in Australia from 1788*, Pergamon Press, 1978, p. 92; also P. H. Curson, *Times of Crisis: Epidemics in Sydney 1788–1900*, SUP, Sydney, 1985, chaps 4–8.
21 Robin Walker, 'The Struggle against Pulmonary Tuberculosis in Australia, 1788–1950', *HS*, 20 (1983), pp. 439–61.
22 The motto with which he prefaced each chapter of *Climate and Health in Australasia*, Street and Co., London, 1886.
23 *W and P*, 1900–1, p. 1014.
24 For an overview of most of these themes see K. S. Inglis, *Hospital and Community: A History of the Royal Melbourne Hospital*, MUP, Melbourne, 1958.
25 C. E. Sayers, *The Women's ... A Social History to mark the 100th Anniversary of The Royal Women's Hospital of Melbourne 1856–1956*, Renwick Pride, Melbourne, 1956; Lyndsay Gardiner, *Royal Children's Hospital Melbourne 1870–1970*, The Royal Children's Hospital,

Melbourne, 1970, chaps I–V; Joan Gillison, *Colonial Doctor and His Town*, Cypress Books, Melbourne, 1974; also Ian Cope, Frank M. C. Forster and Sheila Simpson, *Obstetrics and Gynaecology. Short-title Catalogue of Books published before 1900 and available in Australia together with references to these subjects in Australian and British journals published before 1900*, The Benevolent Society of NSW, Sydney, 1973.

26 Oscar Comettant, *In the Land of Kangaroos and Gold Mines*, Rigby, Adelaide, 1980, p. 183.
27 T. S. Pensabene, *The Rise of the Medical Practitioner in Victoria*, ANU, Canberra, 1980; also Michael Cannon, *Life in the Cities*, Nelson, Melbourne, 1975, chap. 11.
28 *W and P*, 1900–1, p. 1024.
29 F. K. Crowley, 'The British Contribution to the Australian Population: 1860–1919', *University Studies in History*, UWA Press, Nedlands, 1954, vol. 2, pp. 55–88.
30 P. O'Farrell, *The Irish in Australia*, UNSWP, Kensington, 1987; Colm Kiernan (ed.), *Australia and Ireland 1788–1988*, Gill and Macmillan, Dublin, 1986.
31 *The Irish in Australia*, George Robertson and Co., Melbourne, 1888.
32 *ADB*, 5, p. 356.
33 Brian McKinlay, *The First Royal Tour 1867–1868*, Rigby, Adelaide, 1970; A. W. Martin, *Henry Parkes: A Biography*, MUP, Melbourne, 1980, pp. 232–43.
34 John Watson (ed.), *Catalpa 1876. 100 Years ago . . . a special collection of papers on the background and significance of the Fenian escape from Fremantle, Western Australia Easter 1876*, UWA, Nedlands, 1976; Keith Amos, *The Fenians in Australia 1865–1880*, UNSWP, Kensington, 1988.
35 Malcolm Prentis, *The Scots in Australia: A Study of New South Wales, Victoria and Queensland, 1788–1900*, SUP, Sydney, 1983.
36 Janet McCalman, 'Dr Mannix and "Being Irish"', *Meanjin*, 45 (1986), pp. 74–5.
37 D. B. Waterson, 'Thomas McIlwraith: A Colonial Entrepreneur', in D. J. Murphy and R.B. Joyce (eds), *Queensland Political Portraits 1859–1952*, UQP, St Lucia, 1978, pp. 119–42.
38 David S. Macmillan, *Scotland and Australia 1788–1850: Emigration, Commerce and Investment*, OUP, London, 1967.
39 Michael Cannon, *The Land Boomers*, MUP, Melbourne, 1966; Andrew Lemon, *The Young Man from Home: James Balfour 1830–1913*, MUP, Melbourne, 1982.
40 J. Lyng, *Non-Britishers in Australia: Influence on Population and Progress*, Macmillan, Melbourne, 1927, p. 231; Commonwealth of Australia, *Official Year Book*, 1901–8, pp. 157–8.
41 R. B. Walker, 'German-Language Press and People in South Australia, 1848–1900', *JRAHS*, 58 (1972), pp. 121–40.
42 C. A. Price, 'German Settlers in South Australia 1838–1900', *HS*, 7 (1957), pp. 441–51.
43 Jurgen Tampke (ed.), *Wunderbar Country: Germans look at Australia 1850–1914*, Hale and Iremonger, Sydney, 1982, p. 117.
44 ibid., p. 114.
45 A. Brauer, *Under the Southern Cross: History of Evangelical Lutheran Church of Australia*, Lutheran Publishing House, Adelaide, 1956, p. 223.

46 J. Lyng, *The Scandinavians in Australia, New Zealand and the Western Pacific*, MUP, Melbourne, 1939, p. 125.
47 Stanley Raymond (ed.), *Tourist to the Antipodes: William Archer's 'Australian Journey, 1876–77'*, UQP, St Lucia, 1977.
48 Based on Charles A. Price, *Jewish Settlers in Australia*, ANU Press, Canberra, 1964, p. 39.
49 R. B. Walker, *Under Fire: A History of Tobacco Smoking in Australia*, MUP, Melbourne, 1984, p. 20.
50 C. F. Yong, *The New Gold Mountain: The Chinese in Australia 1901–1921*, Raphael Arts Press, Richmond, SA, 1977, p. 275.
51 Charles A. Price, *The Great White Walls Are Built: Restrictive Immigration to North America and Australasia 1836–1888*, ANU Press, Canberra, 1974, p. 277.
52 Yong, *New Gold Mountain*, p. 275.
53 Andrew Markus, 'Divided We Fall: The Chinese and the Melbourne Furniture Trade Union, 1870–1900', *LH*, 26 (1974), pp. 1–10.
54 See A. Curthoys and A. Markus (eds), *Who Are Our Enemies? Racism and the Australian Working Class*, Hale and Iremonger, Sydney, 1978.
55 Andrew Garran (ed.), *Picturesque Atlas of Australasia*, Picturesque Atlas Publishing Co., Sydney, 1886–88, etc., p. 465.
56 ibid., p. 466; P. F. Donovan, *A Land Full of Possibilities: A History of South Australia's Northern Territory*, UQP, St Lucia, 1981, pp. 172–3.
57 Noel Loos, *Invasion and Resistance: Aboriginal–European Relations on the North Queensland Frontier 1861–1897*, ANU Press, Canberra, 1982, p. 84.
58 Kathryn Cronin, *Colonial Casualties: Chinese in Early Victoria*, MUP, Melbourne, 1982.
59 E. M. Andrews, *Australia and China: The Ambiguous Relationship*, MUP, Melbourne, 1985, pp. 15, 24.
60 Kenneth Maddock, *The Australian Aborigines: A Portrait of their Society*, Penguin, Ringwood, Vic., 1975, p. 46, also 2nd edn 1982, chap. 3.
61 Annette Hamilton, 'A Complex Strategical Situation: gender and power in Aboriginal Australia', in Norma Grieve and Patricia Grimshaw (eds), *Australian Women: New Feminist Perspectives*, OUP, Melbourne, 1981, pp. 69–85.
62 Maddock, *The Australian Aborigines*, p. 25.
63 See Broome, *Aboriginal Australians*, chap. 5; Nancy Cato, *Mister Maloga: Daniel Matthews and his Mission, Murray River, 1864–1902*, UQP, St Lucia, 1976.
64 Michael Davitt, *Life and Progress in Australasia*, Methuen and Co., London, 1898, p. 275; D. C. S. Sissons, 'Karayuki-san: Japanese prostitutes in Australia, 1887–1916', *HS*, 17 (1977), pp. 323–41, 474–88.
65 Colin Forster, 'Aspects of Australian Fertility, 1861–1901', *AEHR*, XIV (1974), pp. 105–22.
66 Margaret Grellier, 'The Family: some aspects of its demography and ideology in mind-nineteenth century Western Australia' in C. T. Stannage (ed.), *A New History of Western Australia*, UWA Press, Nedlands, 1981, pp. 473–510.
67 Marnie Bassett, *The Hentys: An Australian Colonial Tapestry*, MUP, Melbourne, 1955; Mary Durack, *Kings in Grass Castles*, Corgi Books, London, 1967; Judith Wright, *The Generations of Men*, OUP, Melbourne, 1959.
68 Some are reproduced in Neville Hicks, *'This Sin and Scandal'. Austra-

lia's *Population Debate 1891–1911*, ANU Press, Canberra, 1978, pp. 94–5.
69 Peter Macdonald and Patricia Quiggin, 'Lifecourse Transitions in Victoria in the 1880s', in Grimshaw, McConville and McEwen (eds), *Families in Colonial Australia*, p. 82.
70 See Hicks, '*This Sin and Scandal*'.
71 See the selections in Ian Turner (ed.), *The Australian Dream*, Sun Books, Melbourne, 1968, pp. 129–56 and L. T. Hergenhan (ed.), *A Colonial City: High and Low Life: Selected journalism of Marcus Clarke*, UQP, St Lucia, 1972, pp. 72–7; also R. E. N. Twopeny, *Town Life in Australia*, Elliot Stock, London, 1883, pp. 82–99.
72 Bryan Gandevia, 'A Comparison of the Heights of Boys Transported to Australia from England, Scotland and Ireland, c. 1840, with Later British and Australian Developments', *Australian Paediatric Journal*, 13 (1977), pp. 95–7.
73 *W and P*, 1900–1, p. 243.
74 Blanche Mitchell, *Blanche, an Australian Diary*, John Ferguson, Sydney, 1980.
75 George W. E. Russell (ed.), *Lady Victoria Buxton: A Memoir with Some Account of her Husband*, Longmans Green, London, 1919, p. 173.
76 Sir Edmund Barton Papers, NLA.
77 Quoted from 'The Imperial Songster' no. 83 (1908) by Graeme Inson and Russel Ward, *The Glorious Years: Of Australia Fair from the Birth of the* Bulletin *to Versailles*, Jacaranda, Brisbane, 1971, p. 74.
78 Macdonald, *Marriage in Australia*, p. 149.
79 J. F. Hogan, 'The coming Australian' from the *Victorian Review*, November 1880, quoted, Turner, *The Australian Dream*, pp. 134–9.
80 Twopeny, *Town Life*, pp. 82–3.
81 ibid., p. 83.
82 e.g. D.E. McConnell, *Australian Etiquette, or the Rules and Usages of the Best Society in the Australian Colonies together with their Sports, Pastimes, Games and Amusements*, D. E. McConnell, Sydney, 1885.
83 Comettant, *In the Land of Gold and Kangaroos*, pp. 54–6.
84 Robert Travers, *Australian Mandarin: The Life and Times of Quong Tart*, Kangaroo Press, Kenthurst, 1981.
85 See Brian Lewis, *Sunday at Kooyong Road*, Hutchinson, Melbourne, 1976.
86 See p. 73.
87 Commonwealth of Australia, *Official Year Book*, 1901–8, pp. 186–8.
88 P. and D. O'Farrell, *Documents in Australian Catholic History*, vol. 1, 1788–1884, Geoffrey Chapman, London, 1969, pp. 419–20.
89 Stephen Garton, 'Bad or Mad: Developments in Incarceration in NSW 1880–1920', in Sydney Labour History Group, *What Rough Beast? The State and Social Order in Australian History*, Allen and Unwin, Sydney, 1982.
90 e.g. Brian Dickey, 'Dependence in South Australia', *Australia 1888*, 8, September 1981, pp. 88–96.
91 S. L. Hateley, 'The Queen's Fund, Melbourne 1887–1900', *Melbourne Historical Journal*, no. 11 (1972), pp. 11–41.
92 R. B. Walker, 'The Ambiguous Experiment—Agricultural Co-operatives in New South Wales, 1893–96', *LH*, 18 (1970), pp. 19–31.
93 Moore's *Australian Almanac and Handbook*, 1861, p. 64.
94 *NSWVP* 1871–72, p. 626.
95 Moore's *Australian Almanac and Handbook*, 1901, p. 28.

96   ibid., pp. 73-7.
97   Brian Dickey, *No Charity There: A Short History of Social Welfare in Australia*, Nelson, Melbourne, 1980, pp. 38-44, 70-80.
98   Raymond Evans, 'The Hidden Colonists: Deviance and Social Control in Colonial Queensland', Jill Roe (ed.), *Social Policy in Australia: Some Perspectives 1901-1975*, Cassell, Sydney, 1976, pp. 74-100; Joan Brown, *'Poverty is not a Crime': The Development of Social Services in Tasmania 1803-1900*, Tasmanian Historical Research Association, Hobart, 1972.
99   Anthony Trollope, *Australia*, UQP, St Lucia, 1967, p. 507.
100 ibid., p. 572; see also Peter Bolger, *Hobart Town*, ANU Press, Canberra, 1973; H. Reynolds, '"That Hated Stain": The aftermath of transportation in Tasmania', *HS*, 14 (1969), pp. 19-31.
101 C. T. Stannage, *The People of Perth: A Social History of Western Australia's Capital City*, Perth City Council, Perth, 1979, p. 99.
102 D. B. Waterson, 'The Remarkable Career of Wm. H. Groom', *JRAHS*, 49 (1963), pp. 38-57; also K. S. Inglis, *The Rehearsal: Australians at War in the Sudan 1885*, Rigby, Adelaide, 1985, pp. 84-5, 153 on the convict ancestry of New South Wales politicians.
103 Marcus Clarke, *His Natural Life*, George Robertson, Melbourne, 1874; Mrs Campbell Praed, *Policy and Passion*, Bentley, London, 1881; Rolf Boldrewood, *Robbery under Arms*, Macmillan, London, 1889; also Barry Andrews on the romance and mythology of convictism in his *Price Warung (William Astley)*, Twayne, Boston, 1976.
104 Trollope, *Australia*, pp. 586-7, 596-9; Rosamond and Florence Hill, *What We Saw in Australia*, Macmillan, London, 1875, p. 36.
105 Stannage, *The People of Perth*, p. 93.
106 Trollope, *Australia*, p. 501.
107 Raymond Evans, 'The Hidden Colonists'.
108 Paul Hasluck, *Black Australians: A Survey of Native Policy in Western Australia, 1829-1897*, MUP, Melbourne, 1942.
109 Raymond Evans, Kay Saunders, and Kathryn Cronin, *Exclusion, Exploitation, and Extermination: Race Relations in Colonial Queensland*, ANZ Book Co., Sydney, 1975, p. 167; also Kay Saunders, 'Pacific Islander women in Queensland: 1863-1907' in Margaret Bevege, Margaret James and Carmel Shute (eds), *Worth Her Salt: Women at Work in Australia*, Hale and Iremonger, Sydney, 1982, p. 17; Clive Moore, *Kanaka: A History of Melanesian Mackay*, New Guinea Press, Boroko, 1985.
110 O. W. Parnaby, *Britain and the Labor Trade in the Southwest Pacific*, Duke University Press, Durham, 1964; Kay Saunders, *Workers in Bondage: The Origins and Bases of Unfree Labour in Queensland 1824-1916*, UQP, St Lucia, 1982.
111 Evans, et al., *Exclusion, Exploitation*, p. 186.
112 Parnaby, *Britain and the Labor Trade*, p. 205; Smith, *The Aboriginal Population*, p. 226.
113 p. 945.
114 Evans, et al., *Exclusion, Exploitation*, p. 189.
115 Peter Corris, *Passage, Port and Plantation: A History of Solomon Islands Labour Migration 1870-1914*, MUP, Melbourne, 1973, chap. 5.
116 H. N. Moseley, *Notes by a Naturalist: An Account of Observations made during the Voyage of H. M. S. Challenger round the world in the Years 1872-1876*, John Murray, London, 1892, p. 305.
117 ibid., p. 309.

118 ibid., pp. 305-7.
119 Loos, *Invasion and Resistance*, p. 113.
120 ibid., p. 114.
121 ibid., p. 182; see also C. D. Rowley, *The Destruction of Aboriginal Society: Aboriginal Policy and Practice*, vol. 1, ANU Press, Canberra, 1970, p. 96; Linda Wilkinson, 'Fractured Families, Squatting and Poverty: the impact of the 1886 "Half-caste" Act on Framglingham Aboriginal Community', *Law and History in Australia*, 2, La Trobe University, Melbourne, 1986.
122 Barry Jones (ed.), *The Penalty is Death: Capital Punishment in the Twentieth Century*, Sun Books, Melbourne, 1968, p. 256; *W and P*, 1900-1, pp. 244-6.
123 Creighton Burns, *The Tait Case*, MUP, Melbourne, 1962, p. 13.
124 *W and P*, 1900-1, p. 206.
125 ibid., pp. 250-1; also Robert Haldane, *The People's Force: A History of the Victoria Police*, MUP, Melbourne, 1986, p. 111; Robin Walker, 'The New South Wales Police Force, 1862-1900', *Journal of Australian Studies*, 15 (1984), pp. 25-38.
126 *W and P*, 1900-1, p. 249.
127 Malcolm A. C. Fraser, *Western Australian Year Book*, 1902-4, Govt. Printer, Perth, 1906, p. 1041.
128 ibid., pp. 1038, 1055; also J. E. Thomas, 'Crime and Society', in Stannage (ed.), *A New History of Western Australia*, pp. 636-51.
129 Craufurd D. W. Goodwin, *The Image of Australia: British perception of the Australian Economy from the Eighteenth to the Twentieth Century*, Duke University Press, Durham, 1974, pp. 132-3.
130 Ted Robert Gurr, Peter N. Grabosky, Richard C. Hula, *The Politics of Crime and Conflict: A Comparative History of Four Cities*, Sage, Beverly Hills, 1977, p. 371.
131 ibid.
132 Kay Daniels (ed.), *So Much Hard Work: Women and Prostitution in Australian History*, Fontana/Collins, Sydney, 1984, chaps 1-3.
133 Peter N. Grabosky, *Sydney in Ferment: Crime, Dissent and Official Reaction 1788-1973*, ANU Press, Canberra, 1977, chap. 7.
134 Helen Jones, *In Her Own Name: Women in South Australian History*, Wakefield Press, Adelaide, 1986, pp. 21-2.
135 Richard P. Davis, *The Tasmanian Gallows*, Cat and Fiddle Press, Hobart, 1974, pp. 29-32.
136 Carmel Harris, 'The "Terror of the Law" as Applied to Black Rapists in Colonial Queensland', *Hecate*, viii (1982), pp. 22-48.
137 David Walker, 'Youth on Trial: The Mt Rennie Case', *LH*, 50 (1986), pp. 28-41.
138 See Leslie Blackwell, *Death Cell at Darlinghurst*, Hutchinson, Melbourne, 1970.
139 R. B. Walker, *On Fire*, p. 42; see also C. W. Dilke, *Problems of Greater Britain*, Macmillan, London, 1890, pp. 605-23.
140 Emile Durkheim, *Suicide. A Study in Sociology*, Routledge Kegan Paul, London, 1952, p. 391; G. H. Knibbs, 'Suicide in Australia. A Statistical Analysis of the Facts', *Journal and Proceedings of the Royal Society of New South Wales*, XLV (1912), pp. 225-46.
141 A. G. Austin (ed.), *The Webbs' Australian Diary 1898*, Pitman, Melbourne, 1965.
142 Henry Demarest Lloyd, *Newest England: Notes of a Democratic Traveller in New Zealand, with some Australian Comparisons*, Doubleday, New York, 1900.

143 Robert Travers, *Murder in the Blue Mountains*, Hutchinson, Melbourne, 1972, pp. 165–6.
144 Mark Finnane, 'The Popular Defence of Chidley', *LH*, 50 (1986), p. 61.
145 e.g. William Pember Reeves, *State Experiments in Australia and New Zealand*, Macmillan, Melbourne, 1969 (1902).

CHAPTER 4: CULTURE

1 See W. S. Ramson, *Australian English: An Historical Study of the Vocabulary 1788–1898*, ANU Press, Canberra, 1966; G. W. Turner, *The English Language in Australia and New Zealand*, Longmans, London, 1966; also the Prefaces to *The Macquarie Dictionary*, Macquarie Library, St Leonards, NSW, 1981.
2 W. S. Jevons 'Social Survey of Australian Cities', quoted by Barry Groom and Warren Wickman, *Sydney—The 1850s. The Lost Collections: Eyewitness Accounts and Early Photographs of Sydney*, The Macleay Museum, University of Sydney, 1982, p. 74.
3 See Beatrice Bligh, *Cherish the Earth: The Story of Gardening in Australia*, Ure Smith, Sydney, 1973; Victor Crittenden, *The Front Garden. The Story of the Cottage Garden in Australia*, Mulini Press, Canberra, 1979.
4 J. A. Froude, *Oceana, or England and Her Colonies*, Longmans, Green, London, 1886, pp. 80–1.
5 Anthony Trollope, *Australia*, UQP, St Lucia, 1967, p. 232.
6 Lionel Gilbert, *The Royal Botanic Gardens, Sydney: A History 1861–1985*, OUP, Melbourne, 1986, chap. 5.
7 Margaret Willis, *By Their Fruit: a Life of Ferdinand von Mueller Botanist and Explorer*, Angus and Robertson, Sydney, 1949; Edward Kynaston, *A Man on Edge: A Life of Baron Sir Ferdinand von Mueller*, Allen Lane, Ringwood, Vic., 1981.
8 R. T. M. Pescott, *W. R. Guilfoyle 1840–1912: The Master of Landscaping*, OUP, Melbourne, 1974.
9 Mrs Lance Rawson, *Australian Enquiry Book of Household and General Information*, Pater and Knapton, Melbourne, 1894; James McEwan and Co.'s *Illustrated Catalogue of Furnishing and General Ironomongery*, a facsimile printing, Hale and Iremonger, Sydney, n.d.; *Australia in the Good Old Days: facsimile pages from 'Lassetters' Commercial Review'*, Ure Smith, Sydney, 1976.
10 L. T. Hergenhan, *A Colonial City: High and Low Life, Selected Journalism of Marcus Clarke*, UQP, St Lucia, 1972 (c. 1874), pp. 328–30.
11 As in Ada Cambridge, *Thirty Years in Australia*, Methuen, London, 1903, pp. 31–3, or Florence and Rosamond Hill, *What We Saw in Australia*, Macmillan, London, 1875, p. 61; also Margaret Kiddle, *Men of Yesterday: A Social History of the Western District of Victoria 1834–1890*, MUP, Melbourne, 1961.
12 H. Mortimer Franklyn, *A Glance at Australia in 1880*, The Victorian Review Publishing Co., Melbourne, 1881, p. xii; Froude, *Oceana*, p. 73.
13 Phillip E. Muskett, *The Art of Living in Australia*, Eyre and Spottiswoode, London, 1893, p. 57.
14 e.g. Francis Adams, *The Melburnians*, Remington, London, 1892, or Mrs Campbell Praed, *Mrs. Tregaskiss: A Novel of Anglo-Australian Life*, Chatto and Windus, London, 1894.
15 Maisy Stapleton and Patricia McDonald, *Christmas in the Colonies*,

David Ell Press, Sydney, 1981.
16 Edmond Marin la Meslée, *The New Australia*, Heinemann Educational, Melbourne, 1979 (1883), p. 67.
17 Andrew Garran (ed.), *The Picturesque Atlas of Australasia*, Picturesque Atlas Publishing Co., Sydney, 1886–88, p. 506.
18 ibid., p. 530.
19 K. S. Inglis, *The Australian Colonists: An Exploration of Social History 1788–1870*, MUP, Melbourne, 1974, chap. 4.
20 Geoffrey Serle, 'The Victorian Government's Campaign for Federation 1883–1889', in A. W. Martin (ed.), *Essays in Australian Federation*, MUP, Melbourne, 1969, p. 34.
21 W. E. Murphy, *History of the Eight Hours' Movement*, Spectator Pub. Co., Melbourne, 1896–1900.
22 A. E. Dingle, '"The Truly Magnificent Thirst": An historical survey of Australian drinking habits', *HS*, 19 (1980), pp. 227–49.
23 John O'Hara, 'The Australian Gambling Tradition', in Richard Cashman and Michael McKernan (eds), *Sport Money Morality and the Media*, UNSWP, Kensington, 1981, pp. 68–85; Ken Inglis, 'Gambling and Culture in Australia', Geoffrey Caldwell *et al.* (eds), *Gambling in Australia*, Croom Helm, Sydney, 1985, pp. 5–17; also Hugh Buggy, *The Real John Wren*, Angus and Robertson, Sydney, 1986, pp. 12–39.
24 R. B. Walker, 'Tobacco Smoking in Australia, 1788–1914', *HS*, 19 (1980), pp. 267–85, also *Under Fire: A History of Tobacco Smoking in Australia*, MUP, Melbourne, 1984.
25 Harold Finch-Hatton, *Advance Australia!: An account of eight years' work, wandering, and amusement, in Queensland, New South Wales, and Victoria*, W. H. Allen and Co., London, 1886, p. 44.
26 H.M. Green, *A History of Australian Literature*, Angus and Robertson, Sydney, 1961, pp. 360–77.
27 Douglas Sladen, *Twenty Years of My Life*, Constable and Co., London, 1915, p. 17.
28 Figures from Commonwealth of Australia, *Official Year Book*, 1901–7, p. 281.
29 Michael Stringer, *Australian Horse Drawn Vehicles*, Rigby, Adelaide, 1980, p. 113.
30 Michael Davitt, *Life and Progress in Australasia*, Methuen, London, 1898, p. 76.
31 D. Barrie, 'Horse-racing', *The Australian Encyclopaedia*, Grolier Society of Australia, Sydney, 5 (1983), pp. 167–74.
32 Beatrice Webb's comment quoted in A. G. Austin (ed.), *The Webbs' Australian Diary 1898*, Pitman, Melbourne, 1965, p. 33.
33 Alexandra Hasluck (ed.), *Audrey Tennyson's Vice-Regal Days: The Australian letters of Audrey Lady Tennyson to her mother Zacyntha Boyle, 1899–1903*, NLA, Canberra, 1978, e.g. p. 275.
34 Austin, *The Webbs' Australian Diary*, p. 32.
35 John Freeman, *Lights and Shadows of Melbourne Life*, Sampson Low etc., London, 1888, p. 55.
36 Keith Dunstan, *Sports*, Cassell, Melbourne, 1973, p. 322.
37 Trollope, *Australia*, p. 667.
38 Dunstan, *Sports*, p. 328–9.
39 Leonie Sandercock and Ian Turner, *Up Where, Cazaly? The Great Australian Game*, Granada, London, 1982, p. 39.
40 *The Australasian*, 18 November 1882, chap. xiii.

41 Hasluck, *Audrey Tennyson*, pp. 312–13.
42 Alfred Searcy, *In Australian Tropics*, Kegan Paul Trench Trubner, London, 1907, chap. XVIII.
43 W. H. Wilde, *Henry Kendall*, Twayne, Boston, 1976, pp. 43–5.
44 A. C. Bicknell, *Travel and Adventures in Northern Queensland*, Longmans Green, London, 1895, p. 66.
45 Quoted Raymond Evans, Kay Saunders and Kathryn Cronin, *Expulsion, Exploitation and Extermination: Race Relations in Colonial Queensland*, ANZ Book Co., Sydney, 1975, p. 78.
46 Clement Semmler, *The Banjo of the Bush: The Life and Times of A. B. Paterson*, UQP, St Lucia, 1974. p. 108.
47 Ray Robinson, 'Cricket', *The Australian Encyclopaedia*, 3 (1983), pp. 91–9.
48 D. J. Mulvaney, *Cricket Walkabout: The Australian Aboriginal Cricketers on Tour 1867–8*, MUP, Melbourne, 1967, p. 4.
49 John A. Daly, *Elysian Fields: Sport, Class and Community in Colonial South Australia, 1836–1890*, Daly, Adelaide, 1982.
50 ibid., p. 110, quoting an address by William Bundey in Adelaide in 1880.
51 'Sport and Society 1890–1940: A foray', in C. T. Stannage (ed.), *A New History of Western Australia*, UWA Press, Nedlands, 1981, pp. 562–76.
52 See W. F. Mandle, 'Games People Played: Cricket and football in England and Victoria in the late nineteenth century', along with Ian Turner's 'Comment', *HS*, 15 (1973), pp. 511–38.
53 W. F. Mandle, 'Cricket and Australian Nationalism in the Nineteenth Century', *JRAHS*, 59 (1973), pp. 225–46; K. S. Inglis, 'Imperial Cricket: Test matches between Australia and England 1877–1900', in Richard Cashman and Michael McKernan (eds), *The Making of Modern Sporting History*, UQP, St Lucia, 1979, pp. 148–79.
54 Philip Derriman, 'Death in Orange: The strange case of Jack Marsh, Australia's great black cricketer', *Sydney Morning Herald Magazine*, 12 January 1985, pp. 22–4.
55 Peter Corris, *Lords of the Ring: A History of Prize Fighting in Australia*, Cassell, Sydney, 1980, chaps 5–7.
56 Jim Fitzpatrick, *The Bicycle and the Bush: Man and Machine in Rural Australia*, OUP, Melbourne, 1980, chap. 5.
57 Scott Bennett, *The Clarence Comet: The Career of Henry Searle 1866–89*, SUP, Sydney, 1973.
58 Semmler, *The Banjo of the Bush*, pp. 49–51.
59 Eric Drayton Davis, *The Life and Times of Steele Rudd*, Lansdowne, Melbourne, 1976, pp. 55–6.
60 Bennett, *The Clarence Comet*, p. 88.
61 See C. M. H. Clark's treatment in *A History of Australia: IV The Earth Abideth Forever 1851–1888*, MUP, Melbourne, 1978, pp. 159–64.
62 Geoffrey Serle, *The Rush to be Rich*, MUP, Melbourne, 1971, pp. 199–201.
63 Rosemary Gill, 'Thomas Joseph Byrnes: The man and the legend', in D. J. Murphy and R. B. Joyce (eds), *Queensland Political Portraits 1859–1952*, UQP, St Lucia, 1978, pp. 190–1.
64 John McQuilton, *The Kelly Outbreak 1878–1880*, MUP, Melbourne, 1979, pp. 202–9.
65 Bennett, *The Clarence Comet*, p. 74, quoting K. S. Inglis 'The Australians at Gallipoli—II', *HS*, 14 (1970), p. 371.

66 Henry Gyles Turner and Alexander Sutherland, *The Development of Australian literature*, George Robertson and Co., Melbourne, 1898, p. 133.
67 Peter Game, *The Music Sellers*, Hawthorn Press, Melbourne, 1976, p. 149.
68 See Richard Cashman, *'Ave a Go. Yer Mug!': Australian Cricket Crowds from Larrikin to Ocker*, Collins, Sydney, 1984, chaps 1–2.
69 See C. Turney, 'William Wilkins—Australia's Kay-Shuttleworth', in C. Turney (ed.), *Pioneers of Australian Education: A Study in the Development of Education in New South Wales in the Nineteenth Century*, SUP, Sydney, 1969, pp. 193–46.
70 Margaret Barbalet, *Far From a Low Gutter Girl: The Forgotten World of State Wards: South Australia 1887–1940*, OUP, Melbourne, 1983, p. 16.
71 Commonwealth of Australia, *Official Year Book*, 1901–7, pp. 748–50.
72 *W and P*, 1900–1, p. 168.
73 Havelock Ellis, *My Life*, Neville Spearman, London, 1967, pp. 102–3.
74 Cecil Hadgraft, *James Brunton Stephens*, UQP, St Lucia, 1969, pp. 54–60.
75 Una Monk (ed.), *New Horizons: A hundred years of women's migration*, HMSO, London, 1963, p. 32; Catherine Helen Spence, *Clara Morison: a tale of South Australia during the gold fever*, J. W. Parker, London, 1854; Gwen Jones, 'Governessing in the Colonies: The Mobility of Middle Class Women in Nineteenth Century Australia', *Australia 1888*, no. 8, September 1981, pp. 13–20.
76 Joan Lindsay, *Picnic at Hanging Rock*, Cheshire, Melbourne, 1967. Marjorie Theobald's *Ruyton Remembers 1878–1978*, Hawthorn Press, Melbourne, 1978 is a good introduction to the world of the girls' private schools. On the convents see Ronald Fogarty, *Catholic Education in Australia 1806–1950*, vol. II, MUP, Melbourne, 1959.
77 F. W. Brenton, *The Professional, Mercantile and Trades Directory Directory of Adelaide*, 1870, p. 27.
78 Turney, 'William Wilkins', *Pioneers of Australian Education*, chap. viii.
79 R. J. Burns and C. Turney, 'A. B. Weigall's Headmastership of Sydney Grammar School', in Turney (ed.), *Pioneers of Australian Education*, chap. v.
80 See C. E. W. Bean, *Here My Son: An Account of Independent and other Corporate Boys' Schools of Australia*, Angus and Robertson, Sydney, 1950.
81 G. E. Saunders, 'J. A. Hartley and the Foundation of the Public School System in South Australia', in C. Turney (ed.), *Pioneers of Australian Education*, vol. 2: *Studies of the Development of Education in the Australian Colonies 1850–1900*, SUP, Sydney, 1972, chap. v.
82 R. J. W. Selleck, 'F. J. Gladman—Trainer of Teachers', in ibid., chap. iii.
83 C. Turney, 'W. Catton Grasby—Harbinger of Reform', in ibid. chap. vi.
84 ibid., p. 203.
85 ibid., p. 223.
86 ibid., p. 205.
87 L. A. Mandelson, 'Norman Selfe and the Beginnings of Technical Education', in ibid., chap. iv.

88 John Tregenza, *Professor of Democracy: The Life of Charles Henry Pearson 1830–1894: Oxford Don and Australian Radical*, MUP, Melbourne, 1968.
89 See Alan Barcan, *A History of Australian Education*, OUP, Melbourne. 1980, pp. 198–239; also Bruce Mitchell, *Teachers, Education, and Politics: A History of Organizations of Public School Teachers in New South Wales*, UQP, St Lucia, 1975, chap. 1; R. J. W. Selleck and Martin Sullivan (eds), *Not So Eminent Victorians*, MUP, Melbourne, 1984.
90 See M. L. Walker, 'The Development of Kindergartens in Australia', unpub. MEd thesis, University of Sydney, 1964.
91 Barcan, *History of Education*, chap. 10; also Ann J. Truscott, 'Primary Teachers Experiences in Rural Victoria 1888', *Australia 1888*, no. 8, September 1981, pp. 28–39.
92 S. G. Firth, 'Social Values in the New South Wales Primary School 1880–1914; an Analysis of School Texts', in R. J. W. Selleck (ed.), *Melbourne Studies in Education*, 1970, p. 137.
93 Barcan, *History of Education*, p. 408.
94 Commonwealth of Australia, *Official Year Book*, 1901–7, pp. 199, 750.
95 Helen Jones, *Nothing Seemed Impossible: Women's Education and Social Change in South Australia 1875–1915*, UQP, St Lucia, 1985; Alison Mackinnon, 'Educating the Mothers of the Nation: The Advanced School for Girls, Adelaide', in Bevege, James and Schute (eds), *Worth Her Salt*, pp. 62–71.
96 J. R. Lawry, 'Charles Lilley and his Vision of the Queensland System', Turney (ed.), *Pioneer of Australian Education*, 2, chap. ii.
97 Roger B. Joyce, *Samuel Walker Griffith*, UQP, St Lucia, 1984, pp. 9–10.
98 R. M. Crawford, *'A Bit of a Rebel': The Life and Work of George Arnold Wood*, SUP, Sydney, 1975, chaps x–xiii.
99 Ernest Scott, *A History of the University of Melbourne*, MUP, Melbourne, 1936, chap. XI; Geoffrey Blainey, *A Centenary History of the University of Melbourne*, MUP, Melbourne, 1957, pp. 114–18.
100 Tom Collins, *Rigby's Romance*, Angus and Robertson, Sydney, 1946, p. 54.
101 Derek Whitelock, *The Great Tradition: A History of Adult Education in Australia*, UQP, St Lucia, 1974, pp. 119, 127.
102 Robert Gittings, *Young Thomas Hardy*, Heinemann Educational Books, London, 1975, p. 52.
103 Walter Murdoch, *Alfred Deakin: A Sketch*, Constable and Co., London, 1923, p. 43.
104 Barcan, *History of Education*, p. 189.
105 William Splatt and Susan Bruce, *Australian Impressionist Painters. A Pictorial History of the Heidelberg School*, Curry O'Neil, Melbourne, 1981, p. 10.
106 Tom Collins, *Rigby's Romance*, p. 36.
107 ibid., p. 37.
108 R. B. Walker, *The Newspaper Press in New South Wales 1803–1920*, SUP, Sydney, 1976, chap. 9; Richard White, *Inventing Australia: Images and Identity 1688–1980*, Allen and Unwin, Sydney, 1981, chap. 6.
109 Commonwealth of Australia, *Official Year Book*, 1901–7, p. 601.
110 Quoted Sylvia Lawson, *The Archibald Paradox: A Strange Case of*

Authorship, Allen Lane, Ringwood, Vic., 1983, pp. 7–8.
111 White, *Inventing Australia*, p. 89.
112 Lurline Stuart, *Nineteenth Century Australian Periodicals: An Annotated Bibliography*, Hale and Iremonger, Sydney, 1979.
113 Walker, *The Newspaper Press in New South Wales 1803–1920*, chap. 12; also H. J. Gibbney, *Labor in Print*, ANU Press, Canberra, 1975.
114 Whitelock, *The Great Tradition*, p. 119.
115 John Holroyd, *George Robertson of Melbourne 1825–1898: Pioneer Bookseller and Publisher*, Robertson and Mullens, Melbourne, 1968.
116 Green, *History of Australian Literature*, p. 292, quoting an early historian and critic of Australian literature, G. B. Barton, Edmund's brother.
117 Cole Turnley, *Cole of the Book Arcade: A Pictorial Biography of E. W. Cole*, Cole Publications, Melbourne, 1974.
118 William Westgarth, *Half a Century of Australasian Progress: A Personal Retrospect*, Sampson Low, London, 1889, pp. 49–50. He also noted that the glue was hopeless.
119 Pauline M. Kirk, 'Colonial Literature for Colonial Readers?', *Australian Literary Studies*, 5 (1971), pp. 133–45.
120 Brian Elliott, *Marcus Clarke*, OUP, Clarendon, 1958, pp. 113–20.
121 Simon Nowell-Smith, *Letters to Macmillan*, Macmillan, London, 1967, p. 210.
122 Green, *History of Australian Literature*, pp. 179–80, 338–40.
123 Graeme Davison, *The Rise and Fall of Marvellous Melbourne*, MUP, Melbourne, 1978, p. 257, concludes in a typically myopic Melbourne manner, 'At the end of the tragic story of "Marvellous Melbourne" we are embarked on the brilliant career of the "Australian legend", and the proud city, with all her marvellous skill and enterprise is shackled to the alien mythology of a nomad tribe'; see also Harry Heseltine, *Vance Palmer*, UQP, St Lucia, 1970, especially p. 4; David Walker, *Dearm and Disillusion: A Search for Australian Cultural Identity*, ANU Press, Canberra, 1976; Craig Munro, *Wild Man of Letters*, MUP, Melbourne, 1984. Lynne Strahan, *Just City and the Mirrors: Meanjin Quarterly and the Intellectual Front 1940–1965*, OUP, Melbourne, 1984, chap. 1.
124 Wilde, *Henry Kendall*, pp. 58–62.
125 See my introduction to the re-issue of Rosa Campbell Praed, *Policy and Passion*, Virago, London, 1988.
126 Brian Elliott, *Marcus Clarke*, chap. VIII.
127 See Alan Brissenden (ed.), *Rolf Boldrewood*, UQP, St Lucia, 1979.
128 Barry Andrews, *Price Warung (William Astley)*, Twayne, Boston, 1976.
129 See Jack Cato, *The Story of the Camera in Australia*, Institute of Australian Photography, 1979; also Alan Davies and Peter Stanbury, *The Mechanical Eye in Australia: Photography 1841–1900*, OUP, Melbourne, 1985.
130 Oscar Comettant, *In the Land of Kangaroos*, Rigby, Adelaide, 1980 (1888), p. 136.
131 R. H. Croll, *Tom Roberts, Father of Australian Landscape Painting*, George Robertson, Melbourne, 1935, pp. 5, 23.
132 See Daniel Thomas, *Australian Art in the 1870s*, Art Gallery of New South Wales, Sydney, 1976; also J. G. Steele, *Conrad Martens in Queensland: The Frontier Travels of a Colonial Artist*, UQP, St Lucia, 1978.

133 Comettant, *In the Land of Kangaroos*, pp. 156–7.
134 Joan Kerr and Hugh Falkus, *From Sydney Cove to Duntroon: A Family Album of Early Life in Australia*, Hutchinson, Melbourne, 1982.
135 Alan McCulloch, *The Golden Age of Australian Painting: Impressionism and the Heidelberg School*, Lansdowne, Melbourne, 1969, pp. 68, 193.
136 Croll, *Tom Roberts*, p. 148; V. Spate, *Tom Roberts*, Lansdowne, Melbourne, 1972, pp. 33–7; also Helen Topliss, *Tom Roberts 1856–1931: A Catalogue Raisonné*, OUP, Melbourne, 1985, pp. 3–77.
137 e.g. James McEwan and Co.'s *Illustrated Catalogue of Furnishing & General Ironmongery*.
138 Julian Ashton, *Now Came Still Evening On*, Angus and Robertson, Sydney, 1941, pp. 30–2; also John McQuilton, *The Kelly Outbreak 1878–1880*, illustrations pp. 114–15.
139 Vane Lindsay, *The Inked-in Image: A Social and Historical Survey of Australian Comic Art*, Hutchinson, Melbourne, 1979, p. 13.
140 ibid., p. 12; George A. Taylor, *'Those Were the Days' Being Reminiscences of Australian Artists and Writers*, Tyrrell's Ltd, Sydney, 1918, p. 26.
141 Michael Cannon, *That Damned Democrat: John Norton, an Australian populist, 1858–1916*, MUP, Melbourne, 1981.
142 Comettant, *In the Land of Kangaroos*, pp. 156–7.
143 Mona Brand, *Australiana: Over 150 Years of Decorative Crafts, Furniture, Jewellery, Pottery, Coins and Bottles*, Ure Smith, Sydney, 1979; Mimmo Cozzolino, *Symbols of Australia*, Penguin Books, Ringwood, Vic., 1980.
144 See his attempt to create an Australian fairy-tale, L. Henry, *Australian Legend: The War-atah*, Neal's English Library, Paris, 1891; also Paul Nixon, *The Waratah*, Kangaroo Press, Kenthurst, 1987.
145 Peter Cuffley, *Chandeliers and Billy Tea: A Catalogue of Australian Life 1880–1940*, Five Mile Press, Hawthorn, Vic., 1984, p. 89.
146 Terry Sturm in Leonie Kramer (ed.), *The Oxford History of Australian Literature*, OUP, Melbourne, 1981, pp. 197–8.
147 Alec Bagot, *Coppin the Great*, MUP, Melbourne, 1965; Eric Irvin, *Gentleman George, King of Melodrama*, UQP, St Lucia, 1980.
148 Ian G. Dicker, *J. C. W.: A Short Biography of James Cassius Williamson*, Elizabeth Tudor Press, Rose Bay, NSW, 1974, pp. 51–2.
149 Eric Irvin, *Australian Melodrama: Eighty Years of Popular Theatre*, Hale and Iremonger, Sydney, 1981, p. 74.
150 Harold Love, *The Golden Age of Australian Opera: W. S. Lyster and his Companies 1861–1880*, Currency Press, Sydney, 1981, chap. 5.
151 Freeman, *Lights and Shadows*, p. 73.
152 Irvin, *Australian Melodrama*, pp. 106–42.
153 ibid., pp. 62–3.
154 See Margaret Williamson, *Australia on the Popular Stage 1829–1929*, OUP, Melbourne, 1983.
155 Andrew Pike and Ross Cooper, *Australian Film 1900–1977: A Guide to Feature Film Production*, OUP, Melbourne, 1980.
156 Lyon and Stowe Family Papers, ML.
157 Love, *Golden Age of Australian Opera*, p. 273.
158 Nancy Bonnin (ed.), *Katie Hume on the Darling Downs: A Colonial Marriage. Letters of a Colonial Lady, 1866–1871*, Darling Downs Institute Press, Toowoomba, 1985, p. 58.
159 ibid., p. 215.
160 P. and D. O'Farrell, *Documents in Australian Catholic History*, vol. 1

1788–1884, Geoffrey Chapman, London, 1969, pp. 336–40.
161 Based on W. Arundel Orchard, *Music in Australia*, Georgian House, Melbourne, 1952, chap. xi.
162 Weston Bate, *Lucky City: The First Generation at Ballarat 1851–1901*, MUP, Melbourne, 1978, p. 230.
163 Orchard, *Music in Australia*, p. 92.
164 See Game, *The Music Sellers*.
165 Thérèse Radic, 'Music of the Centennial Exhibition, Melbourne 1888–9. The Context', *Australia 1888*, no. 7, pp. 59–67.
166 Orchard, *Music in Australia*, p. 118.
167 ibid., p. 124.
168 ibid., p. 138.
169 Roger Covell, *Australia's Music: Themes of a New Society*, Sun Books, Melbourne, 1967, chap. 1.
170 John Bird, *Percy Grainger*, Macmillan, Melbourne, 1977.
171 Orchard, *Music in Australia*, pp. 161–2, 181.
172 ibid., p. 181.
173 ibid., p. 183.
174 Humphrey McQueen, *A New Britannia*, Penguin Books, Ringwood, Vic., 1986, chap. 9.
175 Orchard, *Music in Australia*, p. 172.
176 Comettant, *In the Land of Kangaroos*, p. 232.
177 E. N. Matthews, *Colonial Organs and Organbuilders*, MUP, Melbourne, 1969, p. 5.
178 Eve Keane, *Music for a Hundred Years: The Story of the House of Paling*, Oswald Ziegler Publications, Sydney, 1954.
179 Game, *The Music Sellers*.
180 Comettant, *In the Land of Kangaroos*, p. 138; also Helen Jones, *Nothing Seemed Impossible*, p. 72.
181 *Betty Wayside*, Hodder and Stoughton, London, 1915.
182 *Maurice Guest*, Heinemann, London, 1908.
183 Charles Philips Trevelyan, *Letters from North America and the Pacific 1898*, Chatto and Windus, London, 1969, p. 182; Shirley Andrews, *Take Your Partners: Traditional Dancing in Australia*, Hyland House, Melbourne, 1979.
184 Edward H. Pask, 'Ballet in Australia', *Hemisphere*, 27 (1982), pp. 89–97; also *Enter the Colonies Dancing: A History of Dance in Australia 1835–1940*, OUP, Melbourne, 1979.
185 J. J. Healy, *Literature and the Aborigine in Australia, 1770–1975*, UQP, St Lucia, 1978.
186 C. Mackerras, *The Hebrew Melodist: A Life of Isaac Nathan*, Currawong, Sydney, 1963, pp. 102–11.
187 Baldwin Spencer and F. J. Gillen, *Across Australia*, Macmillan, London, 1912, p. 231.
188 ibid.
189 ibid., p. 502.
190 Bird, *Percy Grainger*, chap. 10.
191 Marcie Muir (ed.), *My Bush Book: K. Langloh Parker's 1890s Story of Outback Station Life*, Rigby, Adelaide, 1982, p. 150.
192 George Nadel, *Australia's Colonial Culture: Ideas, Men and Institutions in Mid-Nineteenth Century Eastern Australia*, F. W. Cheshire, Melbourne, 1957.
193 Arthur W. Jose, *The Romantic Nineties*, Angus and Robertson, Sydney, 1933; Vance Palmer, *The Legend of the Nineties*, MUP,

Melbourne, 1954.
194 G. A. Wilkes, *The Stockyard and the Croquet Lawn: Literary Evidence for Australian Cultural Development*, Edward Arnold (Australia), Melbourne, 1981.
195 Michael Roe, *Nine Australian Progressives: Vitalism in Bourgeois Social Thought 1890–1960*, UQP, St Lucia, 1984.

CHAPTER 5: POWER

1 Christopher Cunneen, *King's Men: Australia's Governors-General*, Allen and Unwin, Sydney, 1983, chap. 1.
2 Based on P. Loveday, A. W. Martin, and R.S. Parker (eds), *The Emergence of the Australian Party System*, Hale and Iremonger, Sydney, 1977, chap. 1.
3 R. B. Joyce, 'Samuel Walker Griffith: A Liberal Lawyer', D. J. Murphy and R. B. Joyce (eds), *Queensland Political Portraits 1859–1952*, UQP, St Lucia, 1978, pp. 166–7.
4 G. N. Hawker, *The Parliament of New South Wales 1856–1965*, Government Printer, Sydney, 1971, p. 15.
5 P. Weller, 'Tasmania', in Loveday, Martin and Parker (eds), *Emergence of the Australian Party System*, p. 355.
6 P. Loveday and A. W. Martin, 'Colonial Politics before 1890', in ibid., p. 10.
7 Helen Jones, *Nothing Seemed Impossible: Women's Education and Social Change in South Australia 1875–1915*, UQP, St Lucia, 1985, p. 151; also Helen Jones, *In Her Own Name: Women in South Australian History*, Wakefield Press, Adelaide, 1986, pp. 115–16.
8 See Bernard Barrett, *The Civic Frontier: The Origin of Local Communities and Local Government in Victoria*, MUP, Melbourne, 1979.
9 Hawker, *Parliament of New South Wales*, p. 15.
10 See p. 129.
11 Government of Queensland, *Our First Half-Century*, Government Printer, Brisbane, 1909, p. 32.
12 Loveday, Martin and Parker (eds), *Emergence of the Australian Party System*, pp. 170, 355.
13 ibid., p. 10.
14 Quoted F. K. Crowley, *Forrest,*, vol. 1, 1847–91: *Apprenticeship to Premiership*, UQP, St Lucia, 1971, p. 224.
15 ibid., p. 258.
16 Colin A. Hughes and B. D. Graham, *A Handbook of Australian Government and Politics 1890–1964*, ANU Press, Canberra, 1968, p. 280; also G. Sawer, 'The Australian Constitution and the Australian Aborigine', *Federal Law Review*, 2 (1966), pp. 17–36.
17 Hawker, *Parliament of New South Wales*, p. 17.
18 Geoffrey Serle, *The Rush to be Rich: A History of the Colony of Victoria, 1883–1889*, MUP, Melbourne, 1971, p. 229.
19 Based on D. B. Waterson, *A Biographical Register of the Queensland Parliament*, ANU Press, Canberra, 1972.
20 Hawker, *Parliament of New South Wales*, p. 16.
21 Serle, *The Rush to be Rich*, p. 153.
22 Howard Coxon, John Playford and Robert Reid, *Biographical Register of the South Australian Parliament 1857–1957*, Wakefield Press, Adelaide, 1985.
23 Hawker, *Parliament of New South Wales*, p. 16.

24 Joy E. Parnaby, 'The Composition of the Victorian Parliament 1856–1881', *HS* Selected Articles Second Series, MUP, Melbourne, 1967, pp. 75–90.
25 Brian Dickey, 'South Australia', D. J. Murphy (ed.), *Labor in Politics: The State Labor Parties in Australia 1880–1920*, UQP, St Lucia, 1975, p. 231; John Hirst, *Adelaide and the Country 1870–1917: Their Social and Political Relationship*, MUP, Melbourne, 1973, pp. 153–71.
26 The term 'squatter' was usually associated with the south and with sheep. Graziers were more respectable later generations of squatters. Pastoralist often meant cattle, though not exclusively.
27 Cane-growers belong to the later part of the period. Planters had plantations, not cane-farms.
28 *ABD*, 7, p. 490; also K. Buckley and K. Klugman, *The History of Burns Philp: The Australian Company in the South Pacific*, Burns, Philp and Co. Ltd, Sydney, 1981.
29 See J. A. La Nauze, 'A Little Bit of Lawyers' Language: The History of "Absolutely Free", 1890–1900', in A. W. Martin (ed.), *Essays in Australian Federation*, MUP, Melbourne, 1969; also J. A. La Nauze, *The Making of the Australian Constitution*, MUP, Melbourne, 1972.
30 Hawker, *Parliament of New South Wales*, pp. 22–3.
31 Roger B. Joyce, *Samuel Walker Griffith*, UQP, St Lucia, 1984, pp. 178–83.
32 Hirst, *Adelaide and the Country 1870–1917*, p. 82.
33 Alan Martin, *Henry Parkes: A Biography*, MUP, Melbourne, 1980, e.g pp. 346–7, 423–4.
34 *ABD*, 4, p. 252.
35 Sir Edmund Barton Papers, Series 3/958, NLA.
36 J. M. Bennett (ed.), *A History of the New South Wales Bar*, Law Book Co. Ltd, Sydney, 1969, p. 96.
37 Sir George Bowen quoted by Stanley Lane-Poole, *Thirty Years of Colonial Government*, Longmans Green, London, 1889, vol. II, p. 155.
38 Michael Davitt, *Life and Progress in Australasia*, Methuen, London, 1898, pp. 73–110.
39 George Black, *A History of the New South Wales Political Labor Party from its Conception until 1917*, George Black, Sydney, 1918, part 2, p. 33.
40 R.B. Walker, *The Newspaper Press in New South Wales 1803–1920*, SUP, Sydney, 1976, p. 232.
41 L. F. Crisp, *Charles Cameron Kingston: Radical Federationist*, ANU, Canberra, 1984, pp. 71–3.
42 See John Rickard, *Class and Politics in New South Wales, Victoria and the Early Commonwealth 1890–1910*, ANU Press, Canberra, 1976.
43 Sarah Campion's trilogy, *Mo Burdekin* (1941), *Bonanza* (1942), and *The Pommy Cow* (1944), all Peter Davies, London, is set mostly on the northern frontier between the 1870s and the First World War.
44 Based on D. J. Murphy, 'Queensland Biographical Index' in *Labor in Politics*, pp. 216–22.
45 Brian McKinlay, *The ALP: A Short History of the Australian Labour Party*, Drummond, Melbourne, 1981, p. 9.
46 ibid., p. 11.
47 Dickey in Murphy (ed), *Labor in Politics*, pp. 282–5.
48 J. A. La Nauze, 'Who are the Fathers?', *HS*, 13 (1968), pp. 341, 350–1.

49 L. F. Crisp, *Federation Prophets without Honour: A. B. Piddington, Tom Price, H. B. Higgins*, ANU, Canberra, 1980, p. 54.
50 Largely as a result of the work of A. W. Martin and P. Loveday in *Parliament Factions and Parties: The First Thirty Years of Responsible Government in New South Wales, 1856–1889*, MUP, Melbourne, 1966.
51 J. B. Dalton, 'An Interpretative Survey: the Queensland Labour Movement', in D. J. Murphy, R. B. Joyce, Colin A. Hughes (eds), *Prelude to Power: The Rise of the Labour Party in Queensland 1885–1915*, Jacaranda Press, Brisbane, 1970, p. 4.
52 C. W. Dilke, *Problems of Greater Britain*, Macmillan, London, 1890, p. 169.
53 J. M. Powell, *The Public Lands of Australia Felix*, OUP, Melbourne, 1970, pp. xix–xx.
54 W. G. McMinn, *A Constitutional History of Australia*, OUP, Melbourne, 1979, p. 75.
55 Geoffrey Serle, 'The Victorian Legislative Council 1856–1950', *HS* Selected Articles, MUP, Melbourne, 1964, p. 138.
56 ibid., pp. 147–8.
57 ibid., p. 141.
58 Hirst, *Adelaide and the Country*, pp. 10–19.
59 John Shaw Neilson, *Autobiography*, NLA, Canberra, 1978, pp. 31–57.
60 David Denholm, *The Colonial Australians*, Penguin Books, Ringwood, Vic., 1979, chap. 9.
61 G. C. Bolton, 'Black and White after 1897', in C. T. Stannage (ed.), *A New History of Western Australia*, UWA Press, Nedlands, 1981, p. 126.
62 A. G. Austin (ed.), *The Webbs' Australian Diary 1898*, Pitman, Melbourne, 1965, p. 53; also John Merritt, *The Making of the AWU*, OUP, Melbourne, 1986.
63 Richard Kennedy, 'The Leongatha Labour Colony: Founding an anti-Utopia', *LH*, 14 (1968), pp. 54–8.
64 Government of Queensland, *Our First Half Century*, pp. 135–40.
65 Gavin Souter, *A Peculiar People: The Australians in Paraguay*, Angus and Robertson, Sydney, 1968.
66 Karl Marx, 'The Eighteenth Brumaire of Louis Bonaparte', Karl Marx and Federick Engels, *Selected Works*, vol. 1, Lawrence and Wishart, London, 1950, pp. 302–8.
67 Gerald E. Caiden, *The Study of Australian Administrative History: With appendix: The Colonial Public Services, 1850–1900*, ANU, Canberra, 1963.
68 See Henry George, *A Perplexed Philosopher*, Henry George Foundation, London, 1937 (1892).
69 D. L. Clark, '"Roasting the Landowner before a Slow Fire": the origins of rating on unimproved land values', in Jill Roe (ed.), *Twentieth Century Sydney: Studies in Urban and Social History*, Hale and Iremonger, Sydney, 1980, pp. 134–47.
70 See A. G. Austin, *Australian Education 1788–1900*, Pitman, Melbourne, 1965; also J. S. Gregory, *Church and State: Changing Government Policies towards Religion in Australia; with Particular Reference to Victoria since Separation*, Cassell, Melbourne, 1973, chaps 3–5.
71 For a neat summary, colony by colony see B. R. Wise, *The Commonwealth of Australia*, Pitman, London, 1909, chap. vi.

72 See p. 73.
73 P. O'Farrell, *Catholic Church and Community*, Nelson, Melbourne, 1977, p. 264.
74 Susan Magarey, *Unbridling the Tongues of Women: A Biography of Catherine Helen Spence*, Hale and Iremonger, Sydney, 1985, p. 162.
75 O'Farrell, *Catholic Church and Community*, p. 264; also Patrick Ford, *Cardinal Moran and the ALP*, MUP, Melbourne, 1966.
76 Harold Finch-Hatton, *Advance Australia!: An Account of eight years' work, wandering, and amusement, in Queensland, New South Wales, and Victoria*, W. H. Allen and Co., London, 1886, p. 225.
77 Quoted by H.V. Evatt, *William Holman: Australian Labour Leader*, Angus and Robertson, Sydney, 1979 edn, pp. 13–14.
78 Ambrose Pratt, *David Syme: The Father of Protection in Australia*, Ward Lock and Co., London, 1908, p. 127.
79 See C. R. Hall. *The Manufacturers: Australian Manufacturing Achievements to 1960*, Angus and Robertson, Sydney, 1971 for an account of the early manufacturers' organizations; David Plowman, 'Industrial Legislation and the Rise of Employer Associations, *Journal of Industrial Relations*, 27 (1985), pp. 283–309.
80 Nellie Stewart, *My Life's Story*, John Sands Ltd, Sydney, 1923, p. 54.
81 L. F. Crisp, *The Unrelenting Penance of Federalist Isaac Isaacs 1897–1947*, ANU, Canberra, 1981, pp. 1–14.
82 R. S. Neale, 'John Stuart Mill on Australia', *HS*, 13 (1968), pp. 239–45.
83 W. Pember Reeves, *State Experiments in Australia and New Zealand*, Macmillan, Melbourne, 1969 (1902), vol. 2, p. 197.
84 K. D. Buckley, *The Amalgamated Engineers in Australia, 1852–1920*, ANU, Canberra, 1970, p. 113.
85 J. H. Portus, *Australian Compulsory Arbitration*, Law Book Co., Sydney, 1979, p. 1.
86 See L. F. Crisp on Higgins, Isaacs (*Federation Prophets*); also John Rickard, *H. B. Higgins: the rebel as judge*, Allen and Unwin, Sydney, 1984.
87 C. C. Kingston quoted Victor S. Clark, *The Labour Movement in Australasia: A Study in Social-Democracy*, Constable, London, 1907, pp. 182–3.
88 Commonwealth of Australia, *Official Year Book*, 1901–7, p. 866.
89 Chap. 8; see also J. S. D. Mellick, *The Passing Guest: A Life of Henry Kingsley*, UQP, St Lucia, 1963, p. 21, and p. 99 for Lane and *The Workingman's Paradise*.
90 Australasian Federation League of Victoria, *Songs of Union to be sung at... Federal Meetings by the Federal Choir and Audience*, A. and W. Bruce, Melbourne, 1899.
91 Gerald Glynn O'Collins SJ (ed.), *Patrick McMahon Glynn: Letters to His Family (1874–1927)*, Polding Press, Melbourne, 1974, p. 23.
92 G. C. Bolton, 'The Idea of a Colonial Gentry', *HS*, 13 (1968), pp. 307–28.
93 H. Mortimer Franklyn, *A Glance at Australia in 1880*, The Victorian Review Publishing Co., Melbourne, 1881, p. 39.
94 Davitt, *Life and Progress in Australasia*, p. 55.
95 Rosa Campbell Praed, *Policy and Passion*, Virago Press, London, 1988, Introduction.
96 J. A. Froude, *Oceana or England and Her Colonies*, Longmans Green, London, 1886, pp. 202–3.

97 Alexandra Hasluck (ed.), *Audrey Tennyson's Vice-Regal Days: The Australian letters of Audrey Lady Tennyson to her mother Zacyntha Boyle, 1889–1903*, NLA, Canberra, 1978.
98 Ethel Castilla, *Centennial Magazine*, vol. 1, pp. 505–7.
99 ibid.
100 Anthony Trollope, *Australia*, UQP, St Lucia, 1967, pp. 652–3.
101 Ada Cambridge, *Thirty Years in Australia*, Methuen, London, 1903, pp. 43–4.
102 Ruth Bedford, *Think of Stephen. A Family Chronicle*, Angus and Robertson, Sydney, 1954, p. 224.
103 In Russel Ward, *The Australian Legend*, OUP, London, 1958.
104 See P. Loveday in Loveday, Martin and Parker (eds), *Emergence of the Australian Party System*, p. 486; also Ronald Lawson, *Brisbane in the 1890s: A Study of an Australian Urban Society*, UQP, St Lucia, 1973, pp. 78–88.
105 Henry Mayer, *Marx, Engels and Australia*, F. W. Cheshire, Melbourne, 1964, pp. 123–4.
106 G. Greenwood (ed.), *Australia. A Social and Political History*, Angus and Robertson, Sydney, 1958, pp. 145–6.
107 'Tasma', *Uncle Piper of Piper's Hill: An Australian Novel*, Trubner, London, 1889.
108 Lane-Poole, *Thirty Years of Colonial Government*, vol. 1, p. 200.
109 Geoffrey Hawker, 'An Investigation of the Civil Service: The South Australian Royal Commission of 1888–1891', *JRAHS*, 65 (1979), p. 47.
110 A. W. Martin, *Henry Parkes: A Biography*, MUP, Melbourne, 1980, p. 379.
111 Joyce, *Griffith*, p. 128.
112 R. S. Neale, *Class and Ideology in the Nineteenth Century*, Routledge and Kegan Paul, London, chap. 4.
113 Andrew Lemon, *The Young Man from Home: James Balfour 1830–1913*, MUP, Melbourne, 1982.
114 Rickard, *H. B. Higgins*, p. 57.
115 Graeme Davison, *The Rise and Fall of Marvellous Melbourne*, MUP, Melbourne, 1978.
116 Bennett, *The NSW Bar*, p. 77.
117 Joyce, *Griffith*, p. 24.
118 Adrian Merritt, 'Forgotten Militants: Use of the New South Wales Master and Servants Acts by and against female employees 1845–1930', *Law and History*, 1 (1982), La Trobe University, pp. 54–104 summarizes the literature on master and servant relations.
119 Quoted Caiden, *The Study of Australian Administrative History*, p. 24.
120 Constance Jane Ellis, *I Seek Adventure: An autobiographical account of pioneering experiences in outback Queensland from 1889–1904*, APCOL, Sydney, 1981.
121 Helen Jones, *Nothing Seemed Impossible*, p. 92.
122 A. M. Nichol, *Half-round the World with General Booth*, Salvation Army Publishing Dept, London, 1892, pp. 221–3, 218–19.
123 J. A. La Nauze, *Alfred Deakin: A Biography*, MUP, Melbourne, 1965, pp. 133–4.
124 Jill Roe, *Beyond Belief: Theosophy in Australia 1879–1939*, UNSWP, Kensington, 1986, p. 90.
125 Hasluck, *Audrey Tennyson's Vice-Regal Days*, p. 300.
126 George R. Parkin, *Imperial Federation: The Problem of National Unity*, Macmillan, London, 1892, p. 199.

127 Dilke, *Problems of Greater Britain*, p. 195.
128 Peter Firkins, *The Australians in Nine Wars: Waikato to Long Tan*, Rigby, Adelaide, 1971, pp. 3–15.
129 Commonwealth of Australia, *Official Year Book*, 1901–7, pp. 886–7.
130 Donald C. Gordon, *The Dominion Partnership in Imperial Defense 1870–1914*, Johns Hopkins Press, Baltimore, 1965, p. 28; also Robert Haldane, *The People's Force: A History of the Victoria Police*, MUP, Melbourne, 1986, pp. 71–7.
131 See K. S. Inglis, *The Rehearsal: Australians at War in the Sudan 1885*, Rigby, Adelaide, 1985.
132 A. W. Martin (ed.), *Letters from Menie: Sir Henry Parkes and his Daughter*, MUP, Melbourne, 1983, p. 109.
133 Geoffrey Serle, 'The Victorian Government's Campaign for Federation 1883–1889', A. W. Martin (ed.), *Essays in Australian Federation*, MUP, Melbourne, 1969; also Roger C. Thompson, *Australian Imperialism in the Pacific: The Expansionist Era 1820–1920*, MUP, Melbourne, 1980, chaps 2 and 3.
134 McMinn, *A Constitutional History of Australia*, p. 90.
135 ibid., p. 84; see also B. A. Knox, 'Imperial Consequences of Constitutional Problems in New South Wales and Victoria 1865–1870', *HS*, 21 (1985), pp. 515–33.
136 L. F. Crisp, *George Richard Dibbs 1834–1904: Premier of New South Wales, Prophet of Unification*, ANU, Canberra, 1980, p. 21.
137 Donald C. Gordon, *The Australian Frontier in New Guinea 1870–1885*, Columbia University Press, New York, 1951, p. 127.
138 W. P. Morrell, *Britain in the Pacific Islands*, OUP, London, 1960, pp. 250–1.
139 ibid., p. 251.
140 Gordon, *Australian Frontier in New Guinea*, p. 190.
141 L. F. Crisp, *The Later Australian Federation Movement 1883–1901. Outline and Bibliography*, ANU, Canberra, 1979, p. 18.
142 Thompson, *Australian Imperialism in the Pacific*, p. 60.
143 See J. D. Legge, *Australian Colonial Policy: A Survey of Native Administration and European Development in Papua*. Angus and Robertson, Sydney, 1956, chap 2 for a brief account; also R. B. Joyce, Australian Interests in New Guinea before 1906', in W. J. Hudson (ed.), *Australia and Papua New Guinea*, SUP, Sydney, 1971, chap. 1.
144 C. A. Bernays, *Queensland Politics During Sixty (1859–1919) Years*, Government Printer, Brisbane, 1919, p. 92.
145 Thompson, *Australian Imperialism in the Pacific*, p. 59.
146 Martin, *Letters from Menie*, p. 148.
147 ibid., p. 150 (25 June 1885).
148 Geoffrey Serle, *The Rush to be Rich: A History of the Colony of Victoria 1883–1889*, MUP, Melbourne, 1971, p. 201.
149 R. B. Walker, 'Violence in Industrial Conflicts in New South Wales in the Late Nineteenth Century', *HS*, 22 (1986), pp. 54–70.
150 See Thomas W. Tanner, *Compulsory Citizen Soldiers*, APCOL, Sydney, 1980, and John Barrett, *Falling In: Australians and 'Boy Conscription' 1911–1915*, Hale and Iremonger, Sydney, 1979.
151 Eric Irvin, *Gentleman George—King of Melodrama: The Theatrical Life and Times of George Darrell 1844–1921*, UQP, St Lucia, 1980, pp. 97–102.
152 Brian Kiernan (ed.), *Henry Lawson*, UQP, St Lucia, 1976, p. 188.
153 Quoted L. E. Skinner, *Police of the Pastoral Frontier: Native Police 1849–59*, UQP, St Lucia, 1975, pp. 369–70.

154 Lorimer Fison and A. W. Howitt, *Kamilaroi and Kurnai*, George Robertson, Melbourne, 1880, p. 182.
155 Noel Loos, *Invasion and Resistance: Aboriginal–European Relations on the North Queensland Frontier 1861–1897*, ANU Press, Canberra, 1982, p. 190.
156 Richard Broome, *Aboriginal Australians: Black Response to White Dominance 1788–1980*, Allen and Unwin, Sydney, pp. 50–1; also Henry Reynolds, *The Other Side of the Frontier*, James Cook University, Townsville, 1981, pp. 163–6 and *Frontier: Aborigines, Settlers and Land*, Allen and Unwin, Sydney, 1987.
157 Arthur Duckworth, quoted Brian Fitzpatrick, *A Short History of the Australian Labor Movement*, Rawson's Bookshop, Melbourne, 1944, p. 68.
158 See p. 101.
159 Kiernan, *Henry Lawson*, pp. 188–9.
160 *Cosmos*, 29 February 1896, quoted Prout, p. 133.
161 Australasian Federation League, *Songs of Union*.
162 G. H. Reid, *My Reminiscences*, Cassell, London, 1917, p. 190.
163 Rickard, *H. B. Higgins*, p. 110.
164 See C. N. Connolly, 'Manufacturing "Spontaneity": The Australian offers of troops for the Boer War', and 'Class, Birthplace, Loyalty: Australian attitudes to the Boer War', *HS*, 18 (1978), pp. 106–17, 210–32.
165 Firkins, *The Australians in Nine Wars*, p. 15.

EPILOGUE

1 *Sydney Morning Herald*, 25 December 1900.
2 ibid., 2 January 1901.
3 John Mansfield Thomson, *A Distant Music: The Life and Times of Alfred Hill 1870–1960*, OUP, Auckland, 1980, p. 71.
4 *Sydney Morning Herald*, 2 January 1901.
5 Nellie Stewart, *My Life's Story*, John Sands Ltd, Sydney, 1923, p. 133.
6 Alfred Deakin, *The Federal Story: An Inner History of the Federal Cause*, MUP, Melbourne, 1963, p. 162; Walter Murdoch, *Alfred Deakin: A Sketch*, Constable, London, 1923, p. 203.
7 As Miles Frankin wrote of *Joseph Furphy: The Legend of a Man and his Book*, Angus and Robertson, Sydney, 1944, p. 148.
8 R. H. Croll (ed.), *Smike to Bulldog: Letters from Sir Arthur Streeton to Tom Roberts*, Ure Smith, Sydney, 1946, p. 18.
9 Quoted Michael Roe, *Nine Australian Progressives: Vitalism in Bourgeois Social Thought 1890–1960*, UQP, St Lucia, 1984, p. 282.
10 R. C. K. Ensor, *England 1870–1914*, Clarendon Press, Oxford, 1936, p. 111.
11 Brian Fitzpatrick, *The Australian People 1788–1945*, MUP, Melbourne, 1946, p. 217.
12 Gordon Greenwood (ed.), *Australia. A Social and Political History*, Angus and Robertson, Sydney, 1955, pp. 146–7.
13 Geoffrey Serle, *From Deserts the Prophets Come: The Creative Spirit in Australia 1788–1972*, Heinemann, Melbourne, 1973, p. 60.

SOURCES OF ILLUSTRATIONS

Page
xiv Andrew Garran (ed.), *Picturesque Atlas of Australasia*, Picturesque Atlas Publishing Co., Sydney, 1886–8, p. 357
22   H. Mortimer Franklyn, *A Glance at Australia in 1880*, The Victorian Review Publishing Co., Melbourne, 1881
23   J. P. Stow, *South Australia: its History, Productions and Natural Resources*, K. Spiller, Adelaide, 1883, p. 89
36   Mortimer Franklyn, *A Glance at Australia*
39   Garran, *Picturesque Atlas*, p. 85
52   *The Centennial Magazine*, vol. 1, 1888–9, p. 394
55   *The Pictorial Australian*, no. 8, vol. xi (new series), August 1885
72   Garran, *Picturesque Atlas*, p. 218
74   *The Illustrated Christian Weekly*, 14 January 1881, p. 172
80   *The Illustrated Australian Phrenological, Physiognomical and Hygienic Magazine*, vol. 1, no. 1, September 1882
93   *The Pictorial Australian*, 1885, p. 156
134 *The Centennial Magazine*, vol. 1, frontispiece, January 1889
139 *The Western Mail*, Christmas 1897, p. 43
151 *The Pictorial Australian*, September 1885, p. 3
156 *The Illustrated Christian Weekly*, 24 June 1881, p. 448
160 Garran, *Picturesque Atlas*, p. 503
169 *The Illustrated Christian Weekly*, September 1880, p. 8
180 Mortimer Franklyn, *A Glance at Australia*

SOURCES OF ILLUSTRATIONS 353

83 E. E. Morris (ed.), *Cassell's Picturesque Australasia*, Cassell and Co., London, 1889, vol. 1, p. 73
87 *The Centennial Magazine*, vol. 1, frontispiece, September 1888
91 *Cassell's Picturesque Australasia*, vol. IV, p. 131
10 *Cassell's Picturesque Australasia*, vol. I, p. 49
38 *The Illustrated Australian News*, 1 July 1891, p. 1
44 *The Illustrated Australian News*, 1 October 1894, p. 12
54 *The Queenslander*, 9 December 1899, p. 1
64 *The Western Mail*, Christmas 1897, p. 95
96 Garran, *Picturesque Atlas*, p. 407
02 Garran, *Picturesque Atlas*, p. 71
06 *The Illustrated Australian News*, 1 September 1890, p. 1
13 Nellie Stewart, *My Life's Story*, John Sands, Sydney, 1923, facing p. 212

BIBLIOGRAPHICAL NOTE

The late nineteenth century saw the beginning of mass education, mass literacy, and modern bureaucracy, so the volume of printed and written evidence which survives from this period is enormous. I can still only claim to be acquainted with a small fraction of it. Likewise, historical research on this period has been extensive, and intelligent use of indexes and bibliographies will almost certainly turn up a book or article on any given subject. I have tried to refer to as many of the major or standard works as possible in the notes, so that used in conjunction with the index the reader at least has a starting point for further inquiry. It should also be noted that for late nineteenth-century history, the most recent work does not necessarily mean the most useful or the best. Modern perceptions of productivity and scholarship as well as the costs of publication do not often permit reworking of sources on the heroic scale which established many a subject in earlier times. The approach of some older works may seem unfashionable, but the content is frequently more detailed. There are also some areas which have ceased to be of interest to modern scholars, either because they no longer seem relevant, or because they offer nothing new. They are nonetheless part of a history which itself now has quite a history.

Two main sources have been heavily utilized for this work, the vast contemporary statistical compilations represented in nineteenth-century censuses, blue books and yearbooks, and that great modern historical enterprise, the *Australian Dictionary of Biography*. Most individuals referred to in the text will be found in the *ADB*, although this has only been noted in the case of specific quotation. There is now also a biographical dictionary for each colonial parliament in this period.

In addition there are journals which can be expected to yield detailed recent work. *Australia 1888* (1979–1986) was devoted entirely to late nineteenth-century research, while *Labour History*, *Historical Studies*, *Australian Economic History Review*, *Journal of Religious History*, and *Australian Literary Studies* all regularly publish late nineteenth-century material.

As a result of the most recent round of historical celebrations, there are now series of sesquicentenary volumes for Western Australia, Victoria, and South Australia, and of course there is the Australian bicentenary series. It has not been possible yet to assess the usefulness of all these volumes, but if the fate of their 1888 equivalents is any guide, they are certain to be valued for what they say about us now.

INDEX

Aboriginal Protection Act (Queensland 1897) 167
Aboriginals 3, 14, 60, 61, 70, 192
 assimilation 110
 art 235
 cricket 193–4, 195
 crime rate 176
 economy 138–9
 exclusion as voters 246
 labour 5, 9
 missions 130
 music 234
 religion 62–4
 social structure 137–41
 status 265, 266
 tobacco smoking 186–7
 troopers 302
 wars 304
 welfare 166–7
Aborigines Protection Act (Victoria 1886) 110
accidents 121, 172
acclimatization societies 177–8, 191
Adelaide 5, 27, 30, 32, 138, 177, 230–1, 291
Adelaide Hills 183, 204
Adelaide Liedertafel 227
agents general 239
agnostics 87
Agra and Mastermans Bank 40
agricultural machinery 19
agriculture 18–28
Albany 161, 291
Albury 297
Albury, Louisa 131
Allen, George L. 228, 229, 233
Allgemeiner Deutscher Verein 100

America 5, 255
 democracy 259, 302
 feminism 104
 theatre 225
 see also USA
Anglicanism 66, 246, 255
Anniversary Day 184
Anrep-Elmpt, Count Reinhold von 17
Antarctic Committee 17
anthropology 60, 63, 64–5, 130
arbitration 54, 275–6
Archdall, Mervyn 68–9
Arafura Sea 33
Archer family 131
Archer, William Henry 108, 109
Archibald, J. F. 90
Argentina 5
Armidale 214, 247
Arnold, Matthew 217
Arnold, Thomas 202
Arnotts biscuit making 36
art galleries 94, 211–12
artesian water 4, 21
artists 218–22
Ashton, G. R. 221
Ashton, Julian 220–1
Asian trade 6
atheists 87, 88
Atherton 166
audiences, theatre 223–4
Austin, Thomas 190
Australasian Association for the Advancement of Science 41, 91
Australasian New Hebrides Co. 17
Australian Art Association Ltd 219–20
Australian boy 146
Australian Economist 90

357

358 INDEX

Australian girl 118, 146–7, 207
Australian Insurance and Banking Record 90
Australian Jockey Club Derby 189
Australian Labor Party 101, 169
Australian Medical Journal 90
Australian Natives Society (Association) 112, 184
Australian Protestant Defence Association 69
Australian Socialist League 100
Australian Turf Register 189
'Australian type' 94, 106, 123, 146–8, 173
Aveling, Edward 101
AWU (Australian Workers' Union) 266

babies 148–9
Baden Powell, George S. 17
Badham, Charles 91, 211
Balfour, A. J. 127, 288
Ballan 294
Ballarat 13, 227
ballet 234
Balmain 68, 90, 226, 244
Bancroft, Dr Thomas 165
banks 7, 29, 30, 38, 40, 45, 48, 49
  closures 11, 49, 50
Banks, Mary Mcleod 74
Baptists 86
Barbalet, Margaret 200
Barcaldine 99, 248
Barker, Frederic 68
Barker, Possie 51, 52
Bartho, Catherine 234
Barton, Edmund 147–8, 251, 252, 291
Barwon Park 190
Basedow, Friedrich 128
Basilisk 14
Bastow, Henry 209
Bates, Daisy 61
Batho, Thomas 99
Bayley, Arthur and Ford, William 13–14
Baynton, Barbara 236
Beach, Bill 196
Beagle Bay 14
Beale, Octavius 229, 232
Bean, C. E. W. 202
Bean, Edwin 202
Bedford, Randolph 13
beef cattle 5
Bellamy, Edward 98, 99, 101, 257
Belyando river 3
Bendigo 36
benevolent asylums 51
Bennett, Scott 197
Bent, Thomas 250
Besant, Annie 77, 291
Bicknell, Arthur 192
bicycle riding 195

birth control 77
  see also contraception
birth rate 61
biscuit making 36
Black, George 252
'black gins' 8–9
Blair, David 58
Blumberg 130
Boake, Barcroft 188
Board, Peter 205
Boer War 172, 189, 193, 208, 292, 304, 307–8
Boldrewood, Rolf 51, 159, 188, 216, 217
Bolton, G. C. 265
Bondi 182
Bonwick, James 121–2, 202
books
  sellers 88, 214, 255
  shops 90, 208
  see also reading
Booth, General William 291
Boothby, Mr Justice 295
borrowing 49
  government 38–9
botanical gardens xi, 75, 177–8
Botany 182
boundaries, colonial xii
bourgeoisie 48, 283
Bowen 7
Bowen, Sir George Ferguson xi, xii, 3, 7, 287, 295
Bowen, Lady Diamantina Roma xi, xii
Bowen Downs 6
Boxer Rebellion 292, 304
boxing 195
Brady, E. J. 305
branch offices 48, 248–9, 276, 296
brass bands 228, 303
Brennan, Christopher 90
Bright, Annie (Mrs Charles) 305
Brighton 182
Brisbane xi, xii, 2, 5, 30, 32, 138, 177
Brisbane River 196
Britain 42, 57, 278
  army 8, 292–3, 302
  capital 1, 6, 17, 38, 40, 44
  empire 58, 105–6, 107, 257–8, 290, 293–4, 300, 303
  census 108–9
  government 239, 298–9
  navy 171, 292, 293–4, 299
'British' race 94
Britishness xi, 57, 59, 75, 140–1
Broken Hill 16–17
Bromby, Dr 67
Broome 8
Broome, Richard 304
brotherhood 88–9, 305
Browning, Robert 314
Buchanan, Nat 4

Buddhists 87
building
  industry 31, 45
  labourers 49
  societies 30, 45, 49, 269
Bulletin 106, 212, 216–17, 221
Bulli 15
Bunbury 125, 245
Bundaberg 9, 164
Burdekin River 2, 3
bureaucracy 268
Burke, Edmund 257
Burke, R. O'H. and Wills, W. J. 4, 61, 197
Burns, James 248
Burns Philp 248, 294
Bury, J. B. 57
Butler, Samuel 103
Butlin, N. G. 30, 35
Byrnes, T. J. 197

cabinet formation 250, 251–2
Cairns 7, 136
Cairns, Adam 68
Callide valley 15
Calvinism 127–8
Cambridge, Ada 183, 191–2, 216, 280
camels 26
Campion, Sarah 254
cancer 122
Canoona 3, 303
Cape York 2, 3
capital 38, 40, 249, 268–9
capital punishment 167, 169–70
Carcoar 200
Cardwell 7–8, 17, 136
Carlyle, Thomas 102
Caron, Leon 227, 234
Carpenter, W. R. 294
cartoonists 221
Catalpa 125
Catholic church
  authoritarianism 82–3, 84–5
  impact of Ireland 82
  laity 83–4, 272
  leadership 82, 83, 85
  piety 84, 85
  schools 83, 200, 202
Catholicism 64, 66, 68, 69, 75, 153
Catholics 52, 81–5, 86, 246–7, 255
  see also Irish immigration
cattle 3
Cave, W. R. 231
censuses 46, 85, 108–9, 144, 284–5, 286
centenary 1888 1, 58, 221
Centennial Park 309
ceremonies
  Aboriginal 63–4
  federal 198, 309–11
  funeral 197
  Masonic 89

mealtime 149–50
Melbourne Cup 189–90
religious 64
vice-regal xi, 279
Chamberlain, Joseph 311
Chambers, Joseph 234
Chapman, H. S. 287–8
charitable institutions 155–8
Charters Towers 13, 254
Chester, Henry M. 298
Chidley, William 78, 173
childbirth 77, 90
child-care 142–3
children 10, 19, 35, 59, 146–50, 164, 200–2, 203, 205–6
Chinese 9, 47, 62, 294, 296
  Commission 1887 137
  gardeners 21
  immigrants 133–7
  labour 9
Chisholm, Caroline 20
Christianity 3, 62–3, 64–76, 100, 102, 140, 149–50, 152–3, 158, 166
Christian Science 70
Christmas 182, 184
chocolate consumption 53–4
churches 52, 75–6, 77, 152–3, 154, 173
  adherence 85–7
  attendance 71, 85
  leaders 66–7, 82, 83, 85
  music 226–7
  power 271–2
Church of Christ 86
Church of England 67, 68, 85, 86, 126, 170
Cingalese coolies 9
civilization 3, 8, 57, 59, 61, 62, 78, 133, 220
Clarence River 197
Clark, Andrew Inglis 102
Clarke, Marcus 67, 86, 90, 146, 159, 179, 215–16, 217, 225
Clarke, Sir William 191, 230
class 43–4, 282–90
Claythorpe, Jesse 51
clergymen 70, 73, 88, 152, 280
clerks 48–9
Clifford, W. K. 102
closer settlement 17–19, 266–7
coal 12, 14–16
Cobb, N. A. 23
Coghlan, T. A. 18, 41, 42, 50, 52, 73, 109, 169
Cole, E. W. 88, 208, 214
collectivism 97–9, 102, 269
Collie 15
Collins, Tom *see* Joseph Furphy
Colonial Laws Validity Act 295
Colonial Office 236, 278, 292–6
Colonial Sugar Refining Company 21, 131, 248
Comettant, Oscar 122, 150, 218, 233

'Commonwealth of Australia' 56, 77, 105, 110
Conder, Charles 220
Confucians 87
Congregationalists 52, 69, 86, 89, 255
conservatism 81, 95–8, 260, 265
constitutions
   colonial 237, 239, 240, 294–5
   Commonwealth 56, 87–8, 249, 253, 276
consumption 28, 52–4
contagious diseases legislation 170–1
contraception 144
   see also birth control
convictism 59, 94, 155, 158–62, 217–18
Coogee 182
Coolgardie 14
copper 16–17
Coppin, George 223
Coranderrk 140
Corowa 235
cotton 9–10, 164
corroborees 235
coursing 191
courtship 148
Couvreur, Jessie ('Tasma') 236
Cowan, Frederick 228
cremation 65
Cressbrook 74
cricket 193–5
crime rate 159, 167–70
   Aboriginal 176
Crisp, Christopher 102
Crisp, L. F. 256
crocodiles 7
Cullen, Paul 83

dairying 19, 27, 50, 131
Dalby 233
Dale, R. W. 52
Dalley, W. B. 182, 301
Dalrymple, G. E. 2–3, 6
Daly, John A. 194
Daly Waters 70
Dampier, Alfred 223, 225
dancing 233–4
Danes 131
Dapto 196
Darling Downs 2, 6, 24
Darrell, George 224, 303
Darwin 6, 291
Darwin, Charles 59, 61, 69, 92
Davitt, Michael 102, 126, 189, 252, 279
Dawson, Andrew 253–5
Deakin, Alfred 54, 56, 95, 182, 209–11, 249, 291, 311
Dean, George 171–2
Deane, John 229
death rate 121–2, 165–6, 196–7
defence 280, 291–4, 301–3

democracy 106, 256, 277
Democratic Association of Victoria 100
demography 102–21
Denholm, David 265
Denison, Sir William 260
Depression (1890s) 44–5, 49, 50–1, 54, 94–5, 120, 314–15
Diamantina River 4
Dickens, Charles 193
Dickson, Sir James 254
Diggles, Sylvester 227
Dilke, C. W. 43, 87, 88, 258, 273, 291–2
Dillon, Revd G. F. 153
Dillon, John 126
Disraeli, Benjamin 61
divorce rate 152, 153
domestic service 25, 41, 49, 148–9, 280
Douglas, John 287
Drayton 226
drinking 126, 186
droughts 11
drunkenness 169, 172
Dubbo 34
Duckworth, Arthur 90–1
Duffy, Charles Gavan 125, 247, 259
Duke of Edinburgh 125, 133, 190
Duke of Newcastle xii, 3
Dunsford, John 254
Durack, Paddy 4, 143
Durkheim, Emile 172–3

Earle, Tilly 234
early closing 185, 274–5
Easter 125, 184
Echuca 174, 208
ecology 11, 17–18
economy
   Aboriginal 138–9
   development 4
   management 44, 272–7
   security 42–6, 49–51, 54, 65, 253
Edithburgh 26
editors 102, 213
education 91–4, 199–211, 270–2
Edwards, J. Bevan 301
egalitarianism 106, 198, 236, 270–2, 277–82
Eggleston, F. W. 94–5
Eidsvold 131
eight-hour day 50
eisteddfods 227, 228
Elder, Sir Thomas 230–1
electoral qualifications 239
Ellis, Constance Jane 289
Ellis, Havelock 61, 172, 173, 200, 201
Elsey station 8
Emmaville 13
employers' organizations 275
Engels, Freidrich 283

English language 59–60, 213
Ensor, R. C. K. 314
entrepreneurial spirit 14
environment, perceptions of 60, 175–8
ethnographers 14
eugenics 137
Evans, George Essex 216
Evans, Raymond 162
evolution, theory of 92–3
exclusion 89, 136–7, 246, 289–90
executive council xii
exploration 1–6
exports
  income 12
  statistics 4

Facey, A. B. 19
factions 251
families 8, 10, 45, 59, 65, 77, 93, 141–5, 154–5, 158, 173, 223–4
  fatherless 51
  political 249–50
  rural 28, 142–3
family life 76–7, 149–50
Farina (Government Gums) 25, 26
farmers 192, 247, 264
farming
  machinery 19, 24
  methods 23–6, 27–8
  produce 12
Farr, Dr William 108
Farrer, William 23
Faulconbridge 182
Favenc, Ernest 216, 320
federal conventions
  1891 256, 299
  1897 272
Federal Council 299–300
federation 56, 77, 106, 196, 219–20, 257, 274, 299–300, 305, 307
female franchise 261
femininity 62, 81
feminism 103–4, 153
Fenianism 125–6
Fiji 17, 135, 163, 293–4, 299
film making 225, 235
Fincham, George 232
Finch-Hatton, Harold 186, 272–3
Fink Commission 205
firewood 15, 25, 155
Fisher, Andrew 255
Fisher, Thomas 209
Fitchett, Dr William 303
Fitzgerald, C. B. 255
Fitzpatrick, Brian 314
Flinders river grass 3
folksongs 228, 234–5
food processing 21
Foote, Mary Hannay 216
foreign policy 240
Forrest, John 4, 245

Forsyth, Archibald 96–7
Fox, E. Phillips 95, 219, 220
Franklyn, H. Mortimer 20, 53, 279
Frazer, James 63
free thought 65, 71, 87, 88
free selection 10–11, 18, 28, 96, 259–60, 262–4
free trade 273–4, 299
freemasonry 89, 233, 279
Fremantle 125, 132, 161, 291
Freud, Sigmund 78, 144, 172
friendly societies 49, 89, 112, 185, 279
Froude, J. A. 105, 181, 279
fruit 20–1, 135–7
Furphy, Joseph 67, 88, 174, 208, 212, 236, 312

Galle 291
Gallipoli 197, 314
Galton, Francis 172–3
gambling 186
Gandevia, Bryan 146
gardening 176–8
Gawler 194
Genesis 64, 67
Geographical Society of Australasia 17
George, Henry 98, 102, 266, 268–9, 270
Geraldton (Innisfail) 7
Geraldton (Western Australia) 141
The Germ Growers 70–1
German
  immigrants 128–30, 132, 133, 227
  imperialism 298–300
  socialists 100
Gibney, Bishop Matthew 82
Gilbert and Sullivan 226
Gillen, Francis 63–4, 235
Gilles, Duncan 58, 251
Giorza, Signor 227
Gippsland 15, 130
Gladesville 196
Gladman, F. J. 202
Gladstone Debating Society 95
Glenelg 182
Glen Innes 13
Glynn, P. M. 88, 277–8
God 62–3, 73, 87–8
gold discoveries 2, 3, 5, 12, 14, 17, 262, 303
Goldfields Amendment Bill (Queensland 1876) 137, 294, 296
Golding family 17
Goldstein family 95, 267
Gollan, R. A. 287, 314
Gordon, Adam Lindsay 188, 192, 197
Gordon, Arthur Hamilton 192–3
Gordon, General Charles 197, 301
Goulburn River 267
Gould, Nat 225
governesses 201
Government House 2, 281–2

governors xi, xii, 189, 237–8, 278–9, 281
Goyder, G. W. 8, 23, 25
Grabosky, Peter 170
Grainger, Percy 229, 235
Granville 19, 34
Grasby, William Gatton 203–4, 205
Great Barrier Reef 6
Greek flag xi
Griffith, S. W. 89, 90, 98, 99, 165, 207–8, 241, 250, 267, 288–9
Gronlund, Lawrence 98, 101
Groom, W. H. 159
Guilfoyle, William 178
guilt 78
Gulf Country 3
Gunn, Jeanie 8
Gympie 3, 254, 262

Hagenauer, Friedrich August 130
Hall, W. H. 114–15
Hamburg 129, 131
Hancock, Sir William Keith 96, 98
Hansel, Emil 130
Hardacre, H. F. 255
Hardy, Thomas 209, 216
Hargrave, Lawrence 14
Harpur, Charles 192
Hart, John 250
Hartley, J. A. 202, 203
Hawker, G. N. 241
Hawkesbury Agricultural College 204
Hawkesbury River 26
Hay 208
Hayter, Henry 109
Hazon, Roberto 229
Hearn, W. E. 53, 91
Heales, Richard 271
health 121–3, 165–6
heating 15
heaven 65, 88
Heidelberg 220
Heidenreich, Georg Adam 130
Henley Beach 182
Henniker, Mrs 65
Henning, Rachel 48
Henriques, Lulu 95
Henry, Lucien 222
Henty family 143
Hermannsburg 130
hero worship 197–9, 304
Heussler, J. C. 129, 130
Hicks, Neville 79
Higgins, H. B. 288, 307
Higgs, W. G. 100–1
Higinbotham, George 68, 105, 125, 260–1, 295, 296, 297
Hill, Alfred 311
Hirst, John 250
Historical Society of Australasia 58
Hobart 32
Hodgson brothers 6

Hogan, J. F. 124–5, 148
holidays 181–6
Holt, Bland 223
Holtermann, B. O. 128
home ownership 43–4, 49
homoeroticism 199
Hopetoun, Lord 311
Hopkins, Livingston 221
horses 8, 26, 188–90, 221, 225, 290
hospitals 122, 155, 156, 157
Hougoumont 125
housing 29–31, 45, 176, 178–9, 269
Howe, Renate 69–70
Howitt, A. W. 61, 304
Hubbe, Ulrich 129
Hume, Katie 226–7
Hunter River 17, 36, 290
hunting 190–3, 304
husbands, 73, 118, 186, 280
hydro-electricity 16
hygiene 79–80
Hyndman, H. M. 98, 257
hypocrisy 8, 154

ice works 21
illiteracy 200
immigration 59, 112–13, 123, 246, 281–2
  Chinese 133–7
  Cornish 16, 124
  English 16, 123–4
  German 128–30
  Irish 124–6
  Japanese 141
  Jewish 132–3
  Scandinavian 131–2
  Scots 126–8, 131
income tax 40, 41
India 6, 8, 290, 291–2
  coolies 9
  Mutiny 292
individualism 51, 62, 97
industrial development 14–16
industrialization 36–9
infant mortality 121
Inglis, K. S. 184, 197
inheritance 43–4
insane asylums 155
insurance 49
intercolonial rivalry 256–7, 296–7
International Exhibitions 57–8, 227, 232
Inverell 13
investment 4, 46
Ipswich 15, 129
Ireland 82, 83, 84, 98
Irish 105, 115
  immigration 124–6, 255
  question 67, 128
Ironside, Adelaide 220
irrigation 23, 267
Irvin, Eric 224

INDEX 363

Isaacs, Isaac 95
Ives, Joshua 231

Japanese immigrants 141
Jardine brothers 3
Jeffries, R. T. 229
Jennings, Patrick 247
Jevons, W. S. 176
Jews 62, 95, 132–3
Jose, Arthur 236
Junee 34

Kadina 16
Kalgoorlie 16
Kanaka labour 9, 162–7
Kapunda 194
Kay-Shuttleworth, James 202
Knowlton, Charles 77
Kelly, Ned 146, 192, 197, 220–1, 225, 304
Kelsey, Eliza 241
Kendall, Henry 19, 192, 217, 227, 316
Kennedy, Edmund 2, 3
Kenningham, Charles 311
Kidman, Sidney 265
Kidston, William 255
Kimberleys 4
kindergarten movement 205
Kingsley, Charles 74
Kingsley, Henry 217, 277
Kingston, Charles Cameron 98, 275–6, 311
Kipling, Rudyard 60, 291
Knibbs, G. H. 172
Knibbs-Turner Report 205
Knox, Edward 131
Krefft, Gerard 92
Krichauff, E. H. W. 129, 242

La Nauze, J. A. 256
labour
  and capital 173, 249, 266
  Chinese 136
  colonies 155, 252, 267
  day 185
  force 47
  management 96–7
  movement 249, 256, 273
  settlements 102, 267
ladies xi, 34, 174, 279–80
Lambert, George 221
Lamington, Lord 254
land 2, 4, 258–70
  revenue 40–1, 258–9
Landsborough, William 6
Lane, E. H. (Ernie) 101, 305
Lane, William 99–101, 103–4, 267, 277, 305
Lang, Andrew 235
Lang, Dr John Dunmore 10, 105, 127
Lang, Dr W. H. 235
Lasseter, Harry 4

Launceston 16
laundry work 45
Laurie, Henry 213
Lawson, Henry 16, 33, 90, 106, 131, 172, 303, 305
lawyers 249, 250, 251, 255, 277, 288–9
  in politics 247, 248
Leichhardt, Ludwig 2, 3, 4
leisure 175, 182–99
legislative assemblies xii, 239–58
legislative councils 260–1, 265
Leongatha 267
leprosy 165
liberalism 94–8, 257
libraries 74, 90, 208, 209, 212
life expectancy 121
Lilley, Charles 207
Lindsay, Vane 221
Linger, Carl 227
literacy 206–7, 212–13
Lithgow 15, 36, 132
Liverpool 38
Lloyd, Henry Demarest 173
local government 70, 269
Loftus, Lady 279
Logan River 164, 201
Lombroso, Ceasare 173
London 5, 38, 40, 45, 225, 239, 281
Long, C. R. 205
Longreach 99, 248
Longstaff, John 219
Loos, Noel 166, 304
Lord's Day Observance Society 74
Lukin, Gresley 216
lunatic asylums see insane asylums
Lutherans 86, 128–30
Lyons, Margaret 226
Lyrup 102
Lyster, W. S. 226

Macalister, Arthur 89, 127, 182, 251
Macarthur, James 96
Macassans 141
McCoy, Professor Frederick 92
McIlwraith, Thomas 127, 250, 296, 298, 299, 300
Mackay 6, 131, 164
McKillop, Mary 66
MacKinlay, William 14
Maclean 197
McLean, Hector 231
Macmillan, David 127
Macrae, Tommy 235
Macrossan, J. M. 248
Mahony, Frank 221
Maitland 36, 132
Malays 141
Mallee 24
Maloga 140
Maloney, Dr Patrick 90
Malthus, Thomas 61
Manly 182

Marburg 130
market gardens 133–5
Maori Wars 292
Mann, Charles 262
manufacturing 34–5
Marree 62
Maria 14
Marin la Meslée, Edmond 182
marriage 59, 66, 131, 133, 139
  bureaux 148
  market 117–19
  mixed 82–4, 153
  rate 115–21
  registry offices 152
Marsh, Jack 195
Marshall Hall, G. W. L. 208, 228–9, 248, 277
Martens, Conrad 219
Marx, Karl 99, 100, 101, 257, 268, 282, 283
Maryland run 12–13
masculinity 62, 106, 114–15, 159
Masson, Professor David Orme 91
mateship 88–9, 106, 305
Matthews, Julia 234
May, Phil 221
meat canning 3, 19, 193
  consumption 54
mechanics' institutes 208–9
medical profession 122–3
Medlow Bath 182
Meichel, Pastor 130
Melbourne 28–9, 30, 57, 58, 68, 75, 132, 133, 177–8, 311
Melbourne Cup 189, 190
Melbourne Grammar School 67, 202
Melbourne Philharmonic Society 227
Melbourne Public Library 209, 210, 211
Meredith, George 216
The Messiah 227, 230
Methodism 69–71, 75, 86, 246
Métin, Albert 98
middle classes 277, 283–5
militarism 195, 305, 307–8
Mill, J. S. 93, 257, 288
Miller, Edmund Morris 312
mining 1, 12–16, 38
miners 248
missions ix, 60, 130, 140
Mitchell, Blanche 147
Molony, John 83
Monaro 96, 234
Monash, Sir John 95
Montez, Lola 234
Moonta 16
Moorhouse, Bishop James 68
Moran, Cardinal Patrick Francis 82, 84, 272
Morant, Henry 'Breaker' 188, 307–8
Moresby, Captain 14
Morgan brothers 14

Morpeth 36
Morris, E. E. 202
Morris, William 98, 102, 103
mortgage companies 7
Moseley, H. N. 166
Moss Vale 183
mosque 62
Mount Bischoff 16
Mount Gipps station 13
Mount Macedon 183
Mount Magnet 13
Mount Morgan 1, 2, 289
Mount Rennie 171–2
Mount Victoria 182
Mucke, Dr Carl 128
Mueller, Baron Ferdinand von 128, 178
municipal politics 241, 242
Munro, James 127, 250
Murray Bishop James 83
Murray River 23, 67, 102, 208, 267, 297–8
Murrumbidgee River 23
museums 74, 75, 92, 211
Musgrove, George 311
music 225–33
Muskett, Dr Phillip E. 81

Narrogin 19
'Nasturtium Villa' 179
Nathan, Isaac 234
national spirit 106–7, 195, 303
native-born population 59, 94, 112, 118, 123, 246, 282, 299, 311–12
native fauna 190, 222
  flora 18, 60, 222
native police 292, 302
naturalism 76–81, 93–4
Neilson, John Shaw 90, 264
Nerli, Girolamo 220
New Australia 99, 267
New England 17
New Guinea 14, 17, 23, 60, 162, 296, 298–300
New Hebrides 163, 164
New Zealand 16, 47, 59, 257, 293, 299
Newcastle 15, 16, 17, 36, 132, 148, 227
Newport 34
newspapers 213–16
  German 128, 130
Northcote, Lady 291
Northern Territory xii, 1, 5–9, 136–7, 192
Norton, John 221, 252
Norwegians 131
nursing 122, 148

Ogilvie, Will 188
O'Farrell, Henry James 125
O'Farrell, Patrick 84
old-age pension 118, 157–8
O'Loghlen, Bryan 125

Ooldea waterhole 61
opera 226
opportunity 97–8, 287–90
optimism 4–5
Orara River 19
O'Reilly, John Boyle 125
Ormond, Sir Francis 230
orphans 43, 46, 144, 155
Osburn, Lucy 125
O'Shanassy, John 125, 247
overland telegraph 5, 70, 291

Pacific 17, 178, 290, 293–4
Pacific Islanders 162–7, 186, 257, 294, 299, 300
pacificism 305, 307
paddle steamers 87
Pagans 87
Paling, W. H. 48, 229, 232
Palmer, A. H. 13
Palmer, Vance 236
Palmer River 13
Palmerston 8, 9, 17, 136, 137
 see also Darwin
Pambula 294
Paraguay 99, 267
Parker, K. Langloh 60, 235
Parker Range 13
Parkes, Sir Henry 73, 107, 182, 243–4, 249, 250, 287, 294, 309
Parkes, Menie *see* Thom, Menie
Parnaby, Joy 247
Parramatta 34
party politics 253
pastoralism 1, 3, 4
 industry 4, 5
 in politics 247, 248
 management 7, 10–11
Paterson, A. B. 'Banjo' 188, 192, 193, 196
payment of members, parliamentary 242–3, 250, 261
pearling 8–9, 141
Pearson, Charles Henry 76, 91, 93, 204, 299
Pedley, Ethel 231
periodicals 214–17
Perry, Bishop Charles 272
Perth 32, 132, 138
Phillipini, Rosalie 234
Phillips family 95
Philp, Robert 248
photography 218–19
pianos 48, 231–3, 277
Picnic at Hanging Rock 201
Pine Creek 9, 137
plural voting 241, 261
Polding, Archbishop John Bede 52, 82, 83
police 158, 302
politicians 249
political education 241–3

population
 age distribution 116–19
 birthplaces 118–19, 123–4
 sex ratios 114–15
 statistics 108–13
  Aboriginal 109–10
Port Arthur 158, 160
Port Augusta 25
Port Broughton 26
Port Essington 2
Port Germein 26
Port Pirie 26
possum snaring 45
post offices 29, 212, 215
Potter, Revd. Robert 70–1
potteries 36
poverty 51, 122
Praed, Rosa Campbell 74, 159, 201, 217, 236, 279
Presbyterianism 68, 74, 86, 246, 255, 294
 see also Scots immigration
Prichard, K. S. 49
probate 40
professions 288–9
progress 57–62, 108–9
property 43–4, 49, 268–70
prostitution 141, 170–1
protection 38, 274–5, 299
Protestantism 76, 81, 153
provincialism 90
public
 buildings 29, 31, 45
 utilities 41
 service 288, 289
puritanism 76–81, 99, 104
Pyap 102

Quantong 102
Quart Pot Creek 13
Queen Victoria xii, 59, 106, 128, 290
Queenscliff 182
Queenslander 216–17
Queen's Fund 155
Quilpie 289
Quinn, Bishop James (Brisbane) 82, 83
Quinn, Bishop Matthew (Bathurst) 83
Quong Tart 150

rabbits 11, 45, 190
racism 252
railways 4, 9, 14, 15, 18, 26, 32–5, 44, 46–7, 56, 297
Ramahyuck 130
rape 170–2
Rasp, Charles 13
reading 65, 208, 212–13
Redmond brothers 126
Reeves, William Pember 101–2
refrigeration 4, 21, 27
remoteness 5

Rentoul, Revd. J. A.   88
republicanism   104–5, 198
respectability   48–9
revolution,   101, 305
Reynolds, Henry   304
Richardson, Henry Handel   78, 233
ringbarking   17–18
rites of passage   149, 203–4
Riverina   128
Riverview College   90
roads   33–4
Roberts, Tom   218, 219–20, 312
Robertson, Sir John   182, 259, 260
Robertson, William   102
Rockhampton   2, 3, 303
Rockingham Bay   7
Rogers, Frederic   295
Rome   66, 67, 82, 83
Romilly, H. H.   293
Roper River   4
Rosa, S. A.   101, 173
Roseby, Dr Thomas   71
Roseworthy Agricultural College   129, 204
Ross, Jeannie   148
Rowan, Ellis   91
rowing   195–7
Royal Commission into Accounts and Departments—Southern Side (Tasmania 1863)   289
Royal Society of Victoria   17
Royal Tar   99
Rubinstein, W. D.   43
Rudd, Steele   27, 196
rural employment   19, 49–50, 248
Rusden, Henry Keylock   77
Ruskin, John   102, 220

sabbatarianism   74–6
Sadler, James H.   196
sailing   188
St Kilda   31, 182, 288
St Patrick's Day   184
salt consumption   54
Salvado, Bishop R.   82
Salvation Army   70, 75, 86, 88, 228, 291
sandal wood   8
Sandhurst   13
sanitation   26, 31–3, 121–2
Saunders, Kay   165
savings   31, 49
Scandinavian immigrants   131–2
schools   65
    of arts   107, 208–9, 214
    Catholic   83, 200, 202
    elementary   200, 201–2
    German   128
    for girls   201, 207
    private   202
    secondary   206–7
science   65, 71, 89–90, 91–3
Scotland   38

Scots Church Literary Society   68
Scots   2
    immigration   126–8, 255
Scott brothers   6, 7
Scott, Professor Walter   90
Scott, Walter   188
Scottish Australian Investment Company   6, 45
Scratchley, Peter   293
Searcy, Alfred   8–9, 192
Searle, Harry   196–7
secret ballot   240–1
sectarianism   125, 154–5, 270
secular morality   76–81, 88–9
Select Committee on Transportation (House of Commons 1861)   161
self-government   xiii, 237–77
self-improvement   70
Selfe, Norman   204
Separation Day   184
Serle, Geoffrey   314
Serra, Bishop J. B.   82
service industries   48–9
settlement   4, 5
Seventh-day Adventism   70
sewing   45
sex segregation
    culture   236
    labour   47–8
    Aboriginal society   140
    leisure   198–9
sexuality   77–9, 143–5
    Chinese   136
    instincts   60–1
    morality   173
    repressed   81
shearing   47
sheep   3, 5, 7, 8, 10
shipping   6, 14, 26, 36, 290–1
shooting   190, 304
shop-assistants   49
Sinnett, Frederick   303
Sladen, Douglas   188
Smith, Adam   93, 273
Smith, Bruce   97–8
Smith, Howard   97
Smith, Professor John   91
smog   16
Smythe, Bessye   144
soap manufacturing   36
Social Democratic Federation   100
Social Democratic League   101
social laboratory   173
socialism   85, 98–102
sociology   93–4
Solomon Islands   163
Somerset   6, 8
'Song of Australia'   227
South Brisbane   34
South East Asian trading zone   17
South Melbourne   19
Southern Cross   106, 313

INDEX 367

Sparkes Creek (Scone) 61
speculators 5, 13, 38
speech patterns 174–5
Spence, Catherine Helen 103, 201, 217, 272
Spence, W. G. 88
Spencer, Baldwin 63–4, 91–2, 235
Spencer, Herbert 93, 94
Spiers and Pond 193
spiritualism 88
sport 61, 147–8, 172, 173
squatters 10–11, 264–6
squatting 2, 3
Stagg, William 25–6, 27
standard of living 12, 19–20, 21, 51–4
  Aboriginal 166–7
Stannage, C. T. 159
Stanthorpe 13
state socialism 98–9
statisticians 41–2, 108–9, 167–8
status 73, 174–5, 194–5, 264–6, 279–80, 282–90, 310
  Aboriginal society 138
Stenhouse, David 90
Stephen, Sir Alfred 281
Stephens, A. G. 216
Stephens, James Brunton 59, 200–1, 216
stockmen 49
  Aboriginal 13
Strathalbyn 129, 194
Strehlow, Pastor Carl 130
Stewart, Nellie 226, 274, 311
Stoddardt, Brian 194–5
Stone, Louis 233
Streeton, Arthur 312, 316
strikes 99, 275–6, 304
Strong, Charles 68, 102
Stuart, Sir Alexander 126
Stuart, John McDouall 4
Stuart, Lurline 214
suburban life 179, 181
Sudan campaign 105, 292, 293, 301, 304
Suez Canal 290
sugar 9, 21
  cane varieties 23
  mills 141, 267
suicide 172
Summerhayes, Madame 232
Sunbury 191
Sunday observance 74–6, 150
Sunshine 19
Sutherland, Alexander 58
Sutton Forest 182
sweating 261
Swedes 131
Sydney 30, 57, 75, 132, 133, 138, 177, 296, 309–11
Sydney Grammar School 202
Syme, David 104, 253, 273, 291
Syme, Herbert 291

Tait, Charles 198
Tambaroora 17
Tamrookum 201
Tarcowie 25–6
tariff 47, 56, 273, 295–6
Tasmania 54, 113, 158–9, 183–4, 239
Tasmanian Hydro-Electric Authority 16
Tasmanian Jam and Fruit Preserving Company 21
taste 211
Tate, Frank 205
tax
  income 40, 41
  land 41, 266, 269
  single 268–9
teacher training 202–4
technology 4, 14, 21, 57, 89–90
telegraphs 6, 8, 13, 14
Tennyson, Alfred 188, 217
Tennyson, Lady Audrey 189, 192, 279, 291
Tenterfield School of Arts 107
Thargomindah 16
theatre 222–5
theology 67–9
Thom, Menie 217, 294, 301
Thursday Island 141, 287, 298
timber 4, 17–18, 30
Thompson, Allen 35
tin 12, 13, 21
Tingha 13
tobacco 133, 187
Torrens, Robert 129, 269
Torres Strait 296–7, 300
town halls 29
Towns, Robert 6, 164, 294
Townsville 7
Tozer, Horace 166–7
trade unionism 47–8, 89, 248, 257, 275, 279
  holidays 184–5
  membership 277
  officials 254–5
  records 50
trades and labour councils 241–2
tradition 280
Traill, W. H. 216–17
Trenwith, W. A. 256
Trickett, Ned 196
Trevelyan, Charles 233
Trollope, Anthony 12, 58, 159, 177, 190, 216, 280
Truth 221, 252
tuberculosis 79, 121–2, 165–6
Tully 14
Tully, W. A. 7–8
Turley, Joseph 255
Turner, George 274
Turner, Henry Gyles 38, 68
Turner, Martha 68
tutors 200–1

INDEX

Twopeny, R. E. N.   146, 148, 149
Tyson, James   265

unemployment   46–7, 50–1, 155
Unitarians   68, 69
Union Jack   xi, 59
universities   90, 91, 207–11, 230–1, 290
upper houses   239, 240
upward mobility   73
United States (USA)   10, 33, 38, 88, 100, 120, 124, 125, 163, 164
  see also America
utopian novels   103

Valley of Lagoons   3, 6
Vaughan, Archbishop Roger Bede   82
vegetables   20–1, 49, 53–4
venereal disease   77
Verein Vorwärts   100
Victor Harbour   182
Victoria River   4
Victoria Racing Club   189
Victorian Charity Organisation Society   96
Victorian Common Schools Bill (1862)   271
Victorian Football Association   193
Victorian Sunday Newspapers Act (1889)   74
violence   61, 172, 252, 303, 304
volunteer forces   292–3, 302–3
voters   xii, 245–6

wages
  boards   276
  regulation   275
Wakefield, Edward Gibbon   261–2, 268
Walch, Garnet   223
Wallangarra   34, 297
Wallaroo   26
Walstab, G. A.   216
war   303–4
Warangesda   140
waratah   222
Ward, Ebenezer   262
Ward, Russel   98, 282
Warrego River   2
Warrnambool   90, 213
Warung, Price   159, 218
Watson's Bay   182
wealth   41–4, 95–7, 270, 287–9
Webb, Beatrice   54, 173, 189
Webb, Sidney   173, 189, 266
weddings   73
Weigall, A. B.   202
Welsh immigrants   227
Wesleyans   69, 70
  see also Methodists
Western Australia   8, 54, 114, 117, 161–2, 237, 265
Western Pacific High Commission   293–4
Westgarth, William   1, 18, 57
wheat   20–1, 23–7
White Australia Policy   5, 106, 137
White, Richard   213
widows   43, 46, 155
Wilkins, William   200, 202, 205
Williamson, J. C.   223, 225, 226, 234
Williamstown   34
Wills, T. W. S.   193
Wimmera   24
Windeyer, W. C.   77, 171
wine   129, 180–1
Winter, C. H.   88
Wise, B. R.   61
wives   8, 10, 19, 73, 117–19, 152, 153, 164, 280
  deserted   155
  of governors   xi, 279–80, 281, 291
  leisure of   185
Woman's Christian Temperance Union   104
women
  Aboriginal   138–40
  in the churches   89–90
  and civilization   280
  as dependents   51
  and education   290
  equality of   102, 103, 120
  and medicine   290
  as mothers   81, 118, 264, 289
  musicians   230
  political education   241–2
    exclusion   246
  and power   280
  and science   91, 92
  and sexuality   77–8, 80–1, 119–21
  and status   279–80
  votes for   40, 241
  white   8–9
Wood, G. A.   208
wool   3, 4, 7, 8, 9–10, 11
Woollahra   31
work   9, 49–50, 61–2, 185
  ethic   61
working class   16, 48–9, 54, 85, 87, 145
working man   xi, 28, 174
The Workingman's Paradise   100, 103–4
workingmen's colleges   208–9
Wright family   143
writers   213–18
Wrixson, Henry   96
Wylie, A. C.   90

Yorick Club   90

Zillmere   130
zoological societies   191